GRAPHENE

Energy Storage
and Conversion
Applications

ELECTROCHEMICAL ENERGY STORAGE AND CONVERSION

Series Editor: Jiujun Zhang
National Research Council Institute for Fuel Cell Innovation
Vancouver, British Columbia, Canada

Published Titles

Electrochemical Supercapacitors for Energy Storage and Delivery: Fundamentals and Applications
Aiping Yu, Victor Chabot, and Jiujun Zhang

Proton Exchange Membrane Fuel Cells
Zhigang Qi

Graphene: Energy Storage and Conversion Applications
Zhaoping Liu and Xufeng Zhou

Forthcoming Titles

Electrochemical Polymer Electrolyte Membranes
Yan-Jie Wang, David P. Wilkinson, and Jiujun Zhang

Lithium-Ion Batteries: Fundamentals and Applications
Yuping Wu

Solid Oxide Fuel Cells: From Fundamental Principles to Complete Systems
Radenka Maric

GRAPHENE
Energy Storage and Conversion Applications

Zhaoping Liu

Xufeng Zhou

CRC Press
Taylor & Francis Group
Boca Raton London New York

CRC Press is an imprint of the
Taylor & Francis Group, an **informa** business

CRC Press
Taylor & Francis Group
6000 Broken Sound Parkway NW, Suite 300
Boca Raton, FL 33487-2742

First issued in paperback 2021

ISBN 13: 978-0-367-78371-6 (pbk)
ISBN 13: 978-1-4822-0375-2 (hbk)

Library of Congress Cataloging-in-Publication Data

Liu, Zhaoping (Chemist)
    Graphene : energy storage and conversion applications / author, Zhaoping Liu.
        pages cm. -- (Electrochemical energy storage and conversion ; 6)
    Includes bibliographical references and index.
    ISBN 978-1-4822-0375-2 (hardback)
    1. Electric batteries--Materials. 2. Capacitors--Materials. 3. Solar cells--Materials. 4. Graphene--Electric properties. I. Title.

TK2896.L58 2014
621.31'24240284--dc23   2014042942

Visit the Taylor & Francis Web site at
http://www.taylorandfrancis.com

and the CRC Press Web site at
http://www.crcpress.com

Contents

Series Preface..vii
Preface..ix
Authors..xi

Chapter 1   Graphene Overview...1

   Xufeng Zhou, Wei Wang, and Zhaoping Liu

Chapter 2   Synthesis of Graphene...21

   Wei Wang, Hailiang Cao, Xufeng Zhou, and Zhaoping Liu

Chapter 3   Applications of Graphene in Lithium Ion Batteries.........................65

   Xufeng Zhou and Zhaoping Liu

Chapter 4   Applications of Graphene in New-Concept Batteries......................137

   Xufeng Zhou and Zhaoping Liu

Chapter 5   Applications of Graphene in Supercapacitors.................................171

   Chao Zheng, Xufeng Zhou, Hailiang Cao, and Zhaoping Liu

Chapter 6   Applications of Graphene in Solar Cells...217

   Fuqiang Huang, Dongyun Wan, Hui Bi, and Tianquan Lin

Chapter 7   Applications of Graphene in Fuel Cells..245

   Xuejun Zhou, Jinli Qiao, and Yuyu Liu

Index..295

Series Preface

The goal of the Electrochemical Energy Storage and Conversion book series is to provide comprehensive coverage of the field, with titles focusing on fundamentals, technologies, applications, and the latest developments, including secondary (or rechargeable) batteries, fuel cells, supercapacitors, CO_2 electroreduction to produce low-carbon fuels, electrolysis for hydrogen generation/storage, and photoelectrochemistry for water splitting to produce hydrogen, among others. Each book in this series is self-contained, written by scientists and engineers with strong academic and industrial expertise who are at the top of their fields and on the cutting edge of technology. With a broad view of various electrochemical energy conversion and storage devices, this unique book series provides essential reads for university students, scientists, and engineers and allows them to easily locate the latest information on electrochemical technology, fundamentals, and applications.

Jiujun Zhang
National Research Council of Canada
Richmond, British Colombia

Preface

Graphene is one of the most promising materials in the world and has attracted much attention all over the world. This Nobel Prize–winning material combines multiple outstanding physical and chemical properties within itself, such as ultra-high carrier mobility, excellent electrical and thermal conductivity, high mechanical strength, ultralarge surface, and good optical transparency. In addition, the unique two-dimensional nanostructure of graphene, which significantly differs from conventional three-dimensional materials, is another advantage of this new material for the construction of novel structures to enhance the performance of graphene-based materials or devices. As a result, graphene has exhibited appealing application potentials in various areas that cover a range of different fields in human life.

Clean and high-efficient energy storage and conversion is becoming more and more important for the sustainable development of our society. The emergence of new materials is continuously raising expectations for the performance of energy storage and conversion of devices. Graphene has the potential to be the material that may provide revolutionary advancement in this area. Great progress has been achieved regarding the application of graphene in energy storage and conversion in recent years due to efforts from scientists and engineers all over the world. It is also anticipated that this area may become one of the initial breakthroughs for the practical application of graphene.

Consequently, it is necessary to review the development and progress of the application of graphene in various energy storage and conversion systems in the past few years to promote related research in the future, which is the major aim of this book. No other book has focused specifically on this topic. The application of graphene in several important energy storage and conversion devices, including lithium-ion batteries, supercapacitors, fuel cells, solar cells, lithium sulfur batteries, and lithium-air batteries, are systematically reviewed in this book. In addition, the main synthesis methods of graphene are also briefly introduced, which is important to the design and synthesis of graphene-based materials for energy storage and conversion. This book is suitable for a broad readership, ranging from experts in related areas to beginners who are interested in graphene and its applications. We hope that our book will give you a better understanding of the energy-related applications of graphene and will motivate those who would like to participate in the research.

Authors

Prof. Dr. Zhaoping Liu received his PhD in inorganic chemistry from the University of Science and Technology of China (USTC) in 2004. He then worked as a postdoctoral researcher in the National Institute for Material Science (NIMS), Japan, and the State University of New York (SUNY) at Binghamton, between 2004 and 2008. Since September 2008, he has been working in Ningbo Institute of Materials Technology and Engineering (NIMTE), Chinese Academy of Sciences. He is currently the director and professor of Advanced Li-ion Battery Engineering Lab. His main research interests include Li-ion batteries and graphene materials. He has published more than 70 peer-reviewed papers, with a total citing of more than 3000, and has received more than 100 patents.

Dr. Xufeng Zhou is an associate professor in Ningbo Institute of Materials Technology and Engineering (NIMTE), Chinese Academy of Sciences. He received his PhD in inorganic chemistry from Fudan University, China, in 2008. He then joined NIMTE as a postdoctoral fellow. In 2011, he was promoted to the position of associate professor. His research interests focus on preparation of graphene and its applications in energy storage devices, including Li-ion batteries and supercapacitors. He has published more than 40 research papers in peer-reviewed journals and has applied for more than 20 patents. He received the Lu Jiaxi Young Talent Award from Chinese Academy of Sciences in 2013 and the Excellent Postdoctoral Fellow of Zhejiang Province in 2014.

Dr. Fuqiang Huang is a professor of materials chemistry in Shanghai Institute of Ceramics, Chinese Academy of Sciences. His work concentrates on thin-film semiconductor PV cells (new process and device assembly [CIGS, CZTS, CdTe]), key solar materials (microstructure design and high-performance processing [TCO, AR, etc.]), and new solar materials for new-concept PV cells (materials design, process, and PV applications [graphene, Cu-based intermediate band PV materials, nanostructured semiconductors, etc.]). He has published more than 300 papers and 150 patents. He is a recipient of many awards including China National Funds for Distinguished Young Scholars and 100 Talents Program of CAS. He received his PhD at Beijing Normal University in 1996 and then spent two years as a research associate at the University of Michigan. In 1998, he moved to Northwestern University as a postdoctoral fellow to work on solid-state chemistry. Then he worked as a principal scientist in R&D, Osram Sylvania Inc., from 2000 to 2002 and as a research staff member in the University of Pennsylvania in 2003.

Dr. Jinli Qiao is a professor, PhD supervisor, and disciplines leader of the College of Environmental Science and Engineering, Donghua University, China. She received her PhD in electrochemistry from Yamaguchi University, Japan, in 2004. After that, she joined the National Institute of Advanced Industrial Science and Technology (AIST), Japan, as a research scientist working on both acidic/alkaline polymer electrolyte membranes and non-noble metal catalysts for PEM fuel cells. From 2004 to 2008, as a principal investigator, she carried out seven fuel cell–related projects including two NEDO projects of Japan, and since 2008, she has carried out a total of nine projects funded by the Chinese government including the National Natural Science Foundation of China and the International Academic Cooperation and Exchange Program of Shanghai Science and Technology Committee. Dr. Qiao's research areas include PEM fuel cell catalysts/membranes, CO_2 electroreduction catalysts/membranes, and supercapacitors. As the first author and corresponding author, Dr. Qiao has published more than 100 research papers in peer-reviewed journals, 40 conference and invited oral presentations, 4 coauthored books/book chapters, and 13 patent publications. She is the vice-chairman/vice-president of the International Academy of Energy Science (IAOEES) and an active member of The Electrochemical Society, The Electrochemical Society of Japan, China Association of Hydrogen Energy, and the American Chemical Society.

1 Graphene Overview

Xufeng Zhou, Wei Wang, and Zhaoping Liu

CONTENTS

1.1 Carbon Materials 1
1.2 History of Graphene 5
1.3 Structure and Properties of Graphene 6
    1.3.1 Structure 6
    1.3.2 Properties..8
        1.3.2.1 Electronic Properties 8
        1.3.2.2 Optical Properties 9
        1.3.2.3 Thermal Properties 9
        1.3.2.4 Mechanical Properties 10
1.4 Application of Graphene..10
    1.4.1 Electronics 10
    1.4.2 Photonic 11
    1.4.3 Energy Conversion and Storage...12
    1.4.4 Composite Materials and Coatings...13
    1.4.5 Sensors...14
    1.4.6 Thermal Management..14
    1.4.7 Bioapplications 15
1.5 Challenges and Research Directions 16
References...18

1.1 CARBON MATERIALS

Carbon is the most important element for all living organisms on earth, as most of the organic elements are made out of it. The unique and fascinating carbon materials are formed by the carbon atoms bond with each other in various ways, exhibiting linear, planar, and tetrahedral bonding arrangements. Carbon material is an ancient yet vigorous material; in roughly 5000 BC, our ancestors began to utilize charcoal for heating and cooking food (energy materials) and subsequently, they used charcoal to smelt copper and iron (chemical materials). Today, carbon materials have been developed to form a brand of the inorganic nonmetallic material that occupies an important place in materials science. Carbon materials are widely used in various industries from aerospace to chemical to nuclear. They are ubiquitous in our daily lives, where they are used in metallurgy, machinery, automotive, medical, environmental, and construction applications.

Graphite and diamond are the most common allotrope of carbon materials that exist in nature. By far, diamond is the hardest substance, whereas graphite is one of the softest materials, which is mainly ascribed to the different carbon atomic arrangement. Graphite has a layered, planar structure. In each layer, the carbon atoms are arrangement in a honeycomb lattice with separation of 0.142 nm, the distance between graphene planes is 0.335 nm, and the planes are bonded to each other by weak van der Waals forces (Figure 1.1). The layered structure of graphite allows sliding movement of the parallel graphene plates. Weak bonding between the plates determines the softness and self-lubricating properties of graphite. Graphite is rarely found in the form of monocrystals; most graphite occurs in form of flakes or lumps. Due to the superior electric and thermal conductivity, lubricity, high thermal shock resistance, chemical stability, and high performance of nuclear physics, graphite can be used in battery electrodes, lubricants, pencils, neutron moderators in atomic reactors, and can also be used as the raw material of crucible and synthetic diamonds. In a diamond, the carbon atoms are arranged in a variation of a face-centered cubic crystal structure called a diamond lattice, in which one carbon atom is surrounded by four carbon atoms, the bonding force in every direction is equivalent (Figure 1.1). Diamond is renowned as a material with superlative physical qualities, most of which originate from the strong covalent bonding between its atoms. Because they have the highest hardness and thermal conductivity, diamonds have been widely applied in cutting and polishing tools.

The carbon atoms not only bond with each other by sp^3 hybrid to form a single bond, but also can form double and triple bonds with sp^2 and sp hybrid, respectively. Therefore, apart from the variety of carbon allotropes that exist in nature, scientists have successfully synthesized many carbon materials with different structures and properties. In particular, an important breakthrough has been achieved in the areas of microscopy technologies and nanomaterials in the past few years that has greatly promoted the progress of carbon materials. In 1985, British chemist H.W. Kroto and U.S. scientist R.E. Smalley found a new carbon nanomaterial (named C_{60}) during which they characterized the product of graphite vaporized by laser irradiation using mass a spectrometer [1]. This substance has a nanoscale spherical structure

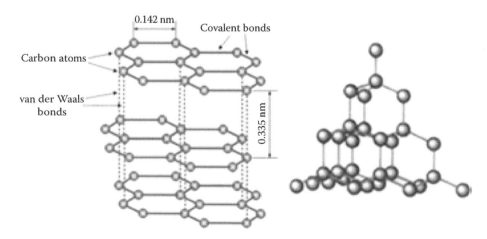

FIGURE 1.1   Atomic structures of graphite and diamond.

known as fullerenes, where one C_{60} molecule consists of 60 carbon atoms arranged as 12 pentagons and 20 hexagons. The discovery of C_{60} has stimulated much activity in chemistry and opened up a new branch of fullerene chemistry which studies new families of molecules that are based on fullerenes. Due to its unique structure and physical and chemical properties, fullerene has had a significant effect on chemistry, physics, and material science, and also has shown to be an attractive prospect regarding potential applications. For this discovery, Kroto and Smalley shared the 1996 Nobel Chemistry Award.

In 1991, after the discovery and verification of fullerenes, tubular carbon structures were first discovered in the soot of the arc-discharge method by Iijima [2]. The nanotubes consisted of up to several dozens of graphitic shells (so-called multiwalled carbon nanotubes [MWCNTs]) with adjacent shell separation of ~0.34 nm, diameters of ~1 nm, and large length/diameter ratio, significantly larger than any other material. Two years later, Iijima and Ichihashi et al. synthesized a single-walled carbon nanotube (CNT), which was formed by wrapping a single-layer graphene sheet. CNTs possess extremely high tensile strengths, high moduli, large aspect ratios, low densities, good chemical and environmental stabilities, and high thermal and electrical conductivities [3]. They are a new type of high-performance carbon nanomaterial and are in demand for various applications, including both large-volume applications (e.g., as components in conductive, electromagnetic, microwave-absorbing, high-strength composites, battery electronic additives, supercapacitors or battery electrodes, fuel cell catalysts, transparent conducting films, field-emission displays, and photovoltaic devices) and limited-volume applications (e.g., as scanning probe tips, drug delivery systems, electronic devices, thermal management systems, and biosensors) [4,5]. CNTs are produced mainly by three techniques: arc discharge, laser ablation, and chemical vapor deposition (CVD). A common feature of the arc discharge and laser ablation methods is high energy input by physical means, such as an arc discharge or a laser beam, to induce the assembly of carbon atoms into CNTs. This leads to a high degree of graphitization in the CNTs. However, these systems require vacuum conditions and continuous graphite target replacement, posing difficulties for continuous large-scale production. In contrast, the CVD process can be operated at mild conditions, for example, atmospheric pressure and moderate temperatures. Furthermore, the CNT structure, such as its diameter, length, and alignment, can be controlled well. Significantly, the mass production of CNTs has been realized by CVD method with low cost [6].

The number of research paper related to CNTs has continually risen in the past years. However, this growth began to decrease in 2009, which is probably ascribed to the appearance of graphene. In 2004, Andrei Geim and Kostya Novoselov successfully obtained single-layered graphene using tape to peel off graphite [7]. This finding of graphene opened up a research boom in carbon namomaterials, which made graphene a rapidly rising star on the horizon of materials science and condensed-matter physics. Graphene is the first two-dimensional (2-D) atomic crystal available. A large number of its material parameters, such as mechanical stiffness, strength and elasticity, very high electrical and thermal conductivity, are supreme. These properties suggest that graphene could become the next disruptive technology, replacing some of the currently used materials and leading to new markets. For these reasons, Geim and Novoselov were awarded the 2010 Nobel Prize in physics for the discovery of graphene and other related 2-D crystal materials.

A summary of the molecular models of different types of sp^2-like hybridized carbon nanostrucutres are shown in Figure 1.2 [8]. These materials have very unique structures and novel physical and chemistry properties, which can be considered as a landmark discovery in the history of carbon materials and will further promote the rapid development of carbon nanomaterials. The carbon family is a magical and colorful world, and while many game-changing discoveries have been achieved during the past several decades, the mystery still remains surrounding the question of how many crystallographic forms, or allotropes, of carbon exist. A variety of novel properties can be observed as a function of topology of carbon leading to metal-insulator transitions, semiconductors, superconductivity, and magnetism.

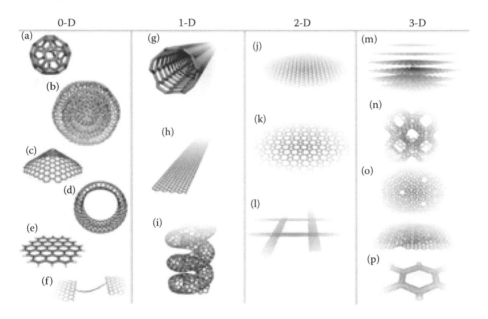

FIGURE 1.2 Molecular models of different types of sp^2-like hybridized carbon nanostructures exhibiting different dimensionalities, 0-D, 1-D, 2-D, and 3-D. (a) C60: Buckminsterfullerene, (b) nested giant fullerenes or graphite onions, (c) nanocones or nanohorns, (d) nanotoroids, (e) graphene cluster, (f) short carbon chains, (g) CNT, (h) GNRs, (i) helicoidal CNTs, (j) graphene surface, (k) Haeckelite surface, (l) nanoribbon 2-D networks, (m) 3-D graphite crystal, (n) 3-D Schwarzite crystals, (o) carbon nanofoams (interconnected graphene surface with channels), and (p) 3-D nanotube networks. (Reprinted from *Nano Today*, 5, Terrones, M., Botello-Méndez, A. R., Campos-Delgado, J. et al., Graphene and graphite nanoribbons: Morphology, properties, synthesis, defects and applications, 351–372, Copyright 2010, with permission from Elsevier.)

1.2 HISTORY OF GRAPHENE

Graphene is defined as a monolayer of carbon atoms that are hexagonally tightly packed into a 2-D lattice. The structure of graphene can also be viewed as the fewest-layer limits of graphite, and graphene can be produced from graphite by complete exfoliation of graphite into monoatomic layers. Graphite as a traditional carbon material has been widely used in our daily lives for centuries. Layered structure is one of the typical characteristics of graphite, which has been successfully utilized in many applications, including the pencil lead for writing, the most universally familiar application of graphite. When you write using a pencil, thin graphite flakes are delaminated from the lead by the mechanical friction between graphite in the pencil lead and paper, and become attached the paper to form writing traces. There is no doubt that monoatomic layer delamination can occur during this process though the possibility may be extremely low. Therefore, it can be speculated that graphene may have already been produced centuries ago by anyone who used a pencil to write or draw, but it has never been realized and taken into serious consideration until the twentieth century.

Graphene as a 2-D unit for the construction of graphite was first explored by theoretical research in 1947 as the starting point to understanding the electronic properties in three-dimensional (3-D) graphite. However, it was theoretically predicted almost 80 years ago that strictly 2-D crystals cannot exist because the thermal fluctuation would destroy long-range order, giving rise to the melt of 2-D lattices [9]. Nevertheless, the experimental research of producing 2-D materials, especially graphene, has never been interrupted. Early attempts to isolate graphene sheets from graphite was conducted by using graphite intercalated compounds as the raw materials, since intercalation of heteroatoms of molecules between graphene layers in graphite can effectively reduce the interlayer interaction and expand interlayer distance [10]. Extremely thin graphite lamellae were also prepared starting with graphite oxide, an oxidized state of graphite. It was discovered that ultrathin carbon films could be obtained by heating or by reduction of graphite oxide in alkaline suspension [11]. This paper reported the ultrasmall thickness of 0.4–2 nm for the exfoliated sheets by analyzing the contrast of the samples in transmission electron microscopy (TEM) images corresponding to mono- and few-layer samples. Although it is now known that it is impossible to distinguish the layer numbers of graphene by simply comparing the contrast, and that the graphite oxide-derived samples have large amounts of structural defects and functional groups, this early attempt does shed light on the chemical preparation of graphene.

Single- or few-layer graphene has been epitaxially grown using gaseous carbonaceous precursors on certain substrates or thermal decomposition of silicon carbide (SiC) in the past few decades, which have high crystallinity and high mobility charge carriers [12,13]. However, the strong interaction between graphene layers and the substrate significantly alters the electronic structure of graphene. Therefore, the measured electronic properties of graphene on the substrates may not truly reflect the properties of intrinsic graphene in a free-standing state.

It was not until the 2004 publication of pioneering work of Geim, Novoselov, and their research group that the curtain of the broad stage for this magic carbon material was finally drawn [7]. Their experiment is quite simple. Its principle is the mechanical exfoliation of graphite, which is actually analogous to the writing on a paper using a pencil but in a more controllable way. The exfoliation process was realized by repeating cleavage of graphite using common Scotch tape. With great patience, ultrathin graphene sheets were finally produced. One of the secrets to their success is the usage of a silicon (Si) wafer with a 300-nm-thick silicon dioxide (SiO_2) layer on top as the substrate to load exfoliated thin graphite flakes. Such a substrate makes it possible to distinguish ultrathin graphitic layers simply by using optical observation, which enormously enhances the efficiency of the search for graphene. The as-prepared graphene is perfect in structure with almost no defects, and therefore it is possible to experimentally measure the intrinsic properties of graphene for the first time. The unique electronic properties of graphene astonished scientists, and soon significant attention was focused on this magic new carbon material all over the world. Due to their remarkable and pioneering research work, the Nobel Prize in Physics 2010 was jointly awarded to Geim and Novoselov for "groundbreaking experiments regarding the two dimensional material graphene" [14]. As of today, tens of thousands of research papers have been published and thousands of patents have been applied based on the materials and techniques related to graphene. It is believed that graphene may become a revolutionary new material that will change the world in the near future.

1.3   STRUCTURE AND PROPERTIES OF GRAPHENE

1.3.1   STRUCTURE

Undoubtedly, graphene, a 2-D atomic crystal, has captured the world's attention due to its extremely significant performance, which greatly benefits from its unique structure [7]. Before the discovery of graphene, most scientists demonstrated that a long-range ordering 2-D structure is impossible to be stable at any finite temperatures [15,16]. This conclusion has been disproved by Novoselov and Geim through the process of repeatedly sticking and tearing graphite with Scotch tape. The graphite was exfoliated when each action occurred. The sheets became thinner and thinner. Finally, one single layer of the graphite was obtained, which coincided with the principle image, as illustrated in Figure 1.3.

The name "graphene" combines the information of "graph" due to the graphite and "ene" due to the carbon–carbon double bonds [17]. Graphene is the first discovered 2-D atomic sheet with hexagonally arranged carbon atoms that are connected to each other with sp^2 bonding. Most recently, it has been rationally defined as "a single-atom-thick sheet of hexagonally arranged, sp^2-bonded carbon atoms that is not an integral part of a carbon material, but is freely suspended or adhered on a foreign substrate" [17]. Because it is one single layer of graphite, the thickness of graphene is around 0.34 nm. It can be considered as the basic structure unit in sp^2-bonding carbon materials. For example, CNTs can be considered as a structure by rolling graphene in certain way. As discussed above, the long-range ordering 2-D structure is unstable in the free state. In fact, graphene has been demonstrated to be not so perfect. Meyer et al. did a very interesting investigation on suspended graphene sheets on a microfabricated scaffold [18]. Broader diffraction peaks were

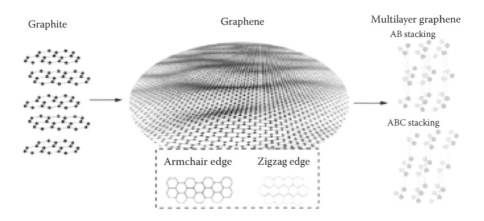

Graphite Graphene Multilayer graphene
AB stacking

ABC stacking

Armchair edge Zigzag edge

FIGURE 1.3 Schematics of the structure of graphene. (Reprinted by permission from Macmillan Publishers Ltd. *Nature*, Ref. 18, copyright 2007.)

observed with increasing tilt angle, which is significantly different from the diffraction behavior in 3-D crystal and hence strongly suggests that the suspended graphene is not perfectly flat. A curved surface is what makes graphene possible, as illustrated in Figure 1.1. This phenomenon becomes much weaker in bilayer graphene, indicating a much flatter surface with increasing layers of graphene.

Besides surface state, edge state had been predicted to play a critical role in the structural property and the corresponding physical performance of graphene [19,20]. Basically, there are two kinds of edge states in graphene: zigzag and armchair configurations, as schemed in Figure 1.3. Zettl systematically investigated edge reconstruction and hole growth under high-energy electron beam in real time [21]. This revealed the long-term stability of zigzag edges in graphene based on the time-average observation and modeling analysis. More importantly, the different edge state in graphene quantum dots (GQDs) and graphene nanoribbons (GNRs) can determine the overall electronic properties. In particular, 7–8-nm GQDs with zigzag-orientated edges are metallic, which is different from other studies on semiconductor behavior. The more zigzag edges in GNRs, the smaller energy gap. Thus, there is great interest to study and hence control the edge states during the growth of graphene. Cheng et al. reported that the edge states can be controlled from zigzag to armchair orientations via changing the reaction precursors and the growth or etching rates for the armchair edge are faster than that of the zigzag edge [22].

Since graphene is the basic building block in graphite, different and significant properties are revealed by stacking graphene in different ways. As two layers of graphene stack together, it is usually found with an AB stacking called a Bernal stacking arrangement [23]. In this case, the vacant centers in one hexagon layer stay at the top of the corner carbon atoms in another hexagon layer, as shown in Figure 1.3. The in-plane lattice constant of graphite is about 0.246 nm and that for the c-axis is about 0.67 nm. Bilayer graphene has been demonstrated to be an effective way to open the bandgap of graphene [24]. When more graphene layers stack together, AB stacking, ABC stacking, or rotationally faulted multilayer graphene appears. The different structures of monolayer, bilayer, trilayer, and other multilayer graphene can be identified in Raman spectrum studies with big differences in I_{2d}/I_G and the shape of I_{2d} and so forth [23].

With the high-speed development of graphene, numerous types of graphene-like materials had been presented, including monolayer/bilayer/trilayer/multilayer graphene, single-crystalline/multicrystalline graphene, GQDs, GNRs, graphene nanomesh, graphene microsheets/nanosheets, graphene oxide (GO), and reduced GO [17]. In each material, the morphologies can be different depending on the fabrication methodologies. For example, bilayer graphene can be constructed by laying monolayer graphene one by one. This will lead to the existence of an unreasonable angle between adjacent layers, which are completely different from the traditional bilayer graphene with AB stacking structure. The Raman spectrum of this kind of bilayer graphene is the same as that of monolayer graphene [25]. This confused situation has attracted the attention of researchers who propose that a rational scientific nomenclature be created to clearly identify these graphene-like structures. It can be expected that many more unknown properties will be discovered as more styles of graphene with different structures will be fabricated in the future.

1.3.2   Properties

1.3.2.1   Electronic Properties

The high quality of 2-D crystal lattice endows graphene with extraordinary electronic properties. The valence and conduction bands in graphene intersect at a single point of zero states, which is called the Dirac point, which makes graphene a zero-gap semiconductor. At the Dirac point, the density of states is zero and the linear dispersion relation results in an effective mass of zero. Therefore, electrons in graphene behave as massless Dirac fermions, giving rise to unprecedentedly high carrier mobility. The carrier mobility measured for a high-quality single-layer graphene sheet prepared by mechanical exfoliation and completely suspended with no influence of the substrate reaches 200,000 cm^2 V^{-1} s^{-1} [26]. With such high-carrier mobility, the charge transport in graphene is actually ballistic on the micrometer scale at room temperature. The charge carriers in graphene behave like relativistic particles. As a result, their behavior should be described using the Dirac equation instead of the Schrodinger equation, generally applied to describe the electronic properties of other materials. Therefore, the emerge of graphene now provides scientists with the possibility to investigate relativity theory on the lab table instead of in the universe, and provides a way to study quantum electrodynamics by measuring graphene's electronic properties.

Another characteristic of graphene is its ambipolar electric field effect. The carriers in graphene can be continuously tuned between holes and electrons by supplying the requisite gate bias. This means that under negative gate bias, the Fermi level drops below the Dirac point, giving rise to a large population of holes in the valence band, while under positive gate bias, the Fermi level was raised above the Dirac point, introducing a large population of electrons in the conduction band. The mobility of charge carriers can exceed 15,000 cm^2 V^{-1} s^{-1} even when their concentration is as high as 10^{13} cm^{-2} at ambient conditions [7,27] and the observed mobility is weakly dependent on temperature.

Due to the atomically thin structure of graphene, electron transport on graphene is strictly confined in the 2-D plane, which creates a so-called 2-D electron gas and induces some unique phenomena, such as a quantum Hall effect even observed at room temperature. Klein tunneling is another feature of graphene, which is derived from the chiral nature of electron transport in graphene. This means that the electrons in graphene have a 100% transmission rate through a potential barrier of any size. This may bring about some difficulties in manipulating graphene-based devices, as square potential barriers usually applied to form the device channel may be ineffective for graphene.

The intrinsic electronic properties of graphene have been extensively studied and will cause not much excitement on their own. However, the interference of graphene with intrinsic or extrinsic scatters, such as doping atoms or substrates, will cause novel transport phenomena, which is a useful way to tune the electron properties of graphene to meet different application purposes.

1.3.2.2 Optical Properties

A constant transparency of about 97.7% of a single layer has been experimentally observed in the visible range, which means that atomic monolayer graphene has an unexpectedly high opacity, absorbing $\pi\alpha \approx 2.3\%$ of white light, where α is the fine-structure constant [28]. This is a consequence of the unusual low-energy electronic structure of monolayer graphene that features electron and hole conical bands meeting each other at the Dirac point. It is noteworthy that the transmittance decreases linearly with the number of layers for multilayer graphene [29], so that the thickness of multilayer graphene with less than 10 layers can be determined using white light illumination on samples supported on a given substrate. However, if the thickness of graphene increases to more than 10 layers, the optical property eventually becomes more like that of graphite. Based on the Slonczewski–Weiss–McClure band model of graphite, the interatomic distance, hopping value, and frequency cancel when optical conductance is calculated using Fresnel equations in the thin-film limit. That is to say, thick graphene platelets with tens of layers will never share the same optical property with the monolayer one.

Considering the practical application, a graphene-based Bragg grating, which is a one-dimensional photonic crystal, has been fabricated and demonstrated its capability for excitation of surface electromagnetic waves in the periodic structure by using a 633-nm helium–neon (He–Ne) laser as the light source.

1.3.2.3 Thermal Properties

Single-layer graphene has the highest intrinsic thermal conductivity that has ever been found in any material. As high as 6000 W m^{-1} K^{-1} [30], the intrinsic thermal conductivity of graphene is even higher than its allotrope-carbon nanotube, which is 3500 W m^{-1} K^{-1} [31,32], and certainly higher than those metals with good thermal performance, such as gold, silver, copper, and aluminum. This excellent thermal performance is born from the unique electronic and topographic features of graphene. However, according to [33], placing graphene on substrates results in serious degradation of thermal conductivity to 600 W mK^{-1}, lower than values obtained for suspended graphene. Since the carrier density of nondoped graphene is relatively low, the electronic contribution to thermal conductivity is negligible according to the Wiedemann–Franz law [34]. The thermal conductivity of graphene is thus dominated by phonon transport, namely diffusive conduction at high temperature and ballistic conduction at sufficiently low temperature. Now we can say that the degradation of thermal conductivity is owing to the phonons leaking across the graphene-support interface and strong interface scattering of flexural modes. In any case, due to the extraordinary thermal conductivity of graphene, graphene functionalized nanocomposites have been studied for various advanced applications including thermal management of advanced electronic chips, thermal pastes, and smart polymers.

1.3.2.4 Mechanical Properties

Graphene is mechanically strong while remaining very flexible, which is attributed to the high strength of the carbon–carbon bond. The elastic properties and intrinsic breaking strength of free-standing monolayer graphene were measured by nanoindentation using an atomic-force microscopy [35]. The experimental data showed that the mechanical properties of graphene measured in experiments have exceeded those obtained in any other material, with some reaching theoretically predicted limits that were investigated by numerical simulations such as molecular dynamics [36]. For example, a single-layer, defect-free graphene was measured with a Young's modulus of 1 TPa and intrinsic fracture strength of 130 GPa. These high values make graphene very strong and rigid and these incredible mechanical performances are benefited from the hexagonal structure of graphene. Interestingly, despite sharing the same honeycomb lattice, bulk graphite is not particularly strong because it shears very easily between layers.

As a derivative of graphene, GO in a paper form was prepared and studied as a ramification of graphene. The average elastic modulus and the highest fracture strength obtained were about 32 GPa and 120 MPa, respectively [37]. The decrease of Young's modulus may be caused by the defects introduced during the chemical reaction. After reduction with hydrazine or annealing, a paper composed of stacked and overlapped graphene platelets was obtained and the stiffness and tensile strength were reinforced [38]. Chemically modified graphene obtained by reducing GO with hydrogen plasma exhibited a mean elastic modulus of 0.25 TPa with a standard deviation of 0.15 TPa.

Just as the Nobel announcement said, a 1-square-meter graphene hammock would support a 4-kg cat, but would weigh only as much as one of the cat's whiskers at 0.77 mg, which is about 0.001% of the weight of 1 m^2 of paper. These outstanding intrinsic mechanical properties make graphene suitable for diverse applications such as pressure sensors, resonators, and engineering components subjected to large mechanical stresses even under extreme conditions.

1.4 APPLICATION OF GRAPHENE

1.4.1 ELECTRONICS

The ultrahigh intrinsic carrier mobility of graphene gives a promising potential for the fast operating speed of graphene-based transistors. In addition, the peak intrinsic average carrier velocity of graphene was theoretically calculated to be four times higher than Si. Therefore, graphene is suitable to be used in high-frequency devices, which is of great importance in the application of communication technologies such as wireless transmission and signal processing. The research group from IBM demonstrated the first experimental study on high-frequency top-gated devices made of graphene transistors. A high cutoff frequency of 26 GHz was obtained with a channel length of 150 nm [39]. Later, the IBM research group further demonstrated that the cutoff frequency could be increased to 100 GHz with a gate length of 240 nm when using graphene epitaxially grown on SiC [40]. Recent research has moved the frequency forward to 280 GHz in 40-nm length channels. Chemical vapor deposition grown graphene has also been fabricated into high-frequency devices. Graphene-based radio-frequency transistors with a gate length of 40 nm could reach a cutoff frequency of 155 GHz [41]. Although exceptionally high switching speed has been realized in graphene-based transistors, the lack of inherent bandgap limits the application of graphene in logic circuits. As a result, there have been numerous efforts to generate and tune the bandgap in graphene to meet different application demands. Methods such as fabrication of nanostructures of graphene using bilayer or trilayer graphene or generating strains in graphene have been developed, and successful formation and control of the bandgap has been achieved, which is of practical importance for graphene in various devices.

Spintronics, which is an approach to logic-based devices using particle spin for signal processing and propagation, is another potential and attractive application area for graphene. Graphene possesses long spin-relaxation time and low spin-orbit coupling, which is important for the use in spin valves and other components in spintronic circuits. It has been demonstrated that GNRs are of special importance in the application of spintronics, as different edge states (zigzag or armchair) can induce different and unique spin states.

1.4.2 PHOTONIC

A photodetector, which is the key component to optoelectronic devices, is one of the potential photonic applications of graphene. In contrast with traditional semiconductors, pristine graphene has no bandgap, which means that is has broader wavelength absorption for broadband photodetection. Meanwhile, ultrahigh carrier mobility makes graphene suitable for ultrafast photodetection. It has been shown by high-frequency measurement that there is no performance degradation in graphene-based photodetectors up to 40 Hz, and the intrinsic bandwidth of graphene can be as high as 500 Hz, comparable with the fastest III–V semiconductor-based ones [42]. However, graphene-based photodetectors usually have low photocurrent responsibility because of the low optical absorption of single-layer graphene. Coupling with plasmonic structures is one effective way to enhance the photocurrent. It has been reported that the photocurrent of gold-decorated graphene was more than one order of magnitude higher than that of pristine graphene [43].

Another potential application of graphene is an optical modulator. Due to its high speed and broad bandwidth, it can be widely used in on-chip optical interconnects to substitute currently used electric interconnects. Compared with traditional semiconductors, graphene shows advantages in high carrier mobility, large optical bandwidth, and strong modulation ability, which makes graphene suitable for high-performance optical modulators. A graphene-based optical modulator has recently been demonstrated [44]. It has an operation frequency as high as 1 GHz. Moreover, its optical bandwidth can range from 1.35 to 1.6 μm, while its footprint is only 25 μm², which was much better than the current modulators.

Graphene can also be used in polarizers, which can convert undefined lights to polarized lights. The polarizing properties can be simply tuned by changing the Fermi level of graphene. In addition, graphene polarizers can cover a much wider bandwidth than current polarizer materials. By integrating graphene within an optical fiber, the device exhibits a strong s-polarization extinction ratio of 27 dB, which is comparable to current metal-based polarizers, but the graphene-based polarizer can work in a much broader wavelength range—from 800 to 1650 nm—indicating the appealing potential of graphene in the application of high-performance polarizers [45].

1.4.3 ENERGY CONVERSION AND STORAGE

In recent years, clean and renewable energy technologies have drawn considerable attention because the combustion of fossil fuels is causing the warming of Earth's climate and having deleterious effects on the environment. Graphene has exhibited appealing potential in the application of energy conversion and storage applications due to its huge theoretical surface area of 2630 m²/g, outstanding thermal and electrical conductivity, and high chemical stability.

Graphene is currently being widely studied in the application of the lithium-ion (Li-ion) battery [46], which is a renewable and clean power source for portable devices, electrical/hybrid vehicles, and miscellaneous power devices. As many potential electrode materials (graphite or transition metal oxide) in a Li-ion battery suffer from slow Li-ion diffusion, poor electron transport, and increased resistance at high charge-discharge rates, graphene-based electrode materials in Li-ion batteries have been proposed as one of the promising alternatives due to graphene's high electrical conductivity and typical 2-D structure [47,48]. Graphene itself can be used as a high-capacity anode material [49]. More importantly, graphene acting as a conducting agent with unique 2-D nanostructure is helpful for fabrication of novel structures with various active materials and enormously improving their electrochemical performance. So far, numerous graphene-based composite cathode and anode materials have been successfully prepared, and their outstanding charge/discharge performance show the broad future of graphene in the application in Li-ion batteries [50,51].

Supercapacitors, also known as electrochemical capacitors or ultracapacitors, exhibit higher power rating, long cycle life, low maintenance cost, and broad thermal operating range. There are two main classifications of supercapacitors based on charge storage mechanism: electrical double-layer capacitance (EDLC) and pseudocapacitance. EDLC is based on electrostatic storage determined by high surface area and a nanoscopic charge separation at the electrode/electrolyte interface, whereas pseudocapacitance is caused by faradaic reactions. Graphene is a potential material for this application, offering good resistance to the oxidative process and defined and accessible pore structure in addition to high temperature stability and intrinsic electrical conductivity [52,53]. In view of the current progress, the characteristics of graphene supercapacitors are very encouraging. However, there are still many challenges before the commercial use of such graphene-based systems come to fruition. Its specific capacitance is still low, which may probably be improved by choosing a better high-voltage electrolyte or increasing effective surface area.

At present, a large body of research has been devoted to graphene solar cell with a view to utilize it for practical applications. Graphene can be used as a transparent electrode material in dye-sensitized or quantum dots solar cell, and the position of Fermi level in graphene can be varied by doping, indicating the electrodes could be used both as hole- and electron-conducting media. In addition, graphene is a promising active material for solar cell. However such devices require plasmonic enhancement or complex interferometry structure [54] because of the low optical absorption of graphene. As well, many groups are investigating the use of graphene as a support material for platinum catalysts in fuel cell applications [10]. The particle size of platinum can be decreased to a nanometer owing to the strong interaction between graphene and platinum, which could increase the catalytic activity. With the cost of graphene prepared by various methods going down, we expect wide use of graphene in solar and fuel cells.

When research on the application of graphene in energy conversion and storage continuously moves forward, the advantages of graphene in terms of both performance and cost have been demonstrated. Therefore, currently used carbon materials in energy-related applications (activated carbon, carbon black, and graphite) may be replaced by graphene in the near future. It can be expected that graphene will achieve great progress in real devices that will be a part of our daily lives.

1.4.4 COMPOSITE MATERIALS AND COATINGS

Graphene has drawn considerable attention for applications in composite materials and coatings because of its outstanding electronic, mechanical, chemical, and barrier properties, as well as its high aspect ratio. Graphene-based fillers have been used in polymer composites to achieve large property enhancement such as elastic modulus, tensile strength, thermal stability, and electrical conductivity [55]. As an additive to a composite matrix, graphene can improve the operating temperature level and reduce moisture uptake, give lightning-strike protection and induce antistatic behavior. Furthermore, these improvements are often observed at low loading owing to its unique 2-D structures, which gives rise to ultralow percolation threshold compared with conventional carbon-based additives. Polymer composites with graphene-based filler include polymethyl methacrylate (PMMA), polyvinyl alcohol, polypropylene (PP), epoxy, and silicone foam and polystyrene [56]. Graphene polymer composites also can be used in making transparent conductive electrode for electrochromic devices and dye-sensitized solar cell.

Graphene is highly inert and can act as a corrosion barrier against oxygen and water diffusion. Graphene can form a protective layer on the surface of almost any metal under certain conditions. In addition, graphene-based coating can be utilized for antistatic, gas barrier, and electromagnetic-interference shielding applications. However, pristine graphene may not have the same adhesion properties to the matrix as that of carbon fiber, so it requires more chemical modification of graphene, which will be developed to control the optical opacity and conductivity of graphene.

We can be optimistic that graphene-based composites and coatings will appear on the market in the near future because some companies have already established such programs.

1.4.5 SENSORS

One of the most important and promising applications of graphene is in sensors, including electrochemical and gas sensors [57]. Charge transfer between graphene and the adsorbed molecules is considered to be responsible for chemical response. The location of adsorption experiences a charge transfer with graphene as an acceptor or donor when molecules adsorb to the surface of graphene, resulting in a change to the Fermi level, carrier density, and electrical resistance of graphene [35]. Some remarkable properties of graphene aid to increase its sensitivity up to molecular- or single-atom-level detection. First, graphene, whose theoretical specific surface area is as large as ~2630 m^2/g, is a typical 2-D material, and almost all carbon atoms can be exposed to the analyte of interest, making graphene extremely sensitive to the environment. Second, graphene has low Johnson noise and high conductivity, so a notable variation of electrical conductivity can be caused by a little change in carrier concentration. Third, graphene is the only crystal with very few defects that can be stretched by 20%. Furthermore, the unique band structure of graphene makes it an ideal material to develop a universal resistance standard based on the quantum Hall effect.

Currently, many graphene-based electrochemical sensors have been reported to detect dopamine, glucose, hydrogen peroxide (H_2O_2), ascorbic acid, antigen, and deoxyribonucleic acid (DNA). Graphene-based materials can not only be developed for environmental electrochemical analysis of some toxic and explosive substances, but can also be used as enhancing material for toxic arsenic removal. In addition, the gas-sensor properties of graphene have been investigated by many groups [58]. Graphene-based gas sensors can be widely used to detect various gases such as nitrogen dioxide (NO_2), ammonia (NH_3), and carbon monoxide (CO). These different kinds of sensors fabricated by graphene have two advantages. One advantage is their multifunctionality—a single device can be utilized in various measurements (e.g., pressure, gas environment, and magnetic field), which offers unique opportunities. The other advantage is that its extremely high sensitivity makes any measurement more precise, which is especially important within the fields of advanced technology. With the development of increasingly interactive consumer electronic devices and the advancement of the technology, such graphene sensors will certainly find their way into a large body of products in the near future.

1.4.6 THERMAL MANAGEMENT

In current electronic and photonic technologies, low operation temperature and efficient heat dissipation are important for life and operation speed of devices. Graphene is proposed as a material for heat removal due to its extraordinarily high intrinsic thermal conductivity. The room temperature thermal conductivity of suspended graphene was determined to range from 3080–5300 W m^{-1} K^{-1}. Moreover, graphene has very high electron mobility and low resistivity. The unique thermal and electrical properties of graphene make it a potentially promising material for thermal management [59,60].

Monolayer graphene synthesized by CVD can be used for the application of heat spreader in electronic packing. The hot-spot temperature is decreased by ~5°C at a heat flux of up to 800 W cm^{-2}. This result can be further improved by optimizing the synthesis parameters and transfer process. In silicon-on-insulator (SOI) metal-oxide-semiconductor field-effect transistor (MOSFET) structure, the buried oxide insulates the active channel from the substrate both electrically and thermally, which results in temperature rise and leads to performance degradation and early thermal breakdown. In order to resolve this problem, graphene can be adopted as the material for lateral heat spreaders, which can lead to great reduction in temperature of the hot spots.

Graphene is also a novel promising candidate for gap-filling thermal interfacial material in electronic and photonic devices. Graphene stacked in a 3-D structure in the electrodes of an electrical device is beneficial for rapid ion dissipation and making full use of a specific surface area. A 3-D vertically aligned graphene architecture between a silicon surface and an exploration has great potential in thermal management of electronic and photonic applications.

1.4.7 BIOAPPLICATIONS

Graphene is also potentially promising in the bioapplications field. For example, graphene can be used for drug delivery due to the large specific surface area, chemical purity, and the possibility of easy functionalization. Also, the combination of conductivity, strength, and ultimate thinness make graphene an ideal support for imaging biomolecules, and its remarkable mechanical property provides opportunity for regenerative medicine and tissue-engineering applications. In addition, based on graphene's capability in fluorescence quenching, chemically functionalized graphene can be fabricated into sensitive, rapid, and cost-effective measurement devices, which could be used for the analysis of a large number of biological molecules such as cholesterol, DNA, glucose, and hemoglobin [61].

The growing demand for the analysis of genomes of many species and cancers, along with the ultimate goal of deciphering individual human genomes, has led to the development of non-Sanger reaction-based technologies toward rapid and inexpensive DNA sequencing. Graphene was considered to be a great potential material for resolving this problem. It is envisioned that porous graphene can be prepared via a certain method and the size of nanopores matched well with the width of DNA, allowing a DNA molecule through the pores. The influences of the four DNA bases (A, C, G, T) on the conductivity of graphene are different, therefore it is easy to determine which base is passing through the pores by measuring the voltage deference during the passage [62]. Some scientists even find that sheets of GO are highly effective at killing bacteria. The sheets with an aerosol rich in *Escherichia coli* (*E. coli*) were sprayed and then placed in an incubator and examined under a microscope. It is interesting that *E. coli* cells were destroyed when they interacted with the GO sheets. In some cases, a toxic graphene derivative could potentially be therapeutic as an antibiotic or anticancer treatment.

However, it is necessary to understand graphene's biodistribution, biocompatibility, and acute and chronic toxicity before large-scale application of graphene is introduced into the biomedical area because outcomes are likely to vary with size, morphology, and chemical structure.

1.5 CHALLENGES AND RESEARCH DIRECTIONS

After 10 years of rapid development of graphene-based materials and applications since the discovery of graphene and its outstanding physical properties in 2004, fruitful achievements and progress have occurred globally within such a short period of time, which has rarely been seen with any other new material. As a novel carbon material with unique structure and multifunction, graphene has been anticipated to lead the next revolution of materials. Although the growing potential of the applications of graphene in a variety of areas has been demonstrated, there are still some big challenges that need to be conquered before this new material can realistically become a part of our daily lives.

Although diverse preparation methods for graphene have been proposed, the mass production of high-quality graphene is still a major challenge. Most of the existing methods only focus on the quality of the graphene product but ignore other important factors for commercial production, such as cost, processing difficulty, and environmental impact. Therefore, only a few among various reported preparation methods are potentially applicable for industrialization. We are encouraged that some pioneering scientists, engineers, investors, and companies have already stepped forward, trying to realize the commercialization of graphene, but this process might take a long time because abundant technical problems still exist. For example, few-layer or multilayer graphene platelets can now be purchased from several companies around the world, but the structures and basic properties of these products differs significantly from one company to another. Even in one product, the distribution of the thickness and the lateral size of graphene platelets is still not uniform enough. Moreover, their cost at the current level is still relatively high compared to the existing competitors. As a result, the spread of these products in commercial applications is not as uniform as possible. Much technical improvement needs to occur to raise the quality of the products and enhance their competitiveness in the future. The production of large-area and transparent graphene film by chemical vapor deposition is even less mature for commercial applications.

Compared to the production of graphene, even bigger challenges exist in its applications. Though the application potential of graphene in diverse areas has been fundamentally verified, it is still far from practical because the unique structure and properties of graphene create some tricky technical problems that have never been encountered in traditional materials. We will look at graphene platelets, for example. Two-dimensional sheetlike morphology is one of the most characteristic features of graphene compared to traditional 3-D materials, which, however, induces the strong tendency of restacking of graphene sheets even after they are completely exfoliated. As a result, dispersing graphene in different matrices is much more difficult than with other nanomaterials. Some new dispersing techniques and matching dispersant specifically for graphene need to be developed. In addition, the properties and hence the application effect of graphene platelets are strongly affected by the structural parameters of individual sheets, including layer numbers, size, defects, and functional groups. Due to the lack of precise control of those parameters using current preparation methods, their influence in most applications is not well understood, which limits the improvement of the application effect of graphene. The application challenge of high-quality graphene film for electronic devices, such as next generation transistors, is even greater, as both the controllable synthesis of qualified samples and fundamental principles are far from maturity to support the change of the current electronic industry to graphene. It might take more than two or three decades to finally replace current Si-based electronics with graphene.

Despite of these big challenges, the bright future of graphene can be optimistically anticipated based on the continuous efforts from scientists and engineers all over the world. Future research on graphene can still be divided into two major directions: the preparation of graphene and the application of graphene.

Chemical exfoliation of graphite to produce graphene platelets and chemical vapor deposition to produce graphene thin film should be the two major areas in which to focus in the future. The main focus on both methods should include improvement of current preparation process and development of new techniques to further improve the quality and consistency of graphene products, raise production efficiency, and lower production cost, in order to meet the requirements for industrialization. Specifically for chemical exfoliation of graphite, the layer number, the lateral size, and structural defects should be more precisely controlled. In addition, a series of graphene platelets with different structural parameters should be designed and their specific preparation procedures should be established for different application purposes. Regarding graphene film derived from chemical vapor deposition, it is urgent to further reduce the surface resistance while still maintaining high transparency in order to raise the technical competitiveness of graphene film in relation to traditional transparent conductive films. On the other hand, large-scale and low-cost preparation techniques, such as roll-to-roll techniques, need to be intensively developed to satisfy manufacturing at the industrial scale.

The application of graphene covers a very wide range within many different areas. According to the characteristics of different application areas and their technical thresholds, they can be categorized into different stages. The most recent application breakthrough of graphene that we expect to be realized in the next few years is as follows: electrochemical energy storage (including Li-ion batteries and supercapacitors), composite materials (including graphene-based composite plastics and rubbers), coatings and inks (including conductive inks, anticorrosive coatings, antistatic coatings, and thermal conductive coatings), and thermal management (including heat dissipation films and thermal conductive pads). Touch screens based on graphene thin films is one of the major breakthroughs in the medium stage that we expect to be commercialized in 5 to 10 years. The application of graphene in sensors and some bio-related aspects may also be realized at this stage. The application of graphene in electronics and optoelectronics, such as graphene-based transistors, is the most remote but most revolutionary aspect, but is one that may need a few decades before it could potentially change the entire electronics industry.

REFERENCES

1. Kroto, H. W., Heath, J. R., O'Brien, S. C. et al. 1985. C_{60}: Buckminsterfullerene. *Nature* 318:162–163.
2. Iijima, S. 1991. Helical microtubules of graphitic carbon. *Nature* 354:56–58.
3. Zhang, Q., Huang, J. Q., Zhao, M. Q. et al. 2011. Carbon nanotube mass production: Principles and processes. *Chem. Sus. Chem.* 4:864–889.
4. Endo, M., Strano, M. S. and Ajayan, P. M. 2008. Potential applications of carbon nanotubes. *Top. Appl. Phys.* 111:13–61.
5. Schnorr, J. M. and Swager, T. M. 2010. Emerging applications of carbon nanotubes. *Chem. Mater.* 23:646–657.
6. Wang, Y., Wei, F., Luo, G. et al. 2002. The large-scale production of carbon nanotubes in a nano-agglomerate fluidized-bed reactor. *Chem. Phys. Lett.* 364:568–572.
7. Novoselov, K. S., Geim, A. K., Morozov, S. et al. 2004. Electric field effect in atomically thin carbon films. *Science* 306:666–669.
8. Terrones, M., Botello-Méndez, A. R., Campos-Delgado, J. et al. 2010. Graphene and graphite nanoribbons: Morphology, properties, synthesis, defects and applications. *Nano Today* 5:351–372.
9. Landau, L. D. 1937. Zur Theorie der phasenumwandlungen II. *Phys. Z. Sowjetunion* 11:26–35.
10. Dresselhaus, M. S. and Dresselhaus, G. 2002. Intercalation compounds of graphite. *Adv. Phys.* 51:1–186.
11. Boehm, H. P., Clauss, A., Fischer, G. et al. 1962. Surface properties of extremely thin graphite lamellae. *Proceedings of the Fifth Conference on Carbon* 73–80.
12. Land, T. A., Michely, T., Behm, R. J. et al. 1992. STM investigation of single layer graphite structures produced on Pt(111) by hydrocarbon decomposition. *Surf. Sci.* 264:261–270.
13. Nagashima, A., Nuka, K., Itoh, H. et al. 1993. Electronic states of monolayer graphite formed on TiC(111) surface. *Surf. Sci.* 291:93–98.
14. The Nobel Prize in Physics. 2010. Nobelprize.org. Nobel Media AB 2014. Web. 23 Sep 2014. Available at http://www.nobelprize.org/nobel_prizes/physics/laureates/2010/.
15. Mermin, N. D. 1968. Crystalline order in two dimensions. *Phys. Rev.* 176:250.
16. Mermin, N. D. and Wagner, H. 1966. Absence of ferromagnetism or antiferromagnetism in one- or two-dimensional isotropic Heisenberg models. *Phys. Rev. Lett.* 17:1133–1136.
17. Bianco, A., Cheng, H.-M., Enoki, T. et al. 2013. All in the graphene family—A recommended nomenclature for two-dimensional carbon materials. *Carbon* 65:1–6.
18. Meyer, J. C., Geim, A. K., Katsnelson, M. et al. 2007. The structure of suspended graphene sheets. *Nature* 446:60–63.
19. Yang, L., Park, C. H., Son, Y. W. et al. 2007. Quasiparticle energies and band gaps in graphene nanoribbons. *Phys. Rev. Lett.* 99:186801.
20. Son, Y.-W., Cohen, M. L. and Louie, S. G. 2006. Half-metallic graphene nanoribbons. *Nature* 444:347–349.
21. Girit, Ç. Ö., Meyer, J. C., Erni, R. et al. 2009. Graphene at the edge: Stability and dynamics. *Science* 323:1705–1708.

22. Ma, T., Ren, W., Zhang, X. et al. 2013. Edge-controlled growth and kinetics of single-crystal graphene domains by chemical vapor deposition. *Proc. Natl. Acad. Sci. U.S.A.* 110:20386–20391.
23. Malard, L., Pimenta, M., Dresselhaus, G. et al. 2009. Raman spectroscopy in graphene. *Phys. Rep.* 473:51–87.
24. Oostinga, J. B., Heersche, H. B., Liu, X. et al. 2008. Gate-induced insulating state in bilayer graphene devices. *Nat. Mater.* 7:151–157.
25. Bae, S., Kim, H., Lee, Y. et al. 2010. Roll-to-roll production of 30-inch graphene films for transparent electrodes. *Nat. Nanotech.* 5:574–578.
26. Bolotin, K. I., Sikes, K. J., Jiang, Z. et al. 2008. Ultrahigh electron mobility in suspended graphene. *Solid State Commun.* 146:351–355.
27. Novoselov, K. S., Jiang, D., Schedin, F. et al. 2005. Two-dimensional atomic crystals. *Proc. Natl. Acad. Sci. U.S.A.* 102:10451–10453.
28. Nair, R. R., Blake, P., Grigorenko, A. N. et al. 2008. Fine structure constant defines visual transparency of graphene. *Science* 6:1308.
29. Kuzmenko, A. B., van Heumen, E., Carbone, F. et al. 2008. Universal optical conductance of graphite. *Phys. Rev. Lett.* 100:117401.
30. Balandin, A. A., Ghosh, S., Bao, W. et al. 2008. Superior thermal conductivity of single-layer graphene. *Nano Lett.* 8:902–907.
31. Pop, E., Mann, D., Wang, Q. et al. 2006. Thermal conductance of an individual single-wall carbon nanotube above room temperature. *Nano Lett.* 6:96–100.
32. Berber, S., Kwon, Y.-K. and Tománek, D. 2000. Unusually high thermal conductivity of carbon nanotubes. *Phys. Rev. Lett.* 84:4613–4616.
33. Yu, C., Shi, L., Yao, Z. et al. 2005. Thermal conductance and thermopower of an individual single-wall carbon nanotube. *Nano Lett.* 5:1842–1846.
34. Seol, J. H., Jo, I., Moore, A. L. et al. 2010. Two-dimensional phonon transport in supported graphene. *Science* 328:213–216.
35. Lee, C., Wei, X., Kysar, J. W. et al. 2008. Measurement of the elastic properties and intrinsic strength of monolayer graphene. *Science* 321:385–388.
36. Van Lier, G., Van Alsenoy, C., Van Doren, V. et al. 2000. Ab initio study of the elastic properties of single-walled carbon nanotubes and graphene. *Chem. Phys. Lett.* 326:181–185.
37. Chen, H., Müller, M. B., Gilmore, K. J. et al. 2008. Mechanically strong, electrically conductive, and biocompatible graphene paper. *Adv. Mater.* 20:3557–3561.
38. Dikin, D. A., Stankovich, S., Zimney, E. J. et al. 2007. Preparation and characterization of graphene oxide paper. *Nature* 448:457–460.
39. Lin, Y. M., Chiu, H. Y., Jenkins, K. A. et al. 2010. Dual-gate graphene FETs with f(T) of 50 GHz. *IEEE Elec. Device Lett.* 31:68–70.
40. Lin, Y. M., Dimitrakopoulos, C., Jenkins, K. A. et al. 2010. 100-GHz transistors from wafer-scale epitaxial graphene. *Science* 327:662–662.
41. Wu, Y. Q., Lin, Y. M., Bol, A. A. et al. 2011. High-frequency, scaled graphene transistors on diamond-like carbon. *Nature* 472:74–78.
42. Xia, F. N., Mueller, T., Lin, Y. M. et al. 2009. Ultrafast graphene photodetector. *Nat. Nanotech.* 4:839–843.
43. Echtermeyer, T. J., Britnell, L., Jasnos, P. K. et al. 2011. Strong plasmonic enhancement of photovoltage in graphene. *Nat. Commun.* 2:5.
44. Liu, M., Yin, X. B., Ulin-Avila, E. et al. 2011. A graphene-based broadband optical modulator. *Nature* 474:64–67.
45. Bao, Q. L., Zhang, H., Wang, B. et al. 2011. Broadband graphene polarizer. *Nat. Photon.* 5:411–415.
46. Pumera, M. 2011. Graphene-based nanomaterials for energy storage. *Energy Environ. Sci.* 4:668.

47. Ji, L., Tan, Z., Kuykendall, T. R. et al. 2011. Fe_3O_4 nanoparticle-integrated graphene sheets for high-performance half and full lithium ion cells. *Phys. Chem. Chem. Phys.* 13:7170–7177.
48. Bhardwaj, T., Antic, A., Pavan, B. et al. 2010. Enhanced electrochemical lithium storage by graphene nanoribbons. *J. Am. Chem. Soc.* 132:12556–12558.
49. Zhao, X., Hayner, C. M., Kung, M. C. et al. 2011. Flexible holey graphene paper electrodes with enhanced rate capability for energy storage applications. *ACS Nano* 5:8739–8749.
50. Zhou, X., Wang, F., Zhu, Y. et al. 2011. Graphene modified $LiFePO_4$ cathode materials for high power lithium ion batteries. *J. Mater. Chem.* 21:3353–3358.
51. Wang, X., Zhou, X., Yao, K. et al. 2011. A SnO_2/graphene composite as a high stability electrode for lithium ion batteries. *Carbon* 49:133–139.
52. Gao, W., Singh, N., Song, L. et al. 2011. Direct laser writing of micro-supercapacitors on hydrated graphite oxide films. *Nat. Nanotech.* 6:496–500.
53. Chen, Y., Zhang, X., Zhang, D. et al. 2011. High performance supercapacitors based on reduced graphene oxide in aqueous and ionic liquid electrolytes. *Carbon* 49:573–580.
54. Echtermeyer, T. J., Britnell, L., Jasnos, P. K. et al. 2011. Strong plasmonic enhancement of photovoltage in graphene. *Nat. Commun.* 2:458.
55. Jang, B. Z. and Zhamu, A. 2008. Processing of nanographene platelets (NGPs) and NGP nanocomposites: A review. *J. Mater. Sci.* 43:5092–5101.
56. Zhu, Y., Murali, S., Cai, W. et al. 2010. Graphene and graphene oxide: Synthesis, properties, and applications. *Adv. Mater.* 22:3906–3924.
57. Choi, W., Lahiri, I., Seelaboyina, R. et al. 2010. Synthesis of graphene and its applications: A review. *Crit. Rev. Solid State* 35:52–71.
58. Huang, X., Yin, Z., Wu, S. et al. 2011. Graphene-based materials: Synthesis, characterization, properties, and applications. *Small* 7:1876–1902.
59. Liang, Q., Yao, X., Wang, W. et al. 2011. A three-dimensional vertically aligned functionalized multilayer graphene architecture: An approach for graphene-based thermal interfacial materials. *ACS Nano* 5:2392–2401.
60. Yan, Z., Liu, G., Khan, J. M. et al. 2012. Graphene quilts for thermal management of high-power GaN transistors. *Nat. Commun.* 3:827.
61. Kuila, T., Bose, S., Khanra, P. et al. 2011. Recent advances in graphene-based biosensors. *Biosen. Bioelec.* 26:4637–4648.
62. Schneider, G. G. F., Kowalczyk, S. W., Calado, V. E. et al. 2010. DNA translocation through graphene nanopores. *Nano Lett.* 10:3163–3167.

2 Synthesis of Graphene

*Wei Wang, Hailiang Cao, Xufeng Zhou,
and Zhaoping Liu*

CONTENTS

2.1   Introduction 21
2.2   Mechanical Exfoliation..23
2.3   Epitaxial Growth on SiC...24
2.4   CVD..25
    2.4.1   Mechanism of Formation of Graphene on Different Substrates.........27
    2.4.2   Toward the Synthesis of High Quality of Graphene Film on Cu
        Substrate 30
    2.4.3   Manipulating the Carbon Precursor for Controlled
        Morphologies of Graphene 31
    2.4.4   Transfer Technique for CVD Graphene Film.....................................33
    2.4.5   Production of Graphene Film in a Roll-to-Roll Way.........................35
2.5   Exfoliation and Reduction of Graphite Oxide...37
    2.5.1   Reduction of GO by Hydrazine 38
    2.5.2   Green Chemical Reduction Methods...40
    2.5.3   Thermal Exfoliation and Reduction of Graphite Oxide 44
    2.5.4   Electrochemical Reduction...47
2.6   Liquid-Phase Exfoliation 49
    2.6.1   Liquid-Phase Exfoliation of Graphite..50
    2.6.2   Liquid-Phase Exfoliation of Graphite Intercalation Compounds.......52
    2.6.3   Solvothermal Exfoliation 53
2.7   Other Methods 54
    2.7.1   Electrochemical Exfoliation 54
    2.7.2   Organic Synthesis...56
    2.7.3   Unzipping of CNTs...57
    2.7.4   Arc Discharge 59
    2.7.5   Solvothermal Method 59
    2.7.6   Igniting Magnesium in Dry Ice 59
References..59

2.1   INTRODUCTION

The attractive application potentials of graphene demand highly efficient, low-cost, and large-scale preparation of this novel carbon material. Therefore, controllable synthesis of graphene has become one of the major tasks in graphene-related research. Since the first successful preparation of single-layer graphene, using a simple mechanical exfoliation method by Geim and his coworkers in 2004, a series of different methods have been developed in the past decade. According to the raw materials and reaction process, the preparation of graphene can be basically divided into two pathways, the top-down process and the bottom-up process. The former mainly employs graphite as the raw material and produces graphene exfoliation of graphite, whereas the latter introduces carbonaceous molecules as the precursor to synthesize graphene through a chemical process.

The mechanical exfoliation of graphite to produce graphene in the pioneering work done by Geim and his research team is a typical top-down process. Although high-quality monolayer graphene with perfect atomic structure can be obtained using this method, the ultralow yield limits its application for large-scale production of graphene. As a result, alternative pathways to achieve higher throughput have been successively developed, which usually adopt chemical-based and liquid phase exfoliation processes. In order to facilitate the exfoliation process of graphite, chemical methods are employed to reduce the interaction between graphene layers in graphite. Intercalation of graphite with heteroatoms or molecules can effectively decouple the π–π interactions between adjacent graphene layers and expand the interlayer distance. The decomposition of intercalated molecules or further intercalation with solvent molecules can finally expand graphite into separated thin sheets. Oxidation of graphite is another effective way to prepare graphene. The oxidized graphite, also known as graphite oxide, no longer has interlayer π–π interactions because of the destruction of conjugated carbon network in graphene layers after chemical oxidation. The oxygen functionalized groups in graphite oxide make it extremely hydrophilic and are gradually intercalated and expanded by water (H_2O) molecules in aqueous solutions. Therefore, graphite oxide can be easily exfoliated into a single-layer state, which is equal to an oxidized state of graphene (GO). Further reduction of GO results in the formation of graphene.

Chemical vapor deposition (CVD) is the most commonly used bottom-up process to synthesize graphene. The reaction between carbonaceous molecules with metal substrate at high temperatures gives rise to the formation of continuous and thin graphene layers closely attached to the substrate. The layer number and size of the graphene film can be tuned by varying the substrate and reaction conditions.

The macroscopic morphology is strongly dependent on the synthesis method. Basically, the top-down process usually produces graphene with small lateral size from hundreds of micrometers to submicrons. The CVD method, however, normally generates large, continuous, and transparent graphene thin film whose size can reach tens of inches. The microscopic structure of graphene is also significantly affected by the preparation methods. Mechanical exfoliation and CVD are capable of producing high-quality graphene with low defects, whereas chemical processes involving exfoliation, especially the GO-mediated process, often induces a certain amount of defects and functional groups. The variation of the structure due to different preparation methods then greatly influences the physical and chemical properties of the graphene products. High carrier mobility and high electrical and thermal conductivity require as little defects as possible, while defects and functional groups are often needed to enhance the dispersing ability of graphene in solvents and other matrices, as well as to further functionalization of graphene for various applications.

Consequently, the structural and physicochemical property requirements of graphene in certain application areas should always be taken into account before the selection of the synthesis method.

Commercial production of graphene is the ultimate goal of graphene preparation, and it is becoming more and more urgent due to the rapidly increasing demand for mass production of graphene for practical applications. Some pioneering companies have already stepped forward and established primary production lines of graphene in recent years. For example, graphene flakes produced by chemical exfoliation of graphite are now commercially available in companies such as Vorbeck Materials and XG Sciences in the United States and Ningbo Morsh Tech. Co. Ltd. in China. The commercialization of graphene film prepared by CVD is also being developed at Samsung, Sony, and other companies. The endeavor to the mass production of graphene will strongly support the broad application of this magic material in the near future.

2.2 MECHANICAL EXFOLIATION

Mechanical exfoliation, just as its name implies, applies mechanical force to exfoliate graphite to obtain graphene. As no chemical process is involved, it retains the perfect honeycomb carbon framework in the graphene sheets during the mechanical exfoliation process.

The first single-layer graphene sample prepared by mechanical exfoliation was achieved in 2004 by Geim's research group [1]. It was a simple process that uses common Scotch tape as the main experimental tool. Highly oriented pyrolytic carbon (HOPG) platelets were used as the raw material in their experiment. Using dry oxygen plasma etching, some 5-μm-deep thick mesas in sizes from 20 μm to 2 mm were prepared on the top of the HOPG platelets. The structured surface was then pressed against a layer of wet photoresist on a glass substrate. After baking, the authors cleaved off the HOPG sample from the substrate, which resulted in the remaining of only mesas strongly attached to the photoresist layer. They then started peeling off graphite flakes from the mesas using Scotch tape. After repeated peeling, the graphite remaining on the photoresist became thinner and thinner. Finally, ultrathin graphite flakes attached to the photoresist was released in acetone by dissolution of the photoresist. An Si wafer with a 300-nm-thick SiO_2 layer on top was dipped into the solution and some flakes were captured by the Si wafer. After ultrasound treatment of the Si wafer to remove relatively thick flakes, ultrathin flakes strongly attached to Si were finally obtained for characterization. Graphite flakes with different thickness on the Si substrate with a SiO_2 layer on top have apparently different colors under optical microscopes. Therefore, it can briefly screen thin exfoliated graphite flakes from thick ones by observation by the eye, although single-layer graphene is almost invisible in optics. Further atomic force microscopy analysis finally identified single-layer graphene with a measured thickness of ~1 nm among numerous graphite flakes. The subsequent characterization of the single-layer sample demonstrated almost magical and surprising physical properties of graphene, a novel carbon material that has now become wildly popular all over the world.

Since mechanical cleavage does not destroy the atomic structure of graphene, it is a useful method to obtain perfect graphene samples for fundamental research. However, the efficiency of mechanical exfoliation is extremely low. Though some attempts of replacing the manual exfoliation process developed by Geim and his coworkers with a machine-based process have been conducted [2], the efficiency is still not much improved, and the yield of high-quality single-layer graphene is still quite low. Therefore, more efficient preparation methods of graphene need to be developed for its large-scale application.

2.3 EPITAXIAL GROWTH ON SiC

Generally, epitaxial growth of graphene on SiC mainly involves the decomposition of SiC hexagonal α-SiC and the subsequent formation of graphene layers on the surface. Because of the epitaxially matching SiC substrate provides the carbon source itself and no metal is involved in the growing process, this method has the advantage that clean and high-quality graphene can be obtained. Actually, the formation of thin graphitic layers on SiC at high temperatures and ultrahigh vacuum (UHV) had already been observed since the 1970s, but this technique was broadly accepted as an effective tool to the production of highly crystalline graphene until Berger and coworkers reported the epitaxial growth of graphene on SiC and measured its transport properties [3,4].

In a typical procedure, the Si (or C)-terminated (0001) face of single-crystal 6H-SiC (or 4H-SiC) was first treated in prepared hydrogen (H_2) in order to produce atomically flat surfaces. It was then heated at ~1000°C in UHV in order to remove any remaining oxide layers. Afterward, the temperature was raised to 1250°C–1450°C for 1–20 min. During this process, Si atoms in SiC sublimate and the surface of SiC undergo several phase changes. Finally, a surface reconstruction is reached and graphene layers are obtained on SiC substrate (Figure 2.1). The formation mechanism of graphene using epitaxial growth on SiC has been systematically investigated. Graphene was observed to nucleate in pits in the SiC substrate that was formed during the high-temperature annealing in vacuum and in which the density of surface steps was high [5]. Desorption of Si from the surface steps mainly controls the nucleation of graphene. Based on this mechanism, the controlled preparation of GNRs by using the steplike facets on SiC substrates was successfully achieved [6].

Theoretically, the size of graphene prepared using epitaxial growth depends on the size of the SiC wafer. However, the large-scale structural quality is usually difficult to achieve due to the lack of continuity and uniformity of the grown graphene film. In order to solve this problem, Emtsev and coworkers modified the growing conditions to finally obtain wafer-size graphene layers with improved quality on SiC [7]. In contrast to the UHV conditions usually employed in the epitaxial growth process, the growth of graphene reported in this paper was carried out in an argon (Ar) atmosphere close to atmospheric pressure. The presence of Ar significantly reduced the evaporation rate of Si, therefore a high annealing temperature of 1650°C was adopted from graphene, which was much higher than the temperature used in the UHV conditions. At such a high temperature, the surface diffusion was enormously enhanced and the reconstruction of the surface was completed before graphene was

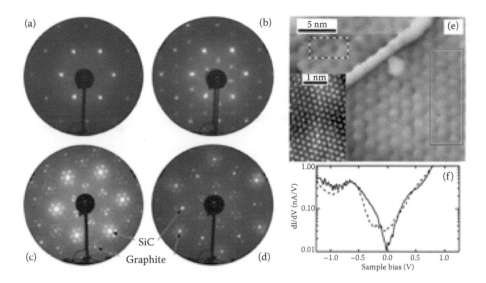

FIGURE 2.1 (a–d) Low-energy electron diffraction patterns from of epitaxially grown thin graphene layers on SiC(0001). (e) STM image of a surface region of the sample described in Figure 2.1d. Inset: Atomically resolved region. (f) dI/dV spectra (log scale) acquired from the regions marked with corresponding line types in the image at the top. (Reprinted with permission from Ref. 3, 19912–19916. Copyright 2004 American Chemical Society.)

formed. Therefore, markedly improved surface morphology was obtained and thus a high-quality and large-size graphene layer was produced.

Because of the insulating nature of SiC, graphene grown on SiC substrate can be directly used without transferring to another insulator substrate. As well, the preparation process of epitaxial growth is close to standard preparation conditions in semiconductor manufacturing. Nevertheless, there exist interactions between the first graphene and the SiC substrate that may influence the electrical characteristics of graphene.

2.4 CVD

The CVD method is a process that the chemical gases or vapors react on the substrate surface to synthesize the coatings or nanostructures, as shown in Figure 2.2 [8]. The CVD method includes the following advantages: uniform deposition, high degree of control, high quality, and low cost. As a successful example, it has been widely used in the fabrication of one of the significant carbon materials, carbon nanotubes (CNTs). The first attempt to grow graphitic layers using the CVD method was carried out in the 1960s [9,10]. However, the short of effective transferring technique and investigating strategy lead to limited understanding of as-grown graphitic layers at that time. In 2009, Hong et al. developed a simple technique to transfer and measure the graphene on the pattern nickel (Ni) films using polydimethylsiloxane (PDMS) films as intermediate adhesive layers, which came to be along with the high-speed development of CVD growth of graphene [11].

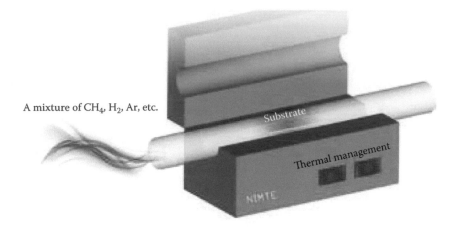

A mixture of CH_4, H_2, Ar, etc.

Substrate

Thermal management

NIMTE

FIGURE 2.2   The schematics of CVD growth of graphene.

However, the graphene precipitated on Ni films was of a few-layer structure so that the corresponding physical properties were quite different from that of monolayer graphene. Due to the ultralow dissolution of carbon, copper (Cu) foil was found to be the most favorable substrate for fabricating monolayer graphene with a self-limiting growth function, which paved an effective route for the preparation of monolayer graphene film using the CVD method and which also attracted extensive attention [12]. One of the concerns was focused on the fabrication of large-area graphene, which is a significant and important issue regarding practical applications. Up to 30-inch continuous graphene film was successfully fabricated through a roll-to-roll process by Hong et al., and the as-grown graphene could be directly used as the conducting electrode in the touch panel for the first time [13]. In a short period of 4 years, CVD growth of graphene has made breakthroughs in characterization, transfer technique, and growth with high quality and large quantity, and hence has become one of the more mature and attractive graphene manufacturing techniques with industry applications.

Compared to other methods, CVD growth of graphene exhibits the following superior features: (1) it is a large-area continuous graphene film, (2) it can be produced in roll-to-roll way, and (3) the method is simple, repeatable, and low-cost. However, it also suffers from the same quality problems that hampered its application in areas such as electronic devices. To meet the requirements for advanced applications in the near future, many efforts are being made in the following areas: (1) a high degree of control of the size of single-crystal graphene and the number of layers of graphene, (2) a mature transferring technique with low defects, low impurity, and low cost in roll-to-roll way, and (3) the manipulation of morphologies and doping. In this chapter, the current status of fabrication approaches, the fundamental mechanism, the manipulating skills, and the transferring technique for CVD growth of graphene will be introduced in detail. The challenges and possible solutions will be discussed to give researchers a better understanding in order to promote the development of graphene and its related technologies.

2.4.1 MECHANISM OF FORMATION OF GRAPHENE ON DIFFERENT SUBSTRATES

Basically, the growth of graphene film via CVD process can be divided into two catalogs depending on what kind of substrate is used: (1) the segregation or precipitation on metal with high solubility of carbon (>0.1 atomic %), such as Ni [14], and (2) the absorption and nucleation growth on the surface of metal, which has a low solubility of carbon (~0.001 atomic %), such as Cu [15]. Besides Cu and Ni, graphene films have also been demonstrated to grow on other metal substrates, including platinum (Pt), ruthenium (Ru), iridium (Ir), cobalt (Co), and palladium (Pd) [16–20], which follow similar rules for Ni and Cu. Considering the cost and the feasibility to etch the substrate for transferring graphene onto other desirable substrates, Ni and Cu foils are widely accepted.

In a typical process for the growth of graphene on Ni substrate, as shown in Figure 2.3a, hydrocarbon gas decomposes at a high temperature as the carbon source that then diffuses and dissolves into the Ni bulk [21]. Usually, the reaction time is kept short for a suitable dissolution of carbon. The substrate is then cooled, and the carbon will get out of the Ni substrate through the segregation and precipitation processes. The amount of carbon precipitated on the Ni surface strongly depends on the cooling rate. An extremely fast or a very slow cooling rate usually leads to no carbon segregation. A moderate cooling rate can generate an appropriate amount of carbon as a result of few-layer graphene. Thus, one of the biggest challenges for the growth of graphene on Ni substrate is how to control the amount of diffusing carbon and hence get the monolayer graphene. This effort is mainly focused on two routes: (1) adjusting the adsorption time or the thickness of Ni to adjust the amount of carbon dissolving into the Ni bulk, and (2) adjusting the cooling rate to adjust the amount of carbon getting out of the Ni substrate. Up to now, this is still quite difficult. Liu et al. exhibited one of the most successful examples on the control of dissociation of hydrocarbon and the dissolution and diffusion of carbon in Ni, which was achieved by manipulating the thickness of the Ni layer, the CVD carburization time, and the H_2 exposure dose [22]. The first two parameters refer to the dissolution of carbon in Ni. Interestingly, the H_2 exposure can not only etch the carbon layer but also promote the sequential segregation process. Then, large-scale graphene films with controllable layers of 1 to 10 can be obtained as demonstrated in Figure 2.3b,c, and d, and the resulting graphene films show excellent optical and electrical properties. Graphene films with 3–4 layers can deliver a sheet resistance of 100 ohm/sq with a transmission of >90%.

Considering that the high solubility of carbon in Ni leads to the formation of few-layer graphene, it is worthwhile to investigate the growth of graphene on a substrate that has a relatively low solubility of carbon. It was reported by Ruoff et al. that large-scale monolayer graphene can be fabricated on Cu substrate, which has an extremely low dissolution of carbon (0.001–0.008 weight % at ~1084°C) [12]. A completely different growth process was discovered. Typically, hydrocarbon gases decompose at a high temperature and are absorbed on the Cu surface and follow a classical nucleation-growth process. As graphene grows to a certain size, it will meet to form a complete monolayer graphene. However, hydrocarbon gas molecules can diffuse into the interface between the graphene and the metal substrate. Thus,

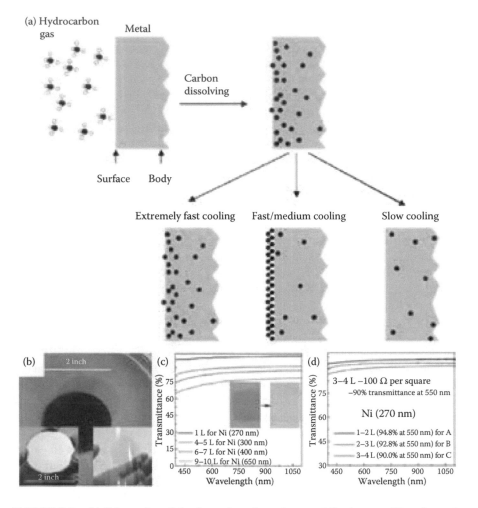

FIGURE 2.3 (a) Schematics of the formation of graphene on Ni substrate; (b) wafer-scale graphene film after etching Ni film; (c) optical transmission of graphene with a controlled layer number; (d) optical transmission of graphene fabricated on thin Ni film with a thickness of 270 nm. (Reprinted with permission from Yu, Q. K., Lian, J., Siriponglert, S. et al., *Appl. Phys. Lett.* 93:113103, Copyright 2008, American Institute of Physics; Gong, Y., Zhang, X., Liu, G. et al., Layer-controlled and wafer-scale synthesis of uniform and high-quality graphene films on a polycrystalline nickel catalyst. *Adv. Funct. Mater.* 22:3153–3159, 2012. Copyright Wiley-VCH Verlag GmbH & Co. KGaA, Weinheim. Reproduced with permission.)

the adlayer starts to grow at the same nucleation center underneath the first grown monolayer graphene film.

The proposed mechanisms have been confirmed by experimental studies through a carbon isotope labeling method. ^{12}C and ^{13}C were used as the carbon sources that were introduced in the CVD process in sequence [23]. As illustrated in Figure 2.4a and b, it is theorized that the segregation and precipitation on the Ni substrate will result in the formation of graphene film with randomly distributed carbon isotope labeling. If the formation of graphene film follows the nucleation-growth mechanism, the isotope labeling carbon will grow at the edges of the previously formed graphene grains. Due to the significantly different frequencies of Raman modes [24], micro-Raman characterization is able to reveal the kinetic growth process. Figure 2.4c demonstrates the uniformity of the ^{12}C and ^{13}C distribution, which is consistent with the proposed scheme in Figure 2.4a. In comparison, Figure 2.4d and e clearly reveal the time evolution of graphene growth on Cu substrate, which is consistent with the surface absorption mechanism in Figure 2.4b. This method provides direct evidence for the different growth mechanisms on the different substrates with significantly different solubility of carbon.

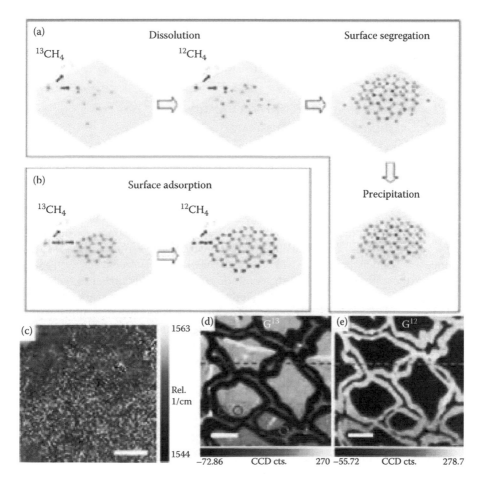

FIGURE 2.4 Schematics of the possible distribution of graphene by carbon isotope labeling based on (a) segregation and precipitation mechanism and (b) surface adsorption; (c) Raman map of G bands on the graphene grown on Ni substrate, and Raman map of (d) G13 and (e) G12 bands on the graphene grown on Cu substrate. (Reprinted with permission from Ref. 23, 4268–4272. Copyright 2009 American Chemical Society.)

For most graphene-based applications, the graphene film has to be transferred onto the dielectric substrates from the metal substrates. The pollution and crack of graphene films happen unavoidably during the transferring process, which results in great degradation of device performance. If graphene can be directly grown on the dielectric substrate, the pollution and cracks can be eliminated. Although graphene film can be grown on SiO_2 and boron nitride (BN) substrates [25,26], compatible dielectric substrates are quite limited, which has a big influence on the carrier mobility of graphene and hence other device properties. Li et al. reported that graphene film can be grown on both the Cu surface and also the interface between the evaporated Cu layer and the substrate [27]. It was suggested that the carbon precursors would diffuse through the grain boundaries during the CVD process. After removing the top graphene and Cu layers, wafer-size graphene film was adhered to the bottom substrate. This method can be easily extended to grow graphene on any kind of dielectric substrates, while the growth condition is the same as that on Cu substrate. Thus, high-quality graphene might be obtained through this method on desirable dielectric substrates and other special structures.

2.4.2 TOWARD THE SYNTHESIS OF HIGH QUALITY OF GRAPHENE FILM ON CU SUBSTRATE

The Cu foils serve as both the catalyst and substrate and hence greatly influence the quality of as-grown graphene film. A good Cu substrate is of low impurity and low roughness because the defects on the Cu surface have a higher surface energy that carbon prefers to nucleate at these sites. Thus, the high nucleation density results in the small sizes of single-crystal graphene domains. The graphene domains will then merge into one continuous film and form the first monolayer graphene due to the self-limiting effect as mentioned above. The higher the nucleation density, the more boundaries form on the surface, which greatly blocks the electron transport [28]. To overcome these problems, one of the common treatments is to anneal the substrates at a high temperature. Higher annealing temperature and longer annealing time lead to bigger Cu grain. Beside this, electrical polishing, evaporating Cu film, electroplating film on flat surfaces, and so forth, were also employed. Recently, Tour et al. reported that annealing at high pressure can be effective in removing sharp wrinkles and defects [29]. After systematic studies on growth condition, single-crystal hexagonal monolayer graphene domain with a size of 2.3 mm was successfully obtained compared to the dense graphene domains with sizes of tens of micrometers without annealing treatment at high pressure. On the other hand, besides the treatment of Cu substrate, researchers have also tried different raw materials. Thinner Cu foil can generate less waste since the Cu foil will be etched in the subsequent transfer process. However, if the Cu foil is too thin, it is not easy to handle, because the hydrogen gas will diffuse into the Cu foil through the grain boundaries and hence the Cu foils become very soft. Thus, the thickness will be balanced depending on the size and application of the final graphene film. It has to be mentioned that the thickness of Cu foil has no obvious influence on the quality of graphene. But the purity of Cu foil can be easily improved to 99.999%, which is still a broadly used commercial material.

It has been demonstrated that the improvement of purity can have a certain effect on the morphology adjustment at a certain temperature. More important, the improvement of purity can significantly increase the carrier mobility from around 1000 to about 7300 cm^2 V^{-1} S^{-1} as the purity increases from 99.8% to 99.999% and the formation of multilayer graphene can be greatly suppressed due to the lower amount of impurities in 99.999% Cu foil [30]. The improvement is very important for reaching a higher frequency for a graphene-based electronic device.

Another important factor is the reaction temperature. Low-reaction temperature is preferred in industry production due to the low cost. Usually, the decomposition of a hydrocarbon precursor at low temperature results in the formation of amorphous carbon coating. Plasma enhanced CVD (PE-CVD) is very helpful to activate the formation of graphene at low temperatures [31]. However, the resulting graphene has a relatively high sheet resistance. It has been demonstrated by experiments that a higher reaction temperature can lead to lower nucleation density and larger size of graphene domains, because the higher reaction temperature reduces the concentration of active carbon species by speeding desorption on the Cu surface. It is known that the melting point of pure Cu is 1083°C and that most experiments were conducted below this. Liu et al. did a very interesting study on the liquid Cu surface [32]. Solid Cu was heated to 1120°C so that it would melt and be dispersed on the tungsten (W) substrate. After reacting for a certain time, perfectly ordered 2-D lattice structure of hexagonal graphene flakes were well aligned on the Cu surface and the I_{2D}/I_G intensity ratio reaches 2.5–4. It is worthwhile to mention that pure monolayer graphene was fabricated without the appearance of any adlayer.

High temperature is preferable for getting large size domains. However, it also remains a big challenge for the growth of graphene at high temperature that Cu can be evaporated during this stage. Although Cu has a relatively low vapor pressure (0.05 Pa at 1120°C), there is a very high vacuum evaporation rate of ~4 μm/h at low pressure CVD (LPCVD) [33]. The evaporation of Cu causes an increase of roughness on the Cu surface and hence has a big influence on the growth of graphene, especially in the growth of large-size single-crystal graphene. Ruoff et al. tried to suppress the evaporative loss of Cu during LPCVD for the growth of millimeter-size single-crystal graphene. Three kinds of structures, including the Cu tube, stacked Cu foils, and Cu foil between two quartz slides, were systematically studied. It was found that the inner surface has a much lower (a tenth) roughness compared with that of the outer surface [34]. The graphene domains prefer to grow on the sites of a rough surface, which greatly limits the size of single-crystal graphene. With the establishment of equilibrium between Cu vapor and Cu substrate in the Cu tube, the resulting smooth inner surface favors the growth of large-size single-crystal graphene with a size of 2 mm.

2.4.3 MANIPULATING THE CARBON PRECURSOR FOR CONTROLLED MORPHOLOGIES OF GRAPHENE

Graphene can be produced from various carbonaceous gases, including methane (CH_4), acetylene (C_2H_2), and ethylene (C_2H_4). It is not required to use a specific carbon source. Currently, CH_4 is the mostly preferable carbon precursor. Other gases are also used during this CVD growth of graphene process. Ar is used as a carrier gas because of its high stability and high thermal conductivity. Hydrogen (H_2) is used as a reduction gas to clean the metal oxide on surface, and it also plays a role in balancing the decomposition of hydrocarbon species and modifying the graphene morphology. The interplay of these factors manipulates the structure and morphologies of the graphene under both atmospheric pressure CVD (APCVD) and LPCVD. As shown in Figure 2.5a, a proper CH_4 concentration is needed to activate the nucleation. However, the higher CH_4 concentration leads to a shorter nucleation time and much denser nucleation sites [35]. For growing large-size single-crystal graphene, the flow of CH_4 is usually limited to be less than 1 sccm. On the other hand, Vlassiouk et al. did a systematic study on the growth of graphene with different H_2 partial pressure at APCVD, which demonstrated that H_2 is a required catalyst in the formation of active surface-bound carbon species and also the etching reagent to modify the structure of graphene by etching away the weak carbon-carbon bonds [36]. It can be concluded by looking at several experimental results that the morphology of graphene is closely dominated by the ratio between H_2 and CH_4. By increasing the ratio ($H_2:CH_4$) in Figure 2.5b, the natural growth of dendrite-shaped graphene can be suppressed

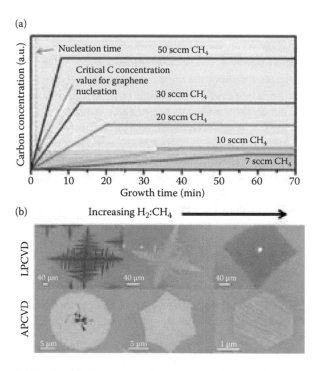

FIGURE 2.5 (a) Relationship between carbon concentration and the growth time. (b) The morphology evolution of graphene with different $H_2:CH_4$ ratio in both APCVD and LPCVD. (From Wu, B., Geng, D., Guo, Y. et al., Equiangular hexagon-shape-controlled synthesis of graphene on copper surface. *Adv. Mater.* 23:3522–3525, 2011. Copyright Wiley-VCH Verlag GmbH & Co. KGaA, Weinheim. Reproduced with permission; Reprinted with permission from Ref. 36, 871–877. Copyright 2012 American Chemical Society.)

so that the graphene islands become smoother to form square or hexagon shapes depending on the growing pressure.

Besides vapor carbon sources, graphene can also be grown from solid carbon sources. Tour et al. casted a thin layer of PMMA or directly placed fluorine and sucrose ground powder on a Cu surface so that the solid carbon could be transferred into graphene film after a heat treatment with a reductive gas flow (H_2/Ar) [37]. The number of layers can be controlled by tuning the reductive gas flow. The quality, as represented by I_D/I_G, is strongly related to the reaction temperature; for example, the ratio of I_D/I_G changes from 0.35 at 750°C to 0.1 at 800°C. Another benefit from solid carbon sources is to grow well-controlled patterned graphene films [38]. Combining the advantages of mature microelectromechanical systems (MEMS) technology, the solid carbon precursor can be easily arranged into desirable patterns that serve as the nucleation center to localize the growth of graphene. Well-ordered arrays of single-crystal graphene domains were fabricated through this process, which might be very useful for assembling desired devices.

2.4.4 Transfer Technique for CVD Graphene Film

The development of a transfer technique opens the door to grow and use CVD-grown graphene. Only after performing this technique can CVD-grown graphene be used in various electronic devices and sensors. Since most CVD-grown graphene films are fabricated on metal substrate, graphene has to be transferred onto a dielectric substrate to build a semiconductor device. In 2009, Hong et al. reported a novel and simple process to transfer patterned graphene for building stretchable transparent electrodes by using PDMS and Ni as intermediate supporting layers, as shown in Figure 2.6a [11]. Graphene film was grown on a Ni substrate that was evaporated on a Si/SiO$_2$ substrate. On the top of the graphene film, a PDMS film was casted and baked, which served as the supporting layer. Subsequently, the Ni layer can be etched by a iron chloride (FeCl$_3$) solution and a graphene/PDMS film floated on the solution, as in Figure 2.6b, which can easily stick to any flat surface due to the van der Waals force as the film is dried. Actually, this transfer technique has been widely used to transfer ultrathin porous anodic alumina (PAA) templates and CNT arrays for a very long time [39]. PDMS film was selected since it has an ultralow viscosity on many surfaces and it can be peeled off as the graphene transferring is completed. Currently, PMMA is more widely used rather than PDMS because it can be dissolved and washed in various organic solutions such as chloroform and acetone. This transfer technique helps researchers to have the first and closest understanding and utilization of CVD-grown graphene film. Using this method, graphene film can be produced into any desired shape for assembling a device, as shown in Figure 2.6c.

It is often observed that wrinkles and cracks in graphene film appear on the new substrate, which is due to the roughness and steps on the metal surface and also the large thermal expansion mismatch between graphene and Cu during the cooling process. In order to improve the quality of graphene on dielectric substrate for better device performance, the cracks must be reduced and eliminated. Ruoff et al. found that the quality of graphene film can be greatly improved by a double PMMA coating method [40]. Typically, after placing PMMA/graphene on the substrate, an

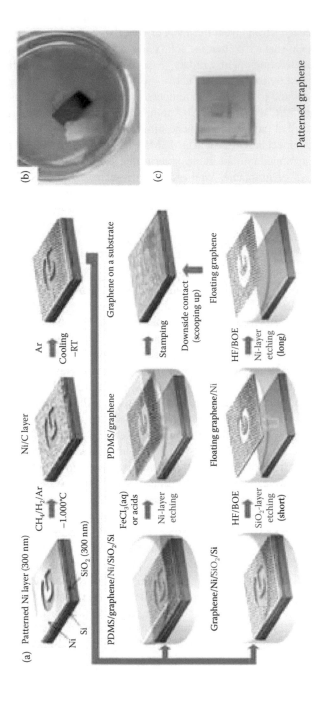

FIGURE 2.6 (a) Schematics of the synthesis of patterned graphene films on thin Ni layers, and optical images of (b) a floating graphene film after removing Ni layer and (c) the patterned graphene film on a SiO$_2$ substrate. (Reprinted by permission from Macmillan Publishers Ltd. *Nature*, Ref. 11, copyright 2009.)

appropriate amount of liquid PMMA solution will be redeposited on the surface to partially or fully dissolve the precoated PMMA. The redeposition of PMMA solution can relax the graphene film underneath, which leads to better electrical and mechanical properties. Resistivity is very stable as the film is bent, and about 2.6% per layer of the attenuation coefficient is achieved, which is much closer to the theoretical value of 2.3%. Besides the influence of cracks, another concern comes from pollution of the polymer, although PMMA and PDMS will be washed out. The as-prepared samples usually have been treated at high temperature to remove the polymer. However, the residual polymer supporting layer is hard to avoid. Li et al. used the Au layer as the supporting layer, which can eliminate contact between the graphene and polymer [41]. Comparing the graphene film before and after transferring by Au and PMMA, the one using Au as an intermediate layer presented lower sheet resistance, lower sheet concentration, and higher mobility.

During the transfer process, metal-based substrate materials have to be removed through the etching process, which presents the following problems: (1) the high cost—Cu is completely etched and wasted, and (2) serious pollution—a large amount of waste solution is produced by etching the Cu, which is not good for large-scale production. Thus, there is an urgent requirement to reduce the cost of substrate in terms of recyclable use of substrates. Loh et al. developed an electrochemical delamination process to peel graphene from Cu foil with little damage to the Cu foil [42]. In a typical process, potassium persulfate ($K_2S_2O_8$) was used as the electrolyte in the electrochemistry process. It was found that small amounts of Cu can be dissolved during this process: $Cu(s) + S_2O_8^{2-}[persulfate](aq) \rightarrow Cu^{2+} + 2SO_4^{2-}[sulfate](aq)$. Subsequently, cuprous oxide (Cu_2O) and cupric oxide (CuO) can be produced by the reaction: $3Cu^{2+}(aq) + 4OH^-(aq) \rightarrow Cu_2O(s) + CuO(s) + 2H_2O(l)$, which served as the passivation layer to further stop the etching of Cu. Using this method, only 40-nm Cu was lost for each growth-delamination-transfer cycle. Thus, a thick Cu foil with thickness of 25 um can be used for hundreds of cycles of repeated growth and delamination. Cheng et al. further improved the delamination process based on platinum substrate. Because the platinum foil is a noble metal, there is no significant loss during the delamination process that the substrate can be reused for longer time [18]. However, the high price of platinum greatly limits the method for practical application.

2.4.5 Production of Graphene Film in a Roll-to-Roll Way

Although much effort has been put into using the transferring technique, the method using polymer as a supporting layer is usually limited to a small area with a size of less than several inches, which is not suitable for scalable production. It is desirable to find an effective approach to manufacture graphene film in a roll-to-roll way. As shown in Figure 2.7a, it was proposed and announced by Samsung and Hong et al. at Sungkyunkwan University for the first time in 2010 [13]. As shown in Figure 2.7b, a Cu foil is wrapped on a large tube furnace in a diameter of 8 inches so that up to a 30-inch graphene film can be obtained after a standard CVD process. The big challenge on the scalable transfer can be overcome by using one kind of thermal release tape that has been widely used in the semiconductor industry to adhere the Si wafer. This tape can stick to the graphene and then easily be released after a heat treatment. The graphene

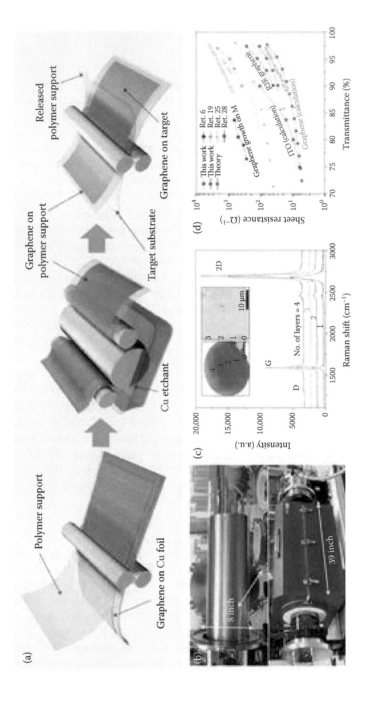

FIGURE 2.7 (a) Schematics of roll-to-roll production of graphene films. (b) Optical image of the growth of graphene on Cu foil in a tube furnace with a diameter of 8 inch. (c) Raman spectra of graphene films with different numbers of stacked layers. (d) Comparison of sheet resistance of the graphene made in a roll-to-roll way and other materials. (Reprinted by permission from Macmillan Publishers Ltd. *Nat. Nanotech.*, Ref. 13, copyright 2010.)

film in a size of up to 30 inches was reported for the first time and was successfully used for the touch panel. As shown in Figure 2.7c and d, the sheet resistance is around 300–400 ohm/sq, and it can be reduced by 60% after a nitric acid (HNO_3) doping process. One layer of as-produced graphene film has a transmission with 2.6% loss, which is very close to the theoretical value of 2.3%. The graphene film can be transferred on the same area repeatedly. With four layers stacked, the sheet resistance can be reduced to 30 ohm/sq. This is the first demonstration of large-area monolayer graphene in a practical application. However, other studies indicate that this transfer technique produces an unstable sheet resistance due to the nonuniform force during the roll-to-roll transferring process. It was found that the sheet resistance can be further improved using a hot-press process, although the size is greatly limited [43].

The above studies demonstrate the revolution of the fabrication and use of CVD-grown graphene films. However, by scaling the Cu foil in a tube furnace, the size of Cu foil is greatly limited due to the finite tube size. It is still very hard to be used in industry applications with a market demand of millions of square meters per year. In 2012, Sony and the National Institute of Advanced Industrial Science and Technology (AIST) announced that graphene film can be manufactured and also transferred to polyethylene terephthalate (PET) substrate in an improved roll-to-roll way [44]. Sony exhibited the graphene films with length of up to 120 m and width of 230 mm. The manufacturing speed reached 10 cm/min. Because the reaction temperature is relatively high, it is not easy to have uniform thermal management. Sony employed a very smart approach where the Cu foil was heated by directly applying a high-level current on the substrate. The as-synthesized graphene film can be transferred into PET substrate with the help of ultraviolet (UV) adhesive, and the Cu substrate was removed by spraying etching solution on surface. The graphene can cover 89%–98% of the Cu foil surface. The sheet resistance reaches 500–600 ohm/sq, and it can be reduced to 150 ohm/sq after chloroauric acid ($HAuCl_4$) doping and finally stabilized at around 250 ohm/sq after a few days. A roll of graphene film with a length of 120 m has been demonstrated. AIST also realized the roll-to-roll production of graphene film using PE-CVD. Using this method, the growth temperature can be decreased to 300°C–400°C. However, the growth speed is relatively low with a value of 1 cm/min, which is 10 times lower than that of the Sony method, and the sheet resistance is higher than 1000 ohm/sq, preventing it from being used in further applications [27].

2.5 EXFOLIATION AND REDUCTION OF GRAPHITE OXIDE

Graphite oxide, which is produced by oxidation of graphite, is an important derivative of graphite. After oxidation, carbon atoms in graphite are functionalized by oxygen-containing groups. As a result, the interaction between adjacent graphene sheets in graphite is enormously reduced when graphite is oxidized, which implies that graphite oxide is much easier to be exfoliated than pristine graphite. The exfoliated graphite oxide, which is also called GO, can be transformed into graphene by reduction. Exfoliation and reduction of graphite oxide has now become one of the most popular methods for the production of graphene, because of its low cost, high yield, and application potential in various areas.

There are three main preparation methods of graphite oxide: the Hummers [45], Brodie [46], and Staudenmaier methods [47]. Strong acids and oxidants are adopted for the oxidation of graphite in all three methods. Among the three methods, the Hummers method using a combination of sulfuric acid and potassium permanganate is more commonly used than the Brodie and Staudenmaier methods, since it is relatively safer compared with the other two. No matter which method is used, three steps are usually needed. First, graphite is converted into graphite oxide through an oxidation process, during which abundant oxygen-containing functional groups are attached onto the basal plane and edges of graphene sheets in graphite, making it highly hydrophilic and stable in water or polar organic solvents. The second step is the exfoliation of graphite oxide into GO nanosheets, and single-layer exfoliation can be easily achieved. The last but most important step is reduction. In order to recover the unique electrical properties of pristine graphene, GO nanosheets need to be reduced for which different reduction methods, such as chemical reduction, thermal reduction, and electrochemical reduction, have been developed. In some cases, exfoliation and reduction may be realized at the same time.

2.5.1 REDUCTION OF GO BY HYDRAZINE

In order to remove the oxygen-containing functional groups on GO and restore the conjugated structure of graphene, relatively strong reduction agents are required. In 2007, Stankovich and his coworkers examined the chemical reduction of GO with several reducing agents and found that hydrazine hydrate was the best one in producing thin graphene sheets [48]. In their research work, GO was produced from natural graphite via the Hummers method. In order to reduce exfoliated GO, 1mL of hydrazine hydrate was added in 100 mL of 1-mg/mL GO dispersion. Graphene was then obtained through heating the mixture in an oil bath at 100°C for 24 h. The reduced GO (RGO) had a highly carbon/oxygen (C/O) atomic ratio of 10.3, much higher than the value of 2.7 for the starting GO, indicating effective removal of oxygen-containing groups by hydrazine hydrate. X-ray photoelectron spectrometer (XPS) and Raman data also strongly support the dramatically decreased O content in the reduced sample. Although the RGO still contains a small amount of oxygen, the conductivity of the material was as high as ~200 S/m. Unfortunately, the RGO was easily aggregated as observed in scanning electron microscope (SEM) images and hard to be redispersed in water or organic solvents because of the hydrophobic nature of RGO after oxygen removal. The specific surface area of the RGO was measured to be 466 m^2/g, which was much lower than the theoretical value of pristine graphene (~2620 m^2/g), also implying the restacking of graphene sheets during the reducing process. In spite of this, it is still a courageous attempt to create graphene using GO as one possible route.

In addition to the reduction of GO, the direct dispersion of GNSs in water without using dispersants is very important for a variety of potential applications, although it has been a great challenge. In order to solve this problem, Li et al. developed a simple process to mass-produce aqueous graphene dispersion without the assistance of surfactant stabilizer through electrostatic stabilization [49]. They also produced graphite oxide via a modified Hummers method. Exfoliation of graphite oxide to GO was conducted by ultrasonication, and they also adopted hydrazine hydrate as the reducing agent. However, they made the improvement that a certain volume of ammonia solution was added to the reaction solution to increase the pH to ~10 and the reaction was carried out in a water bath (~95°C) only for 1 h. Furthermore, they found that the hydrazine (N_2H_4)/GO mass ratio of 7:10 might be an optimal ratio for preparation of stable graphene dispersion. Using the stable suspension of GO in water as the precursor, uniform free-standing graphene paper was obtained by vacuum filtration. Its conductivity was as high as ~7200 S/m and its tensile modulus was up to 35 GPa. This approach with dispersant-free feature makes it attractive for many technological applications, such as electrochemical devices, and in emerging areas, such as transparent/flexible electronics and high-performance composites. Therefore, this work has made a great achievement in the production of stable graphene sheet dispersion.

Ultrasonication is mostly used in the exfoliation of graphite oxide in aqueous solution. Though it is an effective way to delaminate GO sheets within a short period of time, the relatively strong ultrasonic energy could easily tear GO sheets apart into small sheets. As a result, the lateral size of GO or graphene prepared by the solution-phase method is usually in the range of hundreds of nanometers to a few micrometers because of ultrasonication. Such small sheets are not favorable for the application in many areas, such as transparent conductive films, which require graphene sheets to be as large as possible to reduce the contact resistance in the film. In order to prepare large-size GO sheets and graphene, our group adopted a gentle shaking treatment of the aqueous suspension of graphite oxide instead of ultrasonication to delaminate graphite oxide in our lab [50]. As shown in Figure 2.8, using graphite raw materials with a large particle size, most of the lateral size of the obtained GO and RGO sheets was around 100 μm, and some of the sheets even exceeded the size of 200 μm. This was ascribed to the mild shaking process that could maintain the integrity of the GO sheets when delaminating them from graphite oxide. Despite the much milder exfoliation process of this shaking method compared to the ultrasonication process, the exfoliation extent was not affected. More than 95% of the GO sheets had a thickness of ~1 nm, indicating that almost all GO is single-layer. In addition, the size of the graphene sheets could be easily tuned by adjusting the size of the raw graphite. This is an illuminating approach for the production of large graphene, which is very important in practical applications. Meanwhile, we also employed a nonionic polymeric surfactant during the reduction of GO by hydrazine hydrate to the aggregation of graphene sheets. Amphiphilic nonionic surfactants, such as Brij-35, Tween-80, and Triton-X100 were effective in stabilizing hydrophobic graphene sheets in water. When the weight ratio of polymer/GO was fixed at 2, a highly stable aqueous suspension of RGO with a high concentration of 5 mg/mL could be obtained, and there were no sediments or floccules generated after standing for several months. For exploiting more uses out of this stable large graphene dispersion, we further prepared GO foam and paper through a freeze-drying method and vacuum filtration, respectively. After heat treatment in inert atmosphere, GO foam and paper could be transformed to graphene foam and paper. Graphene foam was ultralight and had very high porosity. The conductivity of the graphene paper could be increased up to 10,000 S/m (annealed at 500°C) and 40,000 S/m (annealed at 800°C). This highly conductive graphene foam or paper may be a promising material in Li-ion batteries, supercapacitors, catalytic application, and so on.

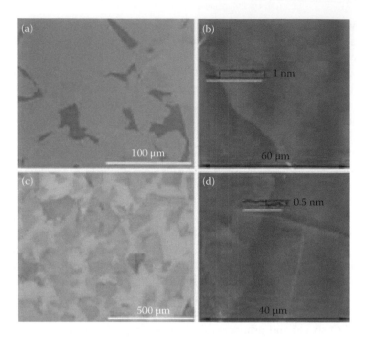

FIGURE 2.8 (a) SEM and (b) AFM images of GO sheets deposited on a Si substrate. (c) SEM and (d) AFM images of reduced graphene sheets deposited on a PEI precoated Si substrate. (From Zhou, X. and Liu, Z., A scalable, solution-phase processing route to graphene oxide and graphene ultralarge sheets. *Chem. Commun.* 46:2611, 2010. Reproduced by permission of The Royal Society of Chemistry.)

Because of its strong reducing capability, hydrazine hydrate has been widely adopted in reducing GO to prepare graphene via a solution-based process. However, it is highly toxic and unstable, which might limit its application in large scale. Moreover, along with effective removal of oxygen functionality, nitrogen from hydrazine hydrate tends to remain covalently bonded to the surface of RGO. Though residual C-N groups might alter the electronic properties of graphene and is favorable for certain applications, the content of N should be minimized or even eliminated if high-quality graphene needs to be produced. Therefore, it is necessary to exploit other effective reducing agents to replace toxic hydrazine hydrate.

2.5.2 GREEN CHEMICAL REDUCTION METHODS

Graphene has attracted more and more attention due to its unique physical and chemical properties. Therefore, it is very important to develop a green method for large-scale production of graphene. Replacing conventionally used toxic hydrazine hydrate with a green reducing agent or developing other green chemical reduction routes are important starts.

Hydrothermal is a classical method for chemical synthesis, which has also been applied to the reduction of GO. Zhou et al. reported a simple hydrothermal route to convert GO to stable graphene dispersion, in which supercritical water plays the role of a reducing agent [51]. A modified Hummers method was used to prepare GO, which was transferred to a Teflon-lined autoclave and heated at 180°C for 6 h. The reduction process was considered to be analogous to the H^+-catalyzed dehydration of alcohol. They also found that the value of pH was critical for the dispersion of the reduced graphene, and the hydrothermal dehydration at a pH of 11 yielded a stable graphene solution. The molar ratio of sp^2 carbon to oxidized sp^3 carbon increased from 1.8 for GO to 5.6 for reduced graphene and the intensity ratio of D band to G band in the Raman spectra decreased to 0.9 after hydrothermal treatment, indicating that hydrothermal reaction was able to reduce/dehydrate the GO and recover the aromatic structure via repairing the defects. Furthermore, the optical performance of reduced graphene could be modified by controlling the hydrothermal treatment temperature. Undoubtedly, the hydrothermal method has potential in preparing graphene-based composite for various applications. In order to prepare much less defective graphene, Wang et al. used solvothermal reduction for graphene with a higher conductivity and lower degree of oxidation through a mild exfoliation-reintercalation-expansion process [52]. Natural graphite was intercalated by oleum and tetrabutylammonium cations and suspended in dimethylformamide (DMF) for solvothermal reduction at 180°C using hydrazine monohydrate as reductant. Raman spectra and electrical transport characterization demonstrated that solvothermally reduced graphene had less defects and oxygen content due to more thorough removal of oxygen functional groups by hydrazine at high temperatures, and a solvothermal condition could also help to heal structure defects. However, more work is still needed in order to prepare high-quality graphene, which is useful in various basic and applied research on graphene-based materials.

In order to explore low-cost and environmentally benign methods for the reduction of GO with high stability and dispersability, Xu et al. used dopamine not only as reductant but also as a capping agent to stabilize and decorate the reduced graphene for further functionalization [53]. During the reaction, dopamine hydrochloride RGO and simultaneously self-polymerized to form polydopamine-capped RGO. The polydopamine adlayer containing catechol group was helpful for further modification of reduced graphene with other organic layers to enhance the dispersity of reduced graphene in different solutions. Taking polyethylene glycol (PEG) for example, the PEG-grafted RGO prepared by Michael addition and Michael addition/Schiff base reaction showed excellent stability in various solvents. By the same token, other thiol- or amine-functionalized polymers or proteins could be grafted on the reduced graphene for production of various graphene-based materials for various applications.

Various methods have been explored to reduce GO; however, the process usually takes a long time, from several hours to days. Therefore, it is important to decrease the reducing time without sacrificing the high quality of graphene for practical applications. Recently, Mohanty et al. reported a simple method to reduce GO and stabilize reduced graphene in a methanol suspension by an ultrafast process that adopted sodium hydride as the reducing agent (Figure 2.9) [54]. The simple hybrid chemistry process could restore the structure of graphene with relatively low defects confirmed by Ranman and UV/Vis spectroscopy, and the whole reduction process could be finished within less than a minute. Meanwhile, the deprotonation

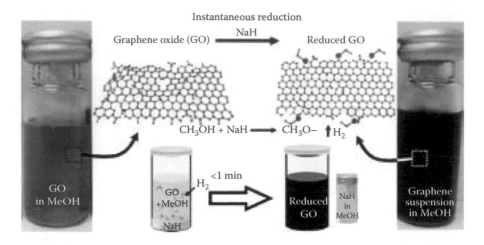

FIGURE 2.9 Schematic diagram of the hydride reduction process. (From Mohanty, N., Nagaraja, A., Armesto, J. et al., High-throughput, ultrafast synthesis of solution-dispersed graphene via a facile hydride chemistry. *Small.* 6:226–231, 2010. Copyright Wiley-VCH Verlag GmbH & Co. KGaA, Weinheim. Reproduced with permission.)

of methanol by sodium hydride (NaH) formed methoxy ions, which could stabilize RGO sheets in methanol. Furthermore, the reduced graphene with one to four layers had high conductivity of ~4500–10,625 S m^{-1}, and electron and hole mobilities of 100–400 and 300–600 cm^2 V^{-1} S^{-1}, respectively.

Sometimes two or more kinds of agents are used together for preparation of graphene with better quality. Sodium borohydride (NaBH$_4$) is an effective reducing agent and can selectively remove different oxygen functional groups with the assistance of different catalysts. Li and his coworkers adopted NaBH$_4$ with anhydrous aluminum chloride (AlCl$_3$) as a mixed reducing agent system to reduce GO that was prepared by a modified Hummers method [55]. The results showed that the electrical conductivity of graphene reduced with anhydrous AlCl$_3$ is improved more than four times, and more oxygen functional group, especially carbonyl, could be removed easily in the presence of anhydrous AlCl$_3$. The C/O atom ratios from GO to the RGO changed from 1.09 to 5.58. In addition, other catalysts may also be used to further improve the reducing ability of NaBH$_4$ and selectively remove other oxygen groups.

Zhao et al. used aluminum iodide (AlI$_3$) synthesized by I$_2$ and Al in ethanol as the reductant to obtain flexible RGO film [56]. AlI$_3$ is a highly efficient reducing agent but less poisonous and explosive than hydrazine. When the AlI$_3$ concentration increased, the conductivity of the GO film also increased intensively. When the concentration of AlI$_3$ reached 0.2 M, the RGO film obtained high electrical conductivity of 5320 S/m. Moreover, the free-standing film served as a counterelectrode showing higher cathodic peak current density than those of a Pt electrode, indicating higher electrochemical activity. At the same time, it is also still attractive to explore inexpensive and eco-friendly material acting as both the reducing agent and the stabilizer for large-scale production graphene, which is helpful for easy preparation of RGO film. Peng et al. developed cellulose, the most abundant polysaccharide in the world, as a green reductant in ionic liquid [57]. Moreover, highly ordered RGO-cellulose composite paper fabricated by vacuum filtration exhibited mechanical flexibility. The conductivity of the paper could reach 700 S/m after enzymatic hydrolysis to remove the cellulose, and may have great potential application in biomedical scaffolds for tissue engineering, medical devices, and so forth.

In terms of further research, how to fabricate flexible graphene devices with high electrical conductivity from GO has received wide attention. Therefore, some reductants were exploited in order to satisfy special applications. Hydroiodic acid is one used to reduce GO film. Pei et al. fabricated free-standing GO film and transparent conductive film assembled on PET substrate via a unique assembly process at a liquid/air interface, and reduced the as-prepared films by immersing into a hydrogen iodide (HI) acid solution [58]. The possible reaction mechanism of GO reduced by HI is illustrated in Figure 2.10. Pei et al. also compared this method with other chemical reducing agents, such as $N_2H_4 \cdot H_2O$ and $NaBH_4$ solution. They found that $N_2H_4 \cdot H_2O$, and $NaBH_4$ solutions could make the films easily break up at higher temperatures, and the thickness of the film, from a cross-section view, showed a shrinkage from 5 to 2.5 µm due to the removal of the oxygen functional groups. However, the film reduced by hydrazine vapor show 10 times an expansion in thickness caused by gas release during the reducing process. Different characterizations demonstrated that GO film reduced in HI acid had a conductivity as high as 298 S cm^{-1} and a C/O ratio above 12. Furthermore, after reduction, the film still maintained good flexibility and its strength and ductility was even improved compared with the original GO film. For transparent conductive film application, Zhao et al. from the same research also investigated the thin GO film fabricated by self-assembly on a liquid/air interface and reduced by HI acid [59]. The difference was that they prepared a large graphene sheet by a modified chemical exfoliated method. They could control the

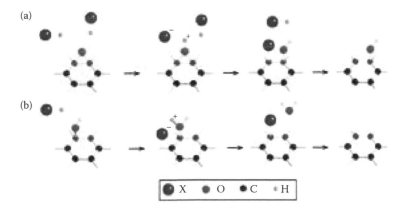

FIGURE 2.10 Possible reaction mechanism of GO reduction by hydrohalic acids. (a) Ring-opening reaction of an epoxy group. (b) Halogenation substitution reaction of a hydroxyl group. The substituted halogen atoms are expected to be easily eliminated from the carbon lattice. (X = iodine or bromine). (Reprinted from *Carbon*, 48, Pei, S., Zhao, J., Du, J. et al. Direct reduction of graphene oxide films into highly conductive and flexible graphene films by hydrohalic acids, 4466–4474, Copyright 2010, with permission from Elsevier.)

area of GO sheets via controlling the C-O content. The largest graphene sheet could reach ~40,000 μm^2 in area. Moreover, the sheet resistance of the reduced GO film decreased with increasing sheet area at the same transmittance due to the decrease in number of intersheet tunneling barriers. A transparent conductive film made from large-area reduced GO sheets by HI acid reduction exhibited a sheet resistance of ~840 Ω/sq at a transmittance of 78%, which could comparable to that of graphene films prepared by CVD method on Ni. Therefore, we can conclude that HI acid is effective in terms of reduction of GO and can retain the flexibility of the as-prepared GO film.

Some new chemical reduction methods are developed at low temperature in both gas and solution phase in order to realize the mass production of high-quality RGO from GO prepatterned on a plastic substrate. Moon et al. adopted hydriodic acid with acetic acid as a novel reducing agent system that allowed for an efficient, one-pot reduction of solution-phased RGO powder and vapor-phased RGO paper and thin film [60]. The conductivity of the RGO increased dramatically in the powder, pape, and thin film form due to the high degree of deoxygenation and high graphitization without nitrogen incorporation. The percentage of carbon in RGO was increased dramatically from 44.56 (GO) to 82.63, and the C/O ratio of the final material was as high as 15.27, indicating that hydriodic acid-acetic acid (HI-AcOH) provides a high level of graphitization of graphene compared with other reducing agents. The conductivity of the RGO pellet prepared under a ~15-ton weight for 10 s was as high as 3.04×10^4 S m^{-1}. This novel approach may be used for the fabrication and processing of various RGO paper and film in the development of new reduced GO on flexible substrate, particularly for a low-temperature process.

Usually, a harsh oxidation method results in small-sized insulting GO sheets with widths in the submicrometer range, which does not facilitate device fabrication. Several research groups have demonstrated exfoliation and intercalation of graphite to produce monolayer graphene sheets. However, all of these methods suffer from low yield and sometimes involve the use of toxic reducing agents. Ang et al. reported an efficient and highly reproducible one-step method of generating monolayer graphene (>90% yield) from weakly oxidized, poorly dispersed GO aggregates [61]. GO sediments were formed after a brief oxidation of graphite via a modified Hummers method. The reduced process is that the 0.5-g dried GO were dispersed in 20 ml of DMF and 2 ml of tetrabutylammonium hydroxide solution (40% water). The mixture was then heated under reflux at 80°C for over to 2 days. Finally, they obtain large-sized conductive graphene sheets (mean sheet area of 330 ± 10 μm^2), which was used to make thin-film field-effect transistors showing high mobility upon nullifying Coulomb scattering by ionic screening. While the reaction time of 2 days is a little longer, this approach is effective in the production of large monolayer graphene.

2.5.3 THERMAL EXFOLIATION AND REDUCTION OF GRAPHITE OXIDE

Chemical reduction is by no means the only method of reducing GO. Thermal reduction is also a common method used to mass-produce graphene. When heating GO at high temperatures, decomposition of the epoxy and hydroxyl groups occurs, which gives rise to the removal of oxygen functionalities, and thus results in reduction of GO. When the decomposition rate exceeds the diffusion rate of the evolved gases, the yielding pressure inside GO can take GO sheets apart to an exfoliated state. Therefore, during thermal treatment, exfoliation and reduction take place simultaneously.

In 2006, Schniepp and coworkers adopted thermal exfoliation and reduction for the first time to prepare graphene [62]. In the experiment, dried graphite oxide was put in a quartz tube and purged with argon. Rapid heating to a high temperature of 1050°C then split graphite oxide into individual sheets. A 500–1000-fold volume expansion was observed for after rapid heat treatment, and the specific surface area of the as-prepared graphene was measured to be ~1000 m_2 g^{-1}. Height analysis of numerous reduced graphene sheets using atomic force microscopy (AFM) revealed a mean thickness of 1.81 ± 0.36 nm, which reflected that the majority of the sheets had 1–3 layers, suggesting quite thorough exfoliation of graphite oxide under the rapid heating process (Figure 2.11). Some wrinkles could be observed on graphene sheets,

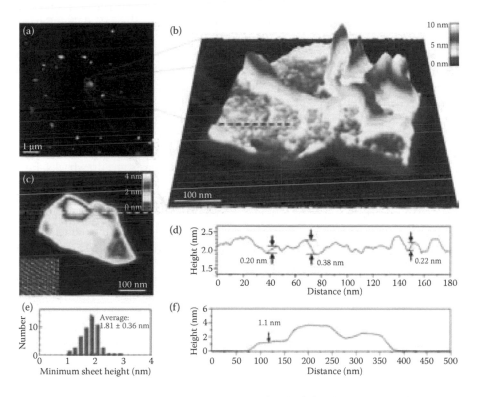

FIGURE 2.11 (a) AFM topography image showing individual thermally exfoliated GO flakes. (b) Pseudo-3-D representation of an AFM scan of an individual graphene sheet showing the wrinkled and rough structure of the surface. (c) Contact-mode AFM scan of a different flake on the same sample. (d) Cross section taken at a rough, unwrinkled area on top of the sheet shown in (b) (position indicated by black dashed line in [b]). (e) Histogram showing the narrow distribution of minimal sheet heights measured on 53 different sheets. (f) Cross section through the sheet shown in (c). (Reprinted with permission from Ref. 61, 8535–8539. Copyright 2006 American Chemical Society.)

which was able to prevent the restacking of thermally exfoliated graphene sheets. The C/O ratio in the thermally exfoliated and reduced sample was 10/1, indicating removal of the majority of oxygen-containing groups in graphite oxide, although some functional groups were still on the graphene sheets. Some small bumps with heights of 0.2–0.4 nm as observed in the AFM images of reduced graphene sheets was related to such remaining oxygen functional groups by the authors. The thermal-induced exfoliation process was associated with the evolution of gases between graphene sheets during heating. A weight loss of ~30% occurred after heat treatment, which was mainly due to the generation of CO_2 by decomposition of oxygen functional groups. The pressure of CO_2 in the sample was estimated to be strong enough to conquer the van der Waals attraction between graphene sheets, which gave rise to highly efficient exfoliation of graphite oxide into single- or few-layer graphene.

Similar experiments were also conducted by McAllister et al. at a later time [63]. Large-volume expansion was also observed at a temperature of 1050°C within a short period of time. After dispersion by ultrasonication in appropriate solvents, statistical analysis by AFM showed that 80% of the observed sheets are single flakes. Surface area measured using the Brunauer–Emmett–Teller (BET) method ranged from 600 to 900 m²/g because the dry RGO was highly agglomerated. They also suggest that a critical temperature of 550°C must be exceeded for exfoliation to occur through comparing of the Arrhenius dependence of the reaction rate against the calculated diffusion coefficient based on Knudsen diffusion. Based on state equation, the pressure ranges from 40 MPa at 200°C to 130 MPa at 1000°C. However, a pressure of only 2.5 MPa is enough to separate two stacked GO sheets.

Reduction and exfoliation of graphite oxide by high-temperature annealing is highly effective, but the thermal annealing method also has some drawbacks. First, high temperature means critical treatment conditions and large energy consumption. Second, the high-temperature condition is not appropriate for some special applications in which a substrate with a low melting-point is needed. Many attempts have been made to prepare exfoliated and reduced graphene at low temperatures. But the obtained samples, in these cases, are only partially exfoliated and still contain extensive domains of staked graphitic layers. However, Lv et al. realized the exfoliation process at a very low temperature, which is far below the proposed critical exfoliation temperature, accompanied with a high vacuum environment [64]. The exfoliation temperature of graphite oxide was as low as 200°C. The successful exfoliation was mainly ascribed to the high vacuum condition that helped to accelerate the expansion of graphene layers by exerting an outward drawing force (Figure 2.12). The x-ray photoelectron spectroscopy measurements showed that the C/O ratio of the low-temperature exfoliated samples was 10:1, which was a slightly lower than that of the samples obtained by the high-temperature approach (C/O ratio, 11:1). Furthermore, the graphene material obtained at low temperatureis of high specific surface area of ~1000 m²/g based on methylene blue dye adsorption and ~400 m²/g based on BET analyses of the nitrogen cry-adsorption method, respectively.

In addition to annealing temperature, the reaction atmosphere also plays an important role in the thermal reduction of GO. Usually annealing reduction is carried out in vacuum or an inert atmosphere. A reducing gas such as H_2 not only consumes the

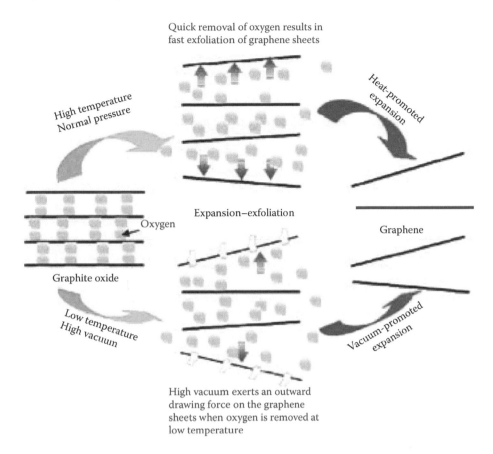

Quick removal of oxygen results in
fast exfoliation of graphene sheets

High temperature
Normal pressure

Heat-promoted
expansion

Expansion–exfoliation

Oxygen

Graphene

Graphite oxide

Low temperature
High vacuum

Vacuum-promoted
expansion

High vacuum exerts an outward
drawing force on the graphene
sheets when oxygen is removed at
low temperature

FIGURE 2.12 Schematic illustration of chemical exfoliation of graphenes. (Top) High-temperature (above 1000°C) exfoliation under an atmospheric pressure; (bottom) low-temperature (as low as 200°C) exfoliation under high vacuum, where the high vacuum introduces a negative pressure surrounding the graphene layers. (Reprinted with permission from Ref. 63, 3730–3736. Copyright 2009 American Chemical Society.)

residual oxygen but also decreases the annealing temperature. Sometimes ammonia (NH_3) or NH_3/Ar is added to produce simultaneous nitrogen doping and reduction of graphite oxide.

2.5.4 ELECTROCHEMICAL REDUCTION

The electrochemical method of reducing GO may open up another way of controlling the reduction of GO and the extent of reduction to obtain highly conducting graphene on electrode materials. Ramesha et al. reduced GO that was assembled on a conducting substrate through a self-assembled monolayer using a layer-by-layer method, using applied DC bias by scanning the potential from 0 to −1 V versus a saturated calomel electrode in an aqueous electrolyte [65]. From the cyclic voltammogram curves the peak observed at −0.75 V suggest that electrochemical reduction takes place, and the reduction of GO to reduced graphene is electrochemically irreversible.

Today, solution-based deposition methods, such as layer-by-layer, spin coating, spray coating, and membrane filtration, are widely utilized to prepare graphene-based films. However, the size, thickness, and uniformity of these films are difficult to control. The electrophoretic deposition (EPD) method is a well-developed and economical method that has many advantages in the preparation of graphene films from charged GO colloidal suspension, including high deposition rate, good uniformity, and thickness controllability. An et al. utilized the EPD method to prepare graphene film deposited on different conductive substrate as illustrated in Figure 2.13 [66]. The typical concentration of GO and applied direct current voltage were 1.5 mg/ml and 10 V, respectively. The conductivity of the air-dried reduced graphene film measured by the van der Pauw method was as high as 10.34×10^2 S m^{-1}, and its C/O atomic ratio was 6.2:1. Moreover, the conductivity and C/O ratio of the as-prepared film after annealing in air at 100°C for 60 min was 1.43×10^4 S m^{-1} and 9.3:1, respectively. Objectively speaking, this approach without harsh and toxic agents has potential for rapid, low-cost, and environmental friendly preparation of graphene films.

Regardless of which method is used to reduce GO for production of graphene, some oxygen functional groups more or less exist on the surface of reduced graphene.

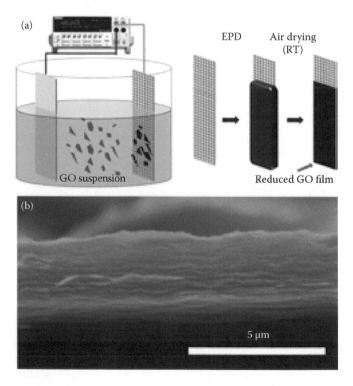

FIGURE 2.13 (a) Schematic diagram of the electrophoretic deposition process and (b) cross-section SEM image of electrochemically reduced GO film. (Reprinted with permission from Ref. 65, 1259–1263. Copyright 2010 American Chemical Society.)

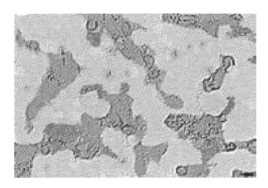

FIGURE 2.14 Atomic resolution, aberration-corrected TEM image of a single-layer RGO sheet. (Reprinted with permission from Ref. 67, 1144–1148. Copyright 2010 American Chemical Society.)

Luo et al. selected six typical reduction methods, including $N_2H_4 \cdot H_2O$, sodium hydroxide (NaOH), $NaBH_4$, solvothermal, high-temperature, and two-step, for preparation of reduced graphene [67]. The reduced materials were systematically compared by four aspects: dipersibility, conductivity, defect repair degree, and reduction degree. Though the reducing degree was significantly dependent on the reducing method, none of the above methods could completely remove all oxygen-containing functional groups. Signals corresponding to C-O and C=O bonds could more or less be observed in the XPS spectra for all six reduced samples. In addition, the reduced graphene samples still possessed a large quantity of defects, as reflected by the existence of D band with large areas in the reduced samples. A careful characterization of the defects in RGO down to the atomic lever was carried out by Gomez-Navarro and his coworkers using high-resolution TEM [68]. As shown in Figure 2.14, five different types of defects highlighted with different colors can be clearly observed a single-layer RGO sheet. The dark gray regions represent contaminated areas on the sheet, the blue regions correspond to disordered carbon network or extended topological defects, the red areas indicated individual ad-atoms or substitutions, the green areas are isolated topological defects, and the yellow areas are large holes existing in the reduced sample. Defect-free graphene areas that have a light gray color only occupied about half of the area of the RGO sheets. Therefore, it has definitely been proved that it is possible to produce defect-free, high-quality graphene sheets by exfoliation and reduction of graphite oxide. The defects and residual functional groups will significantly affect the intrinsic properties of graphene, such as electronic and thermal properties, as well as mechanical properties. However, such defects and functional groups provide RGO with higher chemical activity than pristine graphene, which is favorable for application in many areas, such as composite materials, energy storage, and bioapplications.

2.6 LIQUID-PHASE EXFOLIATION

Though exfoliation and reduction of graphite oxide is an effective and scalable way to prepare graphene, the destruction of the pristine structure of graphene is inevitable in this method, which is not able to obtain high-quality graphene. Therefore, it is important to develop a nondestructive exfoliation method. Direct exfoliation of graphite or graphite intercalated compounds in the liquid phase has been found to be an effective way of producing high-quality graphene. A large body of research has been devoted to this method.

2.6.1 LIQUID-PHASE EXFOLIATION OF GRAPHITE

Hernandez et al. for the first time showed that high-quality graphene sheet dispersions with concentrations up to ~0.01 mg/ml could be successfully obtained by nonchemical, liquid-phase exfoliation of graphite in certain organic solutions [69]. Inspired by the sonication-induced exfoliation of CNTs, they produced a dispersion of sieved graphite powder in N-methylpyrrolidone (NMP) by bath sonication. By analyzing massive TEM images, they found that graphite was extensively exfoliated to yield monolayer or few-layer graphene (Figure 2.15). The monolayer yield was ~1 wt%, which could potentially be improved to 7–12 wt% with sediments recycling. Moreover, high-quality graphene with few defects was confirmed by Raman spectra,

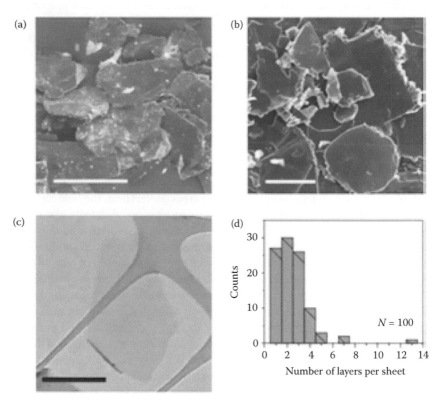

FIGURE 2.15 (a) SEM image of pristine graphite (scale bar: 500 mm). (b) SEM image of sediment after centrifugation (scale bar: 25 mm). (c) TEM image of monolayer graphene deposited from NMP (scale bar: 500 nm). (d) Histogram of the number of graphene flakes as a function of the number of monolayers per flake for NMP dispersions. (Reprinted by permission from Macmillan Publishers Ltd. *Nat. Nanotech.*, Ref. 68, copyright 2008.)

x-ray photoelectron, and infrared (IR) spectroscopy. In order to figure out the optimal solvent for the exfoliation of graphite, the authors repeated the same exfoliation procedure in various solvents with different properties. It was finally concluded that the exfoliation effect was significantly dependent on the surface energy of the solvent. Solvents whose surface energy matches well with that of graphite (~55–90 mJ/m^2) allow for strong interaction with graphene and balancing the energy required for exfoliation, and thus can improve the effect of exfoliation. NMP with a surface energy of ~40 mJ/m^2 was found to be one of the best solvents of graphite exfoliation. The authors also illustrated the application potentials of as-prepared graphene sheets by producing semitransparent conducting films whose electrical conductivity reached ~6500 S/m, and conducting composites with conductivity of ~100 S/m. This work opens up a whole new vista in the preparation of defect-free graphene sheets, although the yield still needs to be improved.

As mentioned above, the interaction between solvents and graphene plays an important role in the exfoliation process. It is thus necessary to understand more details about the interaction, which will greatly facilitate the exfoliation of graphite for wider applications. Hernandez and his coworkers, on the basis of their previous research, identified the Hansen solubility parameters of graphene through measuring its dispersibility in 40 different solvents [70]. Beside NMP, other solvents were discovered to be effective in exfoliation of graphite to form graphene, for example cyclopentanone, acetone, 1,3-dimethyl-2-imidazolinone, N-ethyl-2-pyrrolidone, and N-dodecyl-2-pyrrolidinone. The results demonstrated that the surface tension of a suitable solvent is always close to 40 mJ/m^2, which matched well with their previous work. Furthermore, the authors also demonstrated that good solvents were characterized by a Hildebrand solubility parameter close to 23 MPa$^{1/2}$. After the authors investigated the dispersive, polar, and hydrogen-bonding components of the interaction between the solvent and graphene, they surprisingly found that nonzero values of the polar and hydrogen-bonding Hansen parameters were required for good solvents to disperse graphene. However, it is still a mystery why NMP, the eighth best solvent in the terms of dispersibility, is the best solvent for exfoliation of graphite and stabilizing graphene.

Coleman and coworkers also demonstrated that solvents with the correct surface energy could exfoliate graphite to give reasonable quantities of defect-free graphene by theoretical and experimental studies [71]. The graphene sheets are stabilized either by interaction with the solvents or by the presence of adsorbed surfactant or polymer. Although the solvents mentioned above are effective for exfoliation of graphite, they still have some drawbacks, such as high cost, high boiling points, and requiring special care, which make it difficult to deposit individual monolayers of graphene on surface. Therefore, it is important to exploit safe, user-friendly, low boiling point solvents, preferably water. Lotya and coworkers demonstrated a method of dispersing and exfoliating graphite to produce graphene suspended in a surfactant/water solution [72]. Sodium dodecylbenzenesulfonate was adopted as the surfactant, which made the dispersed graphene flakes stabilize against reaggregation by Coulomb repulsion. By TEM analysis, more than 40% of the graphene flakes had layer numbers less than 5, and ~3% consisted of monolayer. Atomic resolution TEM, Raman, IR, and x-ray photoelectron spectroscopy also indicated that only low levels of defects existed on the graphene basal plane.

In order to prepare graphene dispersion with high concentration and avoid using sonication, which compromise the properties of graphene or reduces its flake size, Behabtu et al. proposed a new method that graphite could be spontaneously exfoliated into monolayer graphene in chlorosulfonic acid with concentrations as high as ~2 mg/ml, which was great progress compared to previous reports [73]. Moreover, this method needed no covalent functionalization, surfactant stabilization, or sonication treatment, which was believed to be a simple and scalable method for producing high-quality graphene. It was found that the acid strength greatly affected the dispersion quality, and when acidity was lowered, the solubility also dropped. Moreover, minimum degree protonation was required for dispersion of graphene at high concentrations, and a liquid-crystalline phase would spontaneously form at high concentrations (~20–30 mg/ml). The high concentrated graphene dispersion, both isotropic and liquid-crystalline, could be promising for making flexible electronics as well as high-performance fibers. Thin transparent graphene films could be produced by filtration of the isotropic dispersion. The sheet resistance of an 80% transparent film was measured to be 1000 Ω/sq. Nontransparent film with a thickness of 8 μm was also produced and showed a high electrical conductivity of 110,000 S/m, which further confirmed the high quality of as-prepared graphene sheets.

2.6.2 LIQUID-PHASE EXFOLIATION OF GRAPHITE INTERCALATION COMPOUNDS

Despite successful exfoliation of pristine graphite in solvents, the yield of graphene is usually low because of the relatively strong interactions between graphene sheets in graphite. It can be expected that better exfoliation can be achieved if the interaction can be reduced. Graphite intercalation compounds (GICs), which introduces heteroatoms or molecules in the interlayer space in graphite, can effectively expand the interlayer distance of graphite, which may enhance the efficiency of graphite exfoliation. Moreover, the chemically active intercalating molecules can serve as a detonator to expand graphite by initiating chemical reactions in the interlayer space. Viculis et al. reported that graphite nanoflakes with a thickness down to 2–10 nm could be prepared via an alkali metal intercalation compound of graphite followed by ethanol exfoliation and microwave drying [74]. Typical procedures are as follows. First, graphite was intercalated with oxidizing acids and exfoliated by rapid heating. Then the graphite is reintercalated with alkali metal to form a first-stage compound, which could be achieved either by using sodium-potassium alloy at room temperature or heating graphite and potassium or cesium at 200°C. Finally, ethanol was used to exfoliate the alkali metal intercalated graphite to form graphene dispersion. In addition, microwave radiation not only dried the sample but also resulted in further expansion of the plates to achieve thickness down to 10 nm.

Gu and his coworkers prepared high-quality graphene sheets by a simple liquid phase exfoliation of GICs [75]. First, graphite flakes were mixed with a mixture of concentrated sulfuric acid and hydrogen peroxide to form GICs. Then, wormlike graphite (WEG) was synthesized by rapidly heating the GIC samples at 900°C for 10 s. Finally, the monolayer or few layers of graphene flakes were obtained by sonication and centrifugation of a 1-methyl-2-pyrrolidinone suspension. The perfect crystalline structure of the exfoliated graphene sheets was confirmed by high-resolution TEM

(HRTEM) characterization and Raman spectra. This approach could potentially lead to the development of new and more effective graphene-based materials.

Recently, Li et al. prepared high-quality GNRs through sonicating thermally exfoliated graphite in a 1,2-dichloroethane solution of poly(m-phenylenevinylene-co-2,5-dichloroethane). But most of the graphene ribbons had two or more layers and the yield was low. To make high-quality graphene, they started by first exfoliating expandable graphite by rapid heating to 1000°C in forming gas (i.e., nitrogen or hydrogen), and then grounded the exfoliated graphite with oleum (fuming sulfuric acid with 20% free SO_3) to obtain reintercalation [76]. Afterward, the oleum-intercalated graphite was further inserted by tetrabutylammonium hydroxide in N,N-dimethylformamide. After sonicating the tetrabutylammonium hydroxide (TBAOH)-inserted oleum-intercalated graphite in the DMF solution of 1,2-distearoyl-sn-glycero-3-phosphoethanolamine-N-[methoxy-(polyethyleneglycol)-5000] for 60 min, a homogeneous suspension was formed, which was then centrifuged to remove non- or not-well-exfoliated pieces (Figure 2.16). The AFM demonstrated that ~90% of the graphene sheets were monolayer with various shape and sizes. The average size of single-layer graphene sheets was 250 nm. TEM and electron diffraction also indicated a well-crystallized, single-layer graphene structure, suggesting high electrical conductivity. Large transparent conducting films on various transparent substrates, including glass and quartz, were made by (LB) assembly of the as-prepared graphene in a layer-by-layer manner. The one-, two-, and three-layer LB films on quarts afforded transparencies of ~93%, 88%, and 83%, and sheet resistances of ~150, 20 and 8 kΩ at room temperature, respectively.

2.6.3 SOLVOTHERMAL EXFOLIATION

The exfoliation of graphite can also be achieved under solvothermal conditions. Qian et al. reported that single or bilayer graphene sheets could be prepared through a solvothermal-assisted exfoliation in acetonitrile, a high polar organic solvent, utilizing expanded graphite as the starting materials [77]. The dipole-induced dipole interactions between acetonitrile and graphene are helpful for the exfoliation and dispersion of graphene sheets. More important, the yield of graphene could be raised to 10–12 wt% via this approach, and the Raman spectroscopy and electron diffraction suggest that the as-prepared graphene flakes are high quality without any obvious defects. Tang and coworkers also prepared high-quality graphene sheets by direct exfoliation of graphite via a solvothermal treatment. It has been reported that graphene could be exfoliated in NMP from graphite by sonication at room temperature [78]. This work further demonstrated that the yield of graphene flakes could be highly improved by increasing the treatment temperature to 200°C. Furthermore, by introducing cyclohexane and water interface, the as-prepared graphene could be easily separated and a graphene film could be obtained by controlling the volume of graphene/NMP solution.

In a word, liquid-phase exfoliation of graphite or its derivatives is very effective for preparation of high-quality graphene with few defects. Though its yield of single-layer graphene is not high enough to be practically applicable, the high throughput preparation of few-layer of multilayer graphene still makes this method attractive for

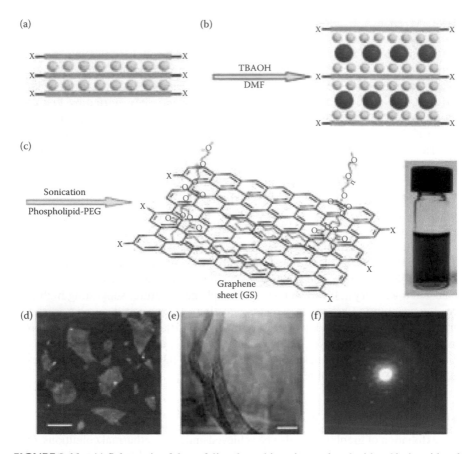

FIGURE 2.16 (a) Schematic of the exfoliated graphite reintercalated with sulfuric acid molecules (teal spheres) between the layers. (b) Schematic of TBAOH (blue spheres) inserted into the intercalated graphite. (c) Schematic of graphene coated with DSPE-mPEG molecules and a photograph of a DSPE-mPEG/DMF solution of graphene. (d) AFM image of a typical graphene sheet. The scale bar is 300 nm. (e) Low-magnification TEM images of a typical graphene sheet. The scale bar is 100 nm. (f) Electron diffraction pattern of a graphene sheet as in (e). (Reprinted by permission from Macmillan Publishers Ltd. *Nat. Nanotech.*, Ref. 75, copyright 2008.)

mass production of graphene. The reproducible fabrication of high-quality graphene in large scale is important for any eventual applications, but for now there is still considerable work to be done.

2.7   OTHER METHODS

2.7.1   ELECTROCHEMICAL EXFOLIATION

Liu et al. demonstrated that functionalized graphene nanosheets could be simply prepared via a one-step electrochemical approach with the assistance of ionic liquid in water at room temperature [79]. It was found that the ratio of the ionic liquid to water was the key factor that influenced the properties of the graphene nanosheets. Those functionalized graphene nanosheets could be separately distributed into various polar aprotic solvents such as DMF, dimethyl sulfoxide, and N-methylpyrrolidone solution instead of water after brief ultrasonic treatment. The as-prepared functionalized graphene nanosheets were characterized with an average thickness of 1.1 nm and width of about 500 nm. Furthermore, a type of graphene nanosheet/polystyrene composites were prepared and exhibited a conductivity of 13.84 S m^{-1}, which is 3–15 times higher than that of CNT/polystyrene composites with the same volume fraction. In conclusion, the electrochemical method developed by Liu et al. is simple, fast, and green for the synthesis of functionalized graphene nanosheets.

Wang et al., taking advantage of the principle of lithium rechargeable batteries, demonstrated a solution route for producing few-layer graphene flakes from negative-graphite electrodes [80]. Specifically, during electrochemical charging in a graphite electrode, the cointercalation of the electrolyte with Li-ions would form a ternary graphite intercalation compound at the interlayers of graphite. The stress caused by those ternary graphite intercalation compounds lead to its fragmentation eventually. The schematic diagram is shown in Figure 2.17. The AFM images indicated that the thickness of those graphene flakes were about 1.5 nm, which corresponded to bilayer graphene and was confirmed by the Raman spectra.

In another work, Huang et al. reported that few-layer graphene sheets could be easily exfoliated from the electrolysis of graphite cathode in a molten lithium hydroxide (LiOH) medium [81]. An exfoliation mechanism of a graphite cathode into graphene sheets was brought up based on the experimental data. It was believed that the exfoliation process could be expounded as three main procedures: intercalation–expansion–microexplosion. First, lithium ions were intercalated into the spaces between the graphene layers and reduced to form intercalation compounds (LixCy). Afterward, the introduction of those intercalation compounds induced the expansion of interlayer spacing of the graphene layers. Finally, the expanded graphites were disintegrated under ultrasonic conditions because the microexplosion reaction

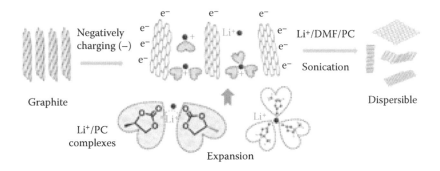

FIGURE 2.17 Schematic illustration of the exfoliation of graphite into few-layer graphene flakes via electrochemical intercalation of Li$^+$ complexes. (Reprinted with permission from Ref. 79, 8888–8891. Copyright 2011 American Chemical Society.)

between lithium and water occurred in the interlayer spaces of graphite and released a large amount of hydrogen. This novel synthesis route is an effective strategy for producing few-layer graphene, which is 2–3 nm in thickness.

2.7.2 Organic Synthesis

In most cases, graphene nanosheets are prepared by exfoliation of graphite or its derivatives typical of a top-down process. It is also possible to synthesize graphene from molecular precursors through organic synthetic protocols, which can be considered as a bottom-up procedure. Mullenmade great contributions in the organic synthesis of graphene. For example, he and his coworkers successfully synthesized linear 2-D GNRs by the designed organic reactions [82]. The overall synthetic strategy of these 2-D GNRs is outlined in Figure 2.18. The scanning tunneling microscope (STM) images reveal that the length of the graphene ribbons ranged from 8 to 12 nm.

Fasel and his coworkers developed an atomically precise bottom-up strategy for fabrication of GNRs [83]. This method can produce atomically precise GNRs of different topologies and widths that are defined by the structure of the precursor monomers. For example, straight and chevron-type GNRs could be synthesized, respectively. The lengths of these two types of GNRs are about 20 to 30 nm and they are atomically precise as confirmed by STM simulations and experimental images.

Jiang et al. developed another bottom-up strategy for synthesizing graphene film at low temperature [84]. To be specific, the C–Br bonds of hexabromobenzene precursors broke and generated hexabromobenzene radicals couple. Subsequently, this

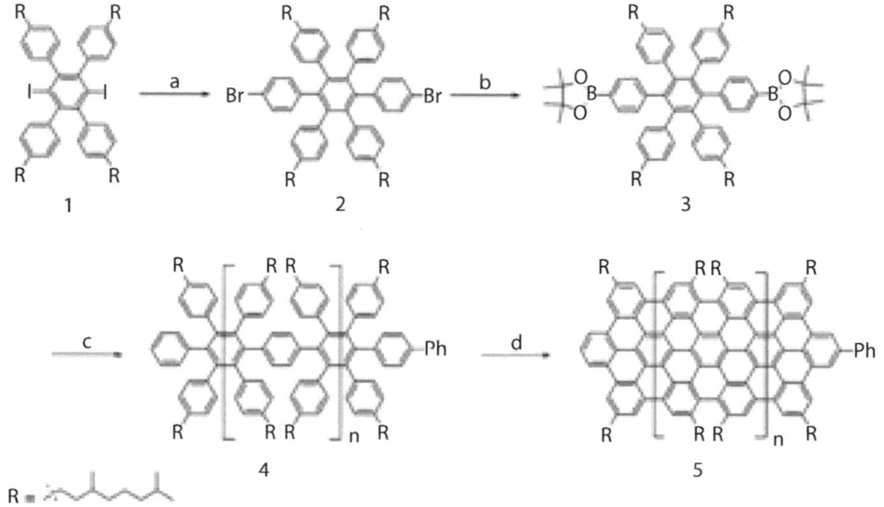

FIGURE 2.18 Reaction process to the synthesis of GNR. (Reprinted with permission from Ref. 81, 4216–4217. Copyright 2008 American Chemical Society.)

radicals couple formed graphene films at a low temperature (220°C to 250°C) efficiently. The thickness of this graphene layer was 1.05 nm, measured by AFM.

Because of the features of the organic synthesis, the atomic structure of as-prepared graphene can be precisely controlled. However, this method has an inevitable limitation that the size of graphene is always limited. It is almost impossible to obtain micrometer-sized graphene sheets using this method. But this method is still attractive in certain applications where nano-sized graphene with controllable structure at the atomic scale is required.

2.7.3 Unzipping of CNTs

In terms of structure, CNTs can be prepared by rolling graphene sheets. Therefore, in a reverse way, graphene can also be prepared by unzipping of CNTs, and it has been experimentally realized recently. In 2009, Tour and his coworkers prepared GNRs by longitudinal unzipping of carbon nanotubes [85]. In a typical procedure, multi-walled CNTs were suspended in concentrated sulfuric acid for a period of 1–12 h. Then, 500 wt% potassium permanganate was added into the reaction mixture and stirred for 1 h at room temperature. After that, the mixture was heated to 55°C to 70°C for another 1 h. Finally, a small amount of hydrogen peroxide was poured into the mixture to quench the reaction. It was observed that the resulting oxidized nanoribbons have a high solubility in water (12 mg ml^{-1}), ethanol, and other polar organic solvents. The obtained oxidized nanoribbons could be easily reduced by hydrazine monohydrate in ammonium hydroxide. Bilayers of these reduced GNRs have field-effect properties with a minimum conductivity at zero gate voltage. A mechanism of opening was developed based on their previous work and is illustrated in Figure 2.19. The unzipping was started at the middle of the CNT instead of the ends. Subsequently, the crack expanded toward the two ends and the tube turned into a nanoribbon.

In 2010, an improved method for the production of GO nanoribbons via longitudinal unzipping of multiwalled CNTs was developed by the same research group [86]. Factors like acid content, time, and temperature were investigated in the improved procedure. As well, a second acid such as trifluoroacetic acid or H_3PO_4 was added into the reaction system in the improved method and the nanoribbons had a higher degree of oxidation with fewer holes on the basal plane at the same time. Electrical measurements revealed that the conductivities of hydrazine-reduced optimized GNRs were 2 to 20 times higher than those of hydrazine-reduced unoptimized ones. What is more, this improved method provided a further understanding of the mechanism raised previously.

Similar work was also reported in 2012. The only difference was that flattened CNTs were used as the starting material in the research by Kang et al. [87]. A mixture of sulfuric acid (H_2SO_4) and potassium permanganate ($KMnO_4$) were employed as the oxidant. Because of the high energy level of the folded edges, cutting was started along the folded edges to yield regular GNRs. As a result, all-bilayer GNRs with narrow width and straight edges were fabricated by the so-called precise unzipping of flattened double-wall CNTs.

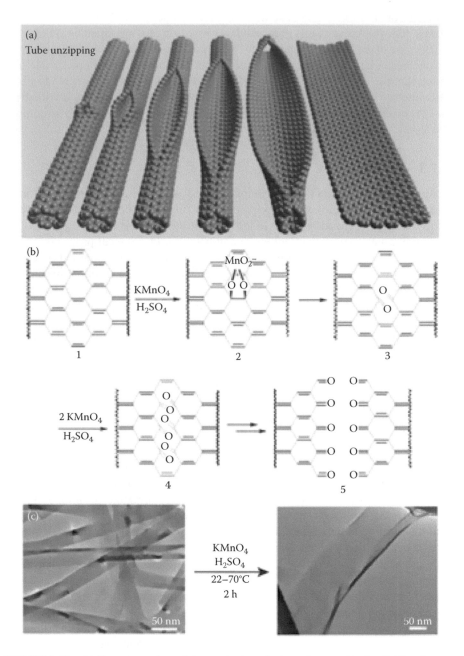

FIGURE 2.19   (a) Representation of the gradual unzipping of one wall of a CNT to form a nanoribbon. (b) The proposed chemical mechanism of nanotube unzipping. (c) TEM images depicting the transformation of multiwalled CNTs into oxidized nanoribbons. (Reprinted by permission from Macmillan Publishers Ltd. *Nature*, Ref. 84, copyright 2009.)

2.7.4 ARC DISCHARGE

Cheng and his coworkers developed a hydrogen arc discharge exfoliation method for the synthesis of graphene sheets using GO as the starting material [88]. The obtained graphene samples exhibited a high electrical conductivity of about $2*10^3$ S cm^{-1} and high thermal stability. They also demonstrated that the high plasma temperature and reducing H_2 played important roles in efficient exfoliation and deoxygenation of GO.

Almost at the same time, Rao et al. reported s similar method for preparing graphene flakes by an arc-discharge strategy [89]. It was demonstrated that graphene flakes generally containing 2 to 4 layers in the inner wall region of the arc chamber were obtained. Surprisingly, this method is eminently suited to dope graphene with boron and nitrogen.

2.7.5 SOLVOTHERMAL METHOD

Stride and coworkers developed a novel bottom-up strategy for producing graphene based on solvothermal synthesis and sonication [90]. In a typical experiment, the graphene precursor was synthesized by solvothermal treatment of sodium and ethanol in a sealed reactor vessel at 220°C for 72 h. Subsequently, the precursor was rapidly pyrolyzed and the remaining product was washed with deionized water. The TEM image displayed an intricate array of folds of graphene sheets and the SEM image clearly showed that the solvothermal product was highly porous but consisted of individual sheets. Furthermore, the definitive evidence was that the step heights measured between the surface of the sheets and the substrate were about 0.4 nm, proving them to be only a single atom thick, which was measured by AFM.

2.7.6 IGNITING MAGNESIUM IN DRY ICE

Hosmane and his coworkers provided an innovative route for conversion of carbon dioxide to few-layer graphene [91]. To be specific, magnesium ribbon was burned inside a dry ice bowl, covered with another dry ice slab. Black products were collected after the burning up of the above-mentioned magnesium ribbon. The chemical equation is formulated as $2Mg + CO_2 = 2MgO + C$. Graphene sheets converted from carbon dioxide have an average length of about several hundred nanometers with the thickness of a few layers. This work paves a new way to the formation of graphene.

REFERENCES

1. Novoselov, K. S. 2004. Electric field effect in atomically thin carbon films. *Science* 306:666–669.
2. Jayasena, B. and Subbiah, S. 2011. A novel mechanical cleavage method for synthesizing few-layer graphenes. *Nanoscale Res. Lett.* 6:1–7.
3. Berger, C., Song, Z. M., Li, T. B. et al. 2004. Ultrathin epitaxial graphite: 2D electron gas properties and a route toward graphene-based nanoelectronics. *J. Phys. Chem. B* 108:19912–19916.
4. Berger, C. 2006. Electronic confinement and coherence in patterned epitaxial graphene. *Science* 312:1191–1196.

5. Hannon, J. B. and Tromp, R. M. 2008. Pit formation during graphene synthesis on SiC(0001): In situ electron microscopy. *Phys. Rev. B* 77:241404.
6. Sprinkle, M., Ruan, M., Hu, Y. et al. 2010. Scalable templated growth of graphene nanoribbons on SiC. *Nat. Nanotech.* 5:727–731.
7. Emtsev, K. V., Bostwick, A., Horn, K. et al. 2009. Towards wafer-size graphene layers by atmospheric pressure graphitization of silicon carbide. *Nat. Mater.* 8:203–207.
8. Choy, K. L. 2003. Chemical vapour deposition of coatings. *Prog. Mater. Sci.* 48:57–170.
9. Tontegode, A. Y. 1991. Carbon on transition metal surfaces. *Prog. Surf. Sci.* 38:201–429.
10. Karu, A. E. and Beer, M. 1966. Pyrolytic formation of highly crystalline graphite films. *J. Appl. Phys.* 37:2179–2181.
11. Kim, K. S., Zhao, Y., Jang, H. et al. 2009. Large-scale pattern growth of graphene films for stretchable transparent electrodes. *Nature* 457:706–710.
12. Li, X., Cai, W., An, J. et al. 2009. Large-area synthesis of high-quality and uniform graphene films on copper foils. *Science* 324:1312–1314.
13. Bae, S., Kim, H., Lee, Y. et al. 2010. Roll-to-roll production of 30-inch graphene films for transparent electrodes. *Nat. Nanotech.* 5:574–578.
14. Singh, V., Joung, D., Zhai, L. et al. 2011. Graphene based materials: Past, present and future. *Prog. Mater. Sci.* 56:1178–1271.
15. Oshima, C. and Nagashima, A. 1997. Ultra-thin epitaxial films of graphite and hexagonal boron nitride on solid surfaces. *J. Phys.: Condens. Matter* 9:1.
16. Sutter, P. W., Flege, J.-I. and Sutter, E. A. 2008. Epitaxial graphene on ruthenium. *Nat. Mater.* 7:406–411.
17. Coraux, J., Engler, M., Busse, C. et al. 2009. Growth of graphene on Ir (111). *New J. Phys.* 11:023006.
18. Gao, L., Ren, W., Xu, H. et al. 2012. Repeated growth and bubbling transfer of graphene with millimetre-size single-crystal grains using platinum. *Nat. Commun.* 3:699.
19. Kwon, S.-Y., Ciobanu, C. V., Petrova, V. et al. 2009. Growth of semiconducting graphene on palladium. *Nano Lett.* 9:3985–3990.
20. Ramón, M. E., Gupta, A., Corbet, C. et al. 2011. CMOS-compatible synthesis of large-area, high-mobility graphene by chemical vapor deposition of acetylene on cobalt thin films. *ACS Nano* 5:7198–7204.
21. Yu, Q. K., Lian, J., Siriponglert, S. et al. 2008. Graphene segregated on Ni surfaces and transferred to insulators. *Appl. Phys. Lett.* 93:113103.
22. Gong, Y., Zhang, X., Liu, G. et al. 2012. Layer-controlled and wafer-scale synthesis of uniform and high-quality graphene films on a polycrystalline nickel catalyst. *Adv. Funct. Mater.* 22:3153–3159.
23. Li, X., Cai, W., Colombo, L. et al. 2009. Evolution of graphene growth on Ni and Cu by carbon isotope labeling. *Nano Lett.* 9:4268–4272.
24. Malard, L., Pimenta, M., Dresselhaus, G. et al. 2009. Raman spectroscopy in graphene. *Phys. Rep.* 473:51–87.
25. Bi, H., Sun, S., Huang, F. et al. 2012. Direct growth of few-layer graphene films on SiO_2 substrates and their photovoltaic applications. *J. Mater. Chem.* 22:411–416.
26. Yang, W., Chen, G., Shi, Z. et al. 2013. Epitaxial growth of single-domain graphene on hexagonal boron nitride. *Nat. Mater.* 12:792–797.
27. Su, C.-Y., Lu, A.-Y., Wu, C.-Y. et al. 2011. Direct formation of wafer scale graphene thin layers on insulating substrates by chemical vapor deposition. *Nano Lett.* 11:3612–3616.
28. Kholmanov, I. N., Magnuson, C. W., Aliev, A. E. et al. 2012. Improved electrical conductivity of graphene films integrated with metal nanowires. *Nano Lett.* 12:5679–5683.
29. Yan, Z., Lin, J., Peng, Z. et al. 2012. Toward the synthesis of wafer-scale single-crystal graphene on copper foils. *ACS Nano* 6:9110–9117.
30. Huang, P. Y., Ruiz-Vargas, C. S., van der Zande, A. M. et al. 2011. Grains and grain boundaries in single-layer graphene atomic patchwork quilts. *Nature* 469:389–392.

31. Yamada, T., Kim, J., Ishihara, M. et al. 2013. Low-temperature graphene synthesis using microwave plasma CVD. *J. Phys. D: Appl. Phys.* 46:063001.
32. Geng, D., Wu, B., Guo, Y. et al. 2012. Uniform hexagonal graphene flakes and films grown on liquid copper surface. *Proc. Natl. Acad. Sci. U.S.A.* 109:7992–7996.
33. Vlassiouk, I., Smirnov, S., Regmi, M. et al. 2013. Graphene nucleation density on copper: Fundamental role of background pressure. *J. Phys. Chem. C* 117:18919–18926.
34. Chen, S., Ji, H., Chou, H. et al. 2013. Millimeter-size single-crystal graphene by suppressing evaporative loss of Cu during low pressure chemical vapor deposition. *Adv. Mater.* 25:2062–2065.
35. Wu, B., Geng, D., Guo, Y. et al. 2011. Equiangular hexagon-shape-controlled synthesis of graphene on copper surface. *Adv. Mater.* 23:3522–3525.
36. Jacobberger, R. M. and Arnold, M. S. 2013. Graphene growth dynamics on epitaxial copper thin films. *Chem. Mater.* 25:871–877.
37. Sun, Z., Yan, Z., Yao, J. et al. 2010. Growth of graphene from solid carbon sources. *Nature* 468:549–552.
38. Wu, W., Jauregui, L. A., Su, Z. et al. 2011. Growth of single crystal graphene arrays by locally controlling nucleation on polycrystalline Cu using chemical vapor deposition. *Adv. Mater.* 23:4898–4903.
39. Chen, Z., Lei, Y., Chew, H. et al. 2004. Synthesis of germanium nanodots on silicon using an anodic alumina membrane mask. *J. Crys. Growth* 268:560–563.
40. Li, X., Zhu, Y., Cai, W. et al. 2009. Transfer of large-area graphene films for high-performance transparent conductive electrodes. *Nano Lett.* 9:4359–4363.
41. Hsu, C.-L., Lin, C.-T., Huang, J.-H. et al. 2012. Layer-by-layer graphene/TCNQ stacked films as conducting anodes for organic solar cells. *ACS Nano* 6:5031–5039.
42. Wang, Y., Zheng, Y., Xu, X. et al. 2011. Electrochemical delamination of CVD-grown graphene film: Toward the recyclable use of copper catalyst. *ACS Nano* 5:9927–9933.
43. Kang, J., Hwang, S., Kim, J. H. et al. 2012. Efficient transfer of large-area graphene films onto rigid substrates by hot pressing. *ACS Nano* 6:5360–5365.
44. Kobayashi, T., Bando, M., Kimura, N. et al. 2013. Production of a 100-m-long high-quality graphene transparent conductive film by roll-to-roll chemical vapor deposition and transfer process. *Appl. Phys. Lett.* 102:023112.
45. Hummers, W. S. and Offeman, R. E. 1958. Preparation of graphitic oxide. *J. Am. Chem. Soc.* 80:1339.
46. Brodie, B. C. 1859. On the atomic weight of graphite. *Philos. Trans. R. Soc. London* 149:249–259.
47. Staudenmaier, L. 1898. Verfahren zur Darstellung der Graphitsaure. *Ber. Dtsch. Chem. Ges.* 31:1481–1487
48. Stankovich, S., Dikin, D. A., Piner, R. D. et al. 2007. Synthesis of graphene-based nanosheets via chemical reduction of exfoliated graphite oxide. *Carbon* 45:1558–1565.
49. Li, D., Müller, M. B., Gilje, S. et al. 2008. Processable aqueous dispersions of graphene nanosheets. *Nat. Nanotech.* 3:101–105.
50. Zhou, X. and Liu, Z. 2010. A scalable, solution-phase processing route to graphene oxide and graphene ultralarge sheets. *Chem. Commun.* 46:2611.
51. Zhou, Y., Bao, Q., Tang, L. A. L. et al. 2009. Hydrothermal dehydration for the "green" reduction of exfoliated graphene oxide to graphene and demonstration of tunable optical limiting properties. *Chem. Mater.* 21:2950–2956.
52. Wang, H. L., Robinson, J. T., Li, X. L. et al. 2009. Solvothermal reduction of chemically exfoliated graphene sheets. *J. Am. Chem. Soc.* 131:9910–9911.
53. Xu, L. Q., Yang, W. J., Neoh, K.-G. et al. 2010. Dopamine-induced reduction and functionalization of graphene oxide nanosheets. *Macromolecules* 43:8336–8339.
54. Mohanty, N., Nagaraja, A., Armesto, J. et al. 2010. High-throughput, ultrafast synthesis of solution-dispersed graphene via a facile hydride chemistry. *Small* 6:226–231.

55. Li, J., Lin, H., Yang, Z. et al. 2011. A method for the catalytic reduction of graphene oxide at temperatures below 150°C. *Carbon* 49:3024–3030.
56. Zhao, X., Lin, H., Li, J. et al. 2012. Low-cost preparation of a conductive and catalytic graphene film from chemical reduction with AlI₃. *Carbon* 50:3497–3502.
57. Peng, H., Meng, L., Niu, L. et al. 2012. Simultaneous reduction and surface functionalization of graphene oxide by natural cellulose with the assistance of the ionic liquid. *J. Phys. Chem. C* 116:16294–16299.
58. Pei, S., Zhao, J., Du, J. et al. 2010. Direct reduction of graphene oxide films into highly conductive and flexible graphene films by hydrohalic acids. *Carbon* 48:4466–4474.
59. Zhao, J. P., Pei, S. F., Ren, W. C. et al. 2010. Efficient preparation of large-area graphene oxide sheets for transparent conductive films. *ACS Nano* 4:5245–5252.
60. Moon, I. K., Lee, J., Ruoff, R. S. et al. 2010. Reduced graphene oxide by chemical graphitization. *Nat. Commun.* 1:1–6.
61. Ang, P. K., Wang, S., Bao, Q. L. et al. 2009. High-throughput synthesis of graphene by intercalation—exfoliation of graphite oxide and study of ionic screening in graphene transistor. *ACS Nano* 3:3587–3594.
62. Schniepp, H. C., Li, J. L., McAllister, M. J. et al. 2006. Functionalized single graphene sheets derived from splitting graphite oxide. *J. Phys. Chem. B* 110:8535–8539.
63. McAllister, M. J., Li, J. L., Adamson, D. H. et al. 2007. Single sheet functionalized graphene by oxidation and thermal expansion of graphite. *Chem. Mater.* 19:4396–4404.
64. Lv, W., Tang, D. M., He, Y. B. et al. 2009. Low-temperature exfoliated graphenes: Vacuum-promoted exfoliation and electrochemical energy storage. *ACS Nano* 3:3730–3736.
65. Ramesha, G. K. and Sampath, S. 2009. Electrochemical reduction of oriented graphene oxide films: An in situ Raman spectroelectrochemical study. *J. Phys. Chem. C* 113:7985–7989.
66. An, S. J., Zhu, Y. W., Lee, S. H. et al. 2010. Thin film fabrication and simultaneous anodic reduction of deposited graphene oxide platelets by electrophoretic deposition. *J. Phys. Chem. Lett.* 1:1259–1263.
67. Luo, D., Zhang, G., Liu, J. et al. 2011. Evaluation criteria for reduced graphene oxide. *J. Phys. Chem. C* 115:11327–11335.
68. Gómez-Navarro, C., Meyer, J. C., Sundaram, R. S. et al. 2010. Atomic structure of reduced graphene oxide. *Nano Lett.* 10:1144–1148.
69. Hernandez, Y., Nicolosi, V., Lotya, M. et al. 2008. High-yield production of graphene by liquid-phase exfoliation of graphite. *Nat. Nanotech.* 3:563–568.
70. Hernandez, Y., Lotya, M., Rickard, D. et al. 2010. Measurement of multicomponent solubility parameters for graphene facilitates solvent discovery. *Langmuir* 26:3208–3213.
71. Coleman, J. N. 2013. Liquid exfoliation of defect-free graphene. *Acc. Chem. Res.* 46: 14–22.
72. Lotya, M., Hernandez, Y., King, P. J. et al. 2009. Liquid phase production of graphene by exfoliation of graphite in surfactant/water solutions. *J. Am. Chem. Soc.* 131:3611–3620.
73. Behabtu, N., Lomeda, J. R., Green, M. J. et al. 2010. Spontaneous high-concentration dispersions and liquid crystals of graphene. *Nat. Nanotech.* 5:406–411.
74. Viculis, L. M., Mack, J. J., Mayer, O. M. et al. 2005. Intercalation and exfoliation routes to graphite nanoplatelets. *J. Mater. Chem.* 15:974–978.
75. Gu, W., Zhang, W., Li, X. et al. 2009. Graphene sheets from worm-like exfoliated graphite. *J. Mater. Chem.* 19:3367–3369.
76. Li, X., Zhang, G., Bai, X. et al. 2008. Highly conducting graphene sheets and Langmuir–Blodgett films. *Nat. Nanotech.* 3:538–542.
77. Qian, W., Hao, R., Hou, Y. et al. 2009. Solvothermal-assisted exfoliation process to produce graphene with high yield and high quality. *Nano Res.* 2:706–712.
78. Tang, Z., Zhuang, J. and Wang, X. 2010. Exfoliation of graphene from graphite and their self-assembly at the oil–water interface. *Langmuir* 26:9045–9049.

79. Liu, N., Luo, F., Wu, H. et al. 2008. One-step ionic-liquid-assisted electrochemical synthesis of ionic-liquid-functionalized graphene sheets directly from graphite. *Adv. Func. Mater.* 18:1518–1525.
80. Wang, J., Manga, K. K., Bao, Q. et al. 2011. High-yield synthesis of few-layer graphene flakes through electrochemical expansion of graphite in propylene carbonate electrolyte. *J. Am. Chem. Soc.* 133:8888–8891.
81. Huang, H., Xia, Y., Tao, X. et al. 2012. Highly efficient electrolytic exfoliation of graphite into graphene sheets based on Li ions intercalation–expansion–microexplosion mechanism. *J. Mater. Chem.* 22:10452–10456.
82. Yang, X. Y., Dou, X., Rouhanipour, A. et al. 2008. Two-dimensional graphene nanoribbons. *J. Am. Chem. Soc.* 130:4216–4217.
83. Cai, J., Ruffieux, P., Jaafar, R. et al. 2010. Atomically precise bottom-up fabrication of graphene nanoribbons. *Nature* 466:470–473.
84. Jiang, L., Niu, T., Lu, X. et al. 2013. Low-temperature, bottom-up synthesis of graphene via a radical-coupling reaction. *J. Am. Chem. Soc.* 135:9050–9054.
85. Kosynkin, D. V., Higginbotham, A. L., Sinitskii, A. et al. 2009. Longitudinal unzipping of carbon nanotubes to form graphene nanoribbons. *Nature* 458:872–876.
86. Higginbotham, A. L., Kosynkin, D. V., Sinitskii, A. et al. 2010. Lower-defect graphene oxide nanoribbons from multiwalled carbon nanotubes. *ACS Nano* 4:2059–2069.
87. Kang, Y.-R., Li, Y.-L. and Deng, M.-Y. 2012. Precise unzipping of flattened carbon nanotubes to regular graphene nanoribbons by acid cutting along the folded edges. *J. Mater. Chem.* 22:16283–16287.
88. Wu, Z. S., Ren, W. C., Gao, L. B. et al. 2009. Synthesis of graphene sheets with high electrical conductivity and good thermal stability by hydrogen arc discharge exfoliation. *ACS Nano* 3:411–417.
89. Subrahmanyam, K. S., Panchakarla, L. S., Govindaraj, A. et al. 2009. Simple method of preparing graphene flakes by an arc-discharge method. *J. Phys. Chem. C* 113:4257–4259.
90. Choucair, M., Thordarson, P. and Stride, J. A. 2008. Gram-scale production of graphene based on solvothermal synthesis and sonication. *Nat. Nanotech.* 4:30–33.
91. Chakrabarti, A., Lu, J., Skrabutenas, J. C. et al. 2011. Conversion of carbon dioxide to few-layer graphene. *J. Mater. Chem.* 21:9491.

3 Applications of Graphene in Lithium Ion Batteries

Xufeng Zhou and Zhaoping Liu

CONTENTS

3.1 Introduction 65
3.2 Graphene Anode Materials.. 66
    3.2.1 Pure Graphene Anode Materials 66
    3.2.2 Doped Graphene Anode Materials... 70
    3.2.3 Graphene Paper as a Flexible Anode... 73
3.3 Graphene-Based Composite Electrode Materials.. 77
    3.3.1 Graphene-Based Composite Cathode Materials............................... 77
        3.3.1.1 Li Metal Phosphate-Graphene Composite Cathode
                 Materials 77
        3.3.1.2 Li-Metal-Oxide Graphene Composite Cathode Materials ... 86
        3.3.1.3 Metal-Oxide-Graphene Composite Cathode Materials....... 90
    3.3.2 Graphene-Based Composite Anode Materials 94
        3.3.2.1 Metal-Graphene Composite Anode Materials.................... 94
        3.3.2.2 Metal Oxide-Graphene Composite Anode Materials........ 105
        3.3.2.3 Lithium Titanate-Graphene Composite Anode Materials... 128
References.. 131

3.1 INTRODUCTION

The Li-ion battery, which was first commercialized by Sony in 1991, is now widely used in many aspects of human life. Due to its relatively high energy density, long cycle life, and good safety, the application areas of Li-ion batteries have expanded rapidly in recent years, ranging from portable and small-scale electronic equipment to electric vehicles and large electricity storage equipment. Technical advancement continuously propels the performance of Li-ion batteries to fulfill increasing market demand. However, practical performance, including energy density, power density, and cycle life at the current levels will be difficult to sustain for the future demand of mass applications of Li-ion batteries in large energy-storage systems, especially in electric cars. Therefore, great effort has been and will continue to be made by scientists and engineers all over the world to improve the performance of Li-ion batteries.

Electrode materials, including cathode and anode materials, are the major components that dominate, to a large extent, the performance of Li-ion batteries. Every big leap in the development of Li-ion batteries is accompanied by a revolution in electrode materials. Consequently, much attention has been paid in research and development to better electrode materials for high-performance Li-ion batteries. Electrode materials with higher capacity, better rate capability, better cycle stability, and better safety are the goals that are pursued.

According to the working principles of Li-ion batteries, during the discharge process, Li-ions inside the anode material are extracted into the electrolyte, and Li-ions in the electrolyte are simultaneously inserted into the cathode material, which is accompanied by the transportation of electrons within the electrode material and generation of electricity in the external circuit. Therefore, performance of Li-ion batteries is significantly affected by the diffusion rate of both Li-ions and electrons in the electrode. However, every existing electrode active material has its own intrinsic Li-ion diffusion coefficient and electrical conductivity, and these two parameters for most active materials are relatively low. As a result, it is usually necessary to modify the electrode materials by either adjusting their morphology or microstructure, or hybridizing with a second phase, in order to achieve faster and matching Li diffusion and electron transportation rate. Hence, modification of electrode materials to fulfill better electrochemical performance is also a major research topic in the area of Li-ion batteries.

Either development of new electrode materials or modification of existing electrode materials is strongly related to material innovation. The emergence of graphene has drawn much attention of scientists who have experience with Li-ion batteries. Numerous research in recent years has demonstrated the appealing application potential of graphene in Li-ion batteries, mainly due to its extremely high electrical conductivity, ultralarge surface area, and unique 2-D nanostructure. Both graphene anode materials and graphene-modified electrode materials with excellent electrochemical performance have been extensively reported on since 2008. In this chapter, the development and some important achievements of graphene-based materials for Li-ion batteries in the last 6 years will be presented.

3.2 GRAPHENE ANODE MATERIALS

3.2.1 PURE GRAPHENE ANODE MATERIALS

Ever since the commercialization of Li-ion batteries, graphite has dominated the field of anode materials, mainly due to its competent capability of insertion and extraction of Li-ions. Although other anode materials may exhibit better performance in certain aspects, no competitor has been able to replace graphite so far with regard to overall performance. However, the electrochemical performance of graphite is still far from the demands on next-generation Li-ion batteries expected to be used for electric vehicles or large energy-storage systems. The relatively low energy density has become one of the major obstacles that have limited the widespread application of Li-ion batteries in the past few years, and maybe also in coming years. Consequently, it is urgent to discover new electrode materials with higher specific capacity. Although graphite anode with a theoretical specific capacity of 372 mAh g^{-1} can still sustain moderate improvement of the energy density, a substitution to graphite with higher capacity is needed after all if the energy density of Li-ion batteries is about to be raised by over 50% or even doubled in the near future. The limited capacity of graphite is mainly attributed to its layered structure, which results in saturated content of inserted Li ions up to a Li/C atomic ratio of 1:6. From a structural point of view, graphene has an identical honeycomb-like atomic structure in the basal plane as that of graphite, which implies that graphene also has a good infinity to Li-ions. However, the huge difference of layer numbers along the c direction between graphene and graphite may give rise to distinct electrochemical reaction mechanisms with Li-ion of two allotropes. For monolayer graphene, the basal planes of this typical 2-D nanomaterial are thoroughly exposed, suggesting no resistance for Li-ion insertion. As for few-layer graphene, the limited layer numbers also enormously reduce the obstruction during Li-ion insertion compared with graphite having numerous graphene layers stacked in the c direction. Therefore, it can be expected that graphene may have more accommodation sites for Li-ions than graphite does, which should result in higher specific capacity. Moreover, the rate capability may also be improved. Consequently, much research on graphene as an anode material has been carried out in the past few years.

The first report of graphene anode was published in 2008 by Honma's group from the National Institute of Advanced Industrial Science and Technology in Japan [1]. Graphene was prepared by exfoliation and reduction of graphite oxide, which has a thickness of 2–5 nm. The charge/discharge profiles indicated a relatively large reversible capacity of 540 mAh g^{-1} for the graphene sample, which was much larger than the theoretical value for graphite. The capacity could even be increased up to 730 and 784 mAh g^{-1} if CNT or fullerene was incorporated with graphene, respectively. It is noted that the charge/discharge profiles for graphene are significantly different from that of graphite (Figure 3.1). Voltage plateaus around ~0.2 V (vs. Li/Li$^+$) in both charge and discharge curves, which corresponds

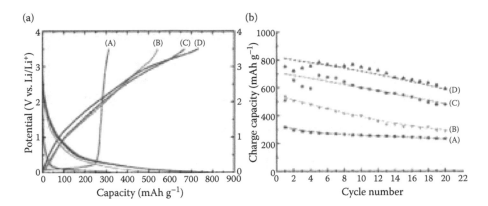

FIGURE 3.1 (a) Charge/discharge profiles and (b) cycle performance of (A) graphite, (B) graphene nanosheet, (C) graphene nanosheet/CNT composite, and (D) graphene nanosheet/C$_{60}$ composite at 0.05 A g^{-1}. (Reprinted with permission from Ref. 1, 2277–2282. Copyright 2008 American Chemical Society.)

to intercalation/deintercalation of Li-ions, can be observed for graphite anode. However, slopes within the voltage range of 0–2.5 V are discerned for graphene, which implies different Li storage sites in graphene compared with graphite. The absence of voltage plateau suggests that graphene sheets are disorderedly stacked, resulting in electrochemically and geometrically nonequivalent Li storage sites, which may be associated with a faradic capacitance on the surface or at the edge of the graphene sheets.

Later, similar research on graphene-based anode materials for Li-ion batteries (LIBs) were successively carried out using graphene from a different preparation method and with different structures. For instance, chemically reduced graphene agglomerated with a flowerlike appearance could deliver an initial discharge capacity of 945 mAh g^{-1} and an initial charge capacity of 650 mAh g^{-1}, and a ~70% retention rate of the specific capacity could be reached after 100 charge/discharge cycles [2]. In another report, graphene was prepared by thermal expansion and reduction of graphite oxide in nitrogen. The sample exhibited ultrahigh reversible capacity of 1264 mAh g^{-1} at the current density of 100 mA g^{-1} and good rate capability (718 mAh g^{-1} at 500 mA g^{-1}) [3]. Unique graphene nanoribbons obtained from unzipping of CNTs using a solution-based oxidative process were also applied as the anode materials [4]. The as-prepared graphene nanoribbons with an oxygen content of 25% presented an initial charge capacity up to ~1400 mAh g^{-1}. However, the Coulombic efficiency for the first cycle was only 53%, which resulted in a largely reduced reversible capacity of ~800 mAh g^{-1} in the first charge process. It was also found that a dramatic decrease of the reversible capacity from ~800 mAh g^{-1} to less than 400 mAh g^{-1} occurred if the GNRs were reduced by hydrogen, which implies that the oxygenous groups are important active sites to generate extra capacity.

The above reports reflect that although graphene has a great advantage over graphite in terms of specific capacity, the measured capacity of graphene in different publications varies in a large range, which is probably due to the fact that structural parameters, such as size, functional groups, and defects vary significantly for graphene samples prepared by different researchers. Consequently, it is important to investigate the critical structural factors that affect the electrochemical performance of graphene, which can help to understand the energy storage mechanism, as well as to rationalize the design of high-performance graphene anode. Pan et al. compared the electrochemical performance of graphene samples prepared under different reduction conditions, including hydrazine reduction, low-temperature pyrolysis, and electron beam irradiation [5]. It was discovered that graphene prepared by pyrolysis and electron beam irradiation showed much higher reversible capacity than the one obtained by hydrazine reduction (Figure 3.2). Raman spectra confirmed that the former two samples had a higher disordered degree than the latter. The enhanced capacity therefore could be ascribed to additional reversible Li-ion storage sites, including edges and other defects in highly disordered graphene sheets. Lee et al. investigated the electrochemical performance of thermally reduced graphene sheets derived from graphene oxide (GO) samples with different oxidation degrees [6]. It was found that as the oxidation degree of GO increased, the reversible capacity of reduced graphene could be enlarged. An ultrahigh charge capacity of 2311 mAh g^{-1} was achieved for the sample oxidized three times. The increased surface area as well as defects and

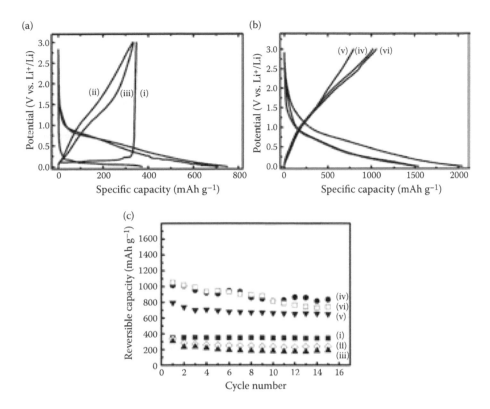

FIGURE 3.2 (a) Charge/discharge profiles of natural graphite (i), GO (ii), and hydrazine reduce GO (iii); (b) charge discharge profiles of (iv) GO pyrolyzed at 300°C, (v) GO pyrolyzed at 600°C, and (vi) electron-beam reduced GO; (c) profile of reversible capacity vs. cycle number at 0.05 A g^{-1}. (Reprinted with permission from Ref. 5, 3136–3142. Copyright 2009 American Chemical Society.)

edges for samples with higher oxidation degrees were believed to be the main cause of the enhanced capacities.

The effect of the layer number on the electrochemical performance of graphene anode was also studied. Pollak and coworkers employed *in situ* Raman spectroscopy combined with electrochemical techniques to investigate the mechanism of Li-carbon (C) interaction of single-layer and few-layer graphene prepared by CVD method during charge/discharge processes [7]. It was discovered that the electrochemical lithiation/delithiation behavior of few-layer graphene resembled that of bulk graphite, while single-layer graphene behaved dramatically differently. It was surprising to find that the LiC$_6$ phase that formed for fully lithiated graphite anode could not form on single-layer graphene. The surface coverage of Li atoms on single-layer graphene was only 5%, which caused much lower capacity of single-layer graphene compared with both few-layer graphene and graphite. This could be mainly ascribed to the lower binding energy of Li to C and strong Coulombic repulsion of Li atoms on the opposite sides of single-layer graphene sheets. It should be pointed out that the sample produced by CVD method has pristine graphene structure with very few defects. Therefore, the Li storage mechanism of CVD-derived graphene should be different from graphene obtained by reduction of GO, as defects are major active sites for Li-ion storage in the latter sample.

Recently, Vargas and coworkers carried out a systematic investigation trying to find out whether the electrochemical performance of graphene can be controlled and predicted [8]. To understand the correlation between the structural properties of graphene and its electrochemical performance, five different graphene samples were prepared in two different ways. Three samples were obtained by thermal exfoliation of graphite oxide at different temperatures, while the other two samples were produced by wet chemical reduction of GO using different reducing agents. Although the samples were carefully characterized by a variety of measurements and their electrochemical properties were analyzed, the attempt to predict the electrochemical performance of graphene anode based on its structural properties still failed. For instance, it was found that higher oxygen (O) content gave rise to higher reversible capacity for graphene prepared by thermal exfoliation; however, the same trend did not work for chemically reduced graphene samples. Other structural properties, such as interlayer spacing measured by x-ray diffraction and structural disorder measured by Raman spectroscopy also failed to explain the reversible capacity observed. The authors also compared their results with previously published graphene anode, and found that the electrochemical performance differed dramatically even between graphene samples prepared in a similar method. Consequently, the authors concluded that the performance of graphene as an anode material in LIBs is difficult to control, since the quality and structure of the graphene samples differ from batch to batch.

3.2.2 Doped Graphene Anode Materials

Though graphene anode has shown attractive electrochemical properties, especially its high reversible capacity, the performance of graphene including both energy density and power density needs further improvement for practical applications. Chemical doping has been successfully used to enhance the electrochemical performance of carbon materials by tailoring their physicochemical properties. Therefore, a similar effect can be expected to be realized in graphene anode.

Ajayan's group published the first report of doped graphene as an anode for LIBs. They grew nitrogen-doped graphene film on Cu current collectors by CVD technique [9]. Using acetonitrile as the precursor, the as-prepared sample had a relatively high nitrogen (N) content of 9%. Electrochemical measurement demonstrated that the reversible capacity of N-doped graphene was almost double than that of pristine graphene. A high percentage of pyridinic N atoms in the sample was believed to be responsible for enhanced electrochemical performance, as they generated large amounts of surface defects on graphene sheets. Though the authors expected that Cu foil supported N-doped graphene could be directly used in flexible thin film batteries, single-layer or few-layer graphene produced by CVD is probably not suitable for large-scale application in LIBs because of its low yield.

Graphene prepared by chemical methods, such as exfoliation and reduction of graphite oxide, has a remarkable advantage over CVD-derived graphene regarding the yield, and therefore doping of chemically derived graphene should be more favorable for practical applications in batteries. The doping process can be realized by thermal treatment of graphene with gaseous precursors of dopants under inert atmosphere, as reported by Wu et al. [10]. Using either NH_3 or boron chloride (BCl_3) as the precursors, N- or boron (B)-doped graphene could be synthesized, respectively. As shown in Figure 3.3, both samples still maintained a typical flexible sheetlike structure of graphene. Elemental mapping using high-angle annular dark-field scanning transmission electron microscopy (STEM) revealed that N and B atoms were homogeneously distributed in the doped graphene sheets. The doping level of N and B was measured to be 3.06% and 0.88%, respectively. XPS spectra indicated that N species in a N-doped graphene sample were mainly pyridinic and pyrrolic N, while B atoms formed BC_3 and BC_2O nanodomains in a B-doped graphene sample. The heteroatoms were bonded to the carbon framework mainly by substituting a carbon atom on edge or defect sites

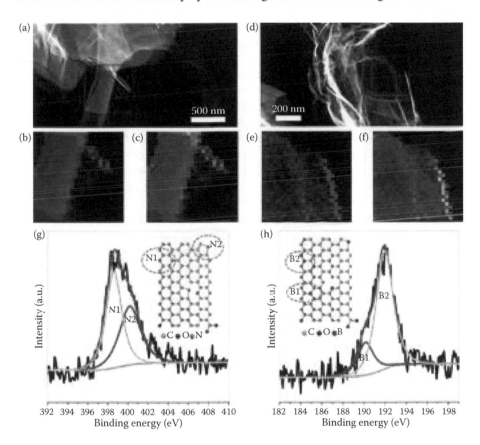

FIGURE 3.3 (a) STEM image of N-doped graphene sheets, (b) C-, and (c) N-elemental mapping of the square region in (a). (d) STEM image of the B-doped graphene sheets, (e) C-, and (f) B-elemental mapping of the square region in (d). (g) N1s XPS spectrum of the N-doped graphene. Inset: schematic structure of the binding conditions of N in a graphene. (h) B1s XPS spectrum of the B-doped graphene. Inset: schematic structure of the binding conditions of B in a graphene. (Reprinted with permission from Ref. 10, 5463–5471. Copyright 2011 American Chemical Society.)

as these carbon atoms are much more active than those perfectly bonded in graphene sheets. Because of the incorporation of heteroatoms, N- and B-doped graphene samples are more disordered than pristine graphene as revealed by Raman spectra. Both samples exhibited excellent electrochemical performance. N- and B-doped graphene have reversible capacity of 1043 and 1549 mAh g^{-1}, respectively, both higher than that of pristine graphene (955 mAh g^{-1}). After 30 charge/discharge cycles at the current rate of 50 mA g^{-1}, the capacity retention for N- and B-doped samples is 83.6% and 79.2%, respectively, also higher than that of pristine graphene (66.8%). The most significant advantage of doped graphene over the pristine one is excellent rate capability. Even at a very high current rate of 25 A g^{-1}, N- and B-doped graphene could still deliver reversible capacity of 235 and 199 mAh g^{-1}, while the capacity of pristine graphene quickly decayed to 100 mAh g^{-1}. Accordingly, high-energy density and high-power density could be achieved simultaneously for both samples. The outstanding electrochemical performance of doped graphene, especially its superior rate capability, could be first ascribed to increased electrical conductivity after doping. Electrochemical impedance spectroscopy measurement showed that doped graphene had much lower electrolyte resistance and charge transfer resistance than those of pristine graphene. In addition, it was discovered that doped graphene sheets have increased disordered surface morphology, such as corrugation and scrolling, which gives rise to higher faradic capacitance on the surface and edge sites of graphene sheets. The topological defects derived from heteroatom doping also favored Li storage. Moreover, chemical doping could enhance hydrophobicity of graphene sheets and their wettability to organic electrolytes, which could further promote Li-ion diffusion in the interior of the electrode.

Similar research was carried out by several other groups. For instance, Li et al. also prepared N-doped graphene by annealing graphene in ammonia gas, and a nitrogen content of 2.8% could be acquired. However, a unique cycling performance of N-doped sample was observed [11]. The reversible capacity of N-doped graphene gradually decayed from 454 mAh g^{-1} to 364 mAh g^{-1} in the initial 17 cycles as normal. Nevertheless, the capacity started growing after the 17th cycle, and a maximum reversible capacity of 684 mAh g^{-1}, which was even higher than the initial one, was reached around the 500th cycle. Then the capacity remained constant. The authors briefly ascribed this phenomenon to defect sites in N-doped graphene; however, deep investigation should be carried out to understand the mechanism. In another report, melamine, a solid nitrogen source, was used instead of a gaseous source [12]. N-doped graphene could be simply prepared by heat treatment of the mixture of pristine graphene and melamine at 700°C for 1 h. Compared with N-doped graphene prepared from a reaction of graphene with NH_3, whose N content normally lies in the range of 2%~3%, the melamine derived sample has a higher N content of 7.04%. The as-prepared N-doped graphene presented a high reversible capacity up to 1123 mAh g^{-1} at the current density of 50 mA g^{-1}, and the capacity of 241 mAh g^{-1} could still be reached at the current density of 20 A g^{-1}. This new approach provides an alternative and an easy way to prepare high-performance N-doped graphene for LIBs by using a simple solid reaction instead of a traditional solid-gas reaction.

Most recently, the preparation of doped graphene was combined with the construction of a porous structure to achieve high capacity and rate capability simultaneously [13]. The porous structure was templated by polystyrene microspheres with the assistance of the surfactant of poly(vinyl pyrrolidone), while decomposition of the templates and surfactants spontaneously doped the graphene sheets with N and S atoms. The doping level of N and S was 4.2% and 0.94%, respectively. Due to the synergetic effect of doping and structure, the electrode could reach a high-energy density 116 kW kg^{-1} and a high-energy density of 322 Wh kg^{-1} at the current density of 80 A g^{-1}, which might bridge the gap between LIBs and supercapacitors.

3.2.3 GRAPHENE PAPER AS A FLEXIBLE ANODE

The 2-D sheetlike nanostructure is a significant structural feature of graphene. Using graphene nanosheets as the building blocks, some macroscopic assemblies with unique morphologies that are not able to be constructed by 3-D or 1-D nanomaterials can be realized. By stacking of graphene sheets along the c direction, a paperlike macrostructure can be easily obtained. The relatively strong interaction and large contact area between adjacent graphene sheets make graphene paper tough enough to be handled with various mechanical forces. At the same time, the structural flexibility of graphene sheets also endows graphene paper with enough flexibility so that it can be bent or rolled repeatedly without generating fractures. Therefore, flexible and self-standing graphene paper is an attractive candidate for many application areas. For the specific application of graphene in LIBs, graphene paper is also expected to be advantageous over powdery graphene as freestanding graphene paper can be directly employed as the anode without any binder or conductive additive, which results in a higher energy density of the electrode. Moreover, graphene paper is especially suitable for the fabrication of thin and flexible LIBs for future applications in next-generation flexible electronic devices.

Abouimrane et al. reported the first graphene paper based anode for LIBs in 2010 [14]. The graphene paper was produced by vacuum filtration of aqueous suspension of GO and the following chemical reduction of GO into graphene by hydrazine. SEM characterization revealed that graphene sheets stacked quite orderly and densely in the paper whose thickness was ~10 μm (Figure 3.4). The average intersheet spacing

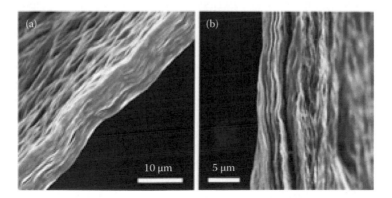

FIGURE 3.4 SEM images showing the undulating surface (a) and a fractured edge (b) of a graphene paper.

was measured to be 0.37 nm, which is ~10% larger than the interlayer distance of graphite. Although the binder-free graphene paper anode could simplify the anode fabrication process compared with graphene powder, the specific capacity of paper electrode was much lower than the latter one. At a current density of 50 mA g^{-1}, only a reversible capacity of 84 mAh g^{-1} was delivered for the graphene paper, while graphene powder exhibited a capacity of 288 mAh g^{-1}, more than three times higher than that of the paper electrode. Even at a very low current density of 10 mA g^{-1}, the graphene paper could still deliver a relatively low capacity of 214 mAh g^{-1}. The poor performance of graphene paper anode is probably because the densely stacked graphene nanosheets in the paper caused a kinetic barrier for the diffusion of Li-ions. Similar limitation of the specific capacity of graphene paper was also discovered by Sun's group [15]. The reversible capacity of ~80 mAh g^{-1} at the current density of 50 mA g^{-1} was also observed for a 10-μm-thick graphene paper. By reducing the thickness of the paper, better electrochemical performance could be obtained. For instance, a 1.5-μm-thick graphene paper presented a reversible capacity of ~200 mAh g^{-1}, and enhanced cycling stability was also observed. However, its capacity is still much lower than powdery samples. Consequently, dense graphene paper produced by simple restacking of graphene sheets is not a good candidate for high performance LIBs. Modification of graphene paper to improve the diffusion of Li-ions is thus important for practical applications.

It is straightforward to anticipate faster Li-ion diffusion by generating pores in graphene paper. To realize this purpose, several different strategies have been implemented and positive results were obtained. One solution is to produce holes in graphene sheets as demonstrated by Zhao et al. [16]. The in-plane pores could be fabricated by combining acid (HNO$_3$) oxidation and ultrasonic vibration of GO in aqueous solutions, and the size and areal density of the pores could be tuned by varying the ratio of acid to GO. By increasing the ratio of HNO$_3$ to GO, the pore size could be enlarged from several nanometers to hundreds of nanometers (Figure 3.5). The holey GO sheets were then vacuum-filtrated to form a continuous film. After thermal reduction in a quartz tube, graphene paper could finally be obtained. It was noted that shrinkage of the pore size occurred if the sample was reduced in H$_2$ atmosphere instead of Ar, which was ascribed to the restoration of the graphene structure aided by H$_2$. The porous structure of the paper was apparently observed under SEM. The surface of porous graphene paper was much rougher than the nonporous one, and in some areas, nano-sized in-plane pores could be directly observed. The porous structure induced apparently improved electrochemical performance of graphene paper as expected. At the current density of 50 mA g^{-1}, porous graphene paper could reach a reversible capacity up to 403 mAh g^{-1}, while the nonporous one could only deliver 336 mAh g^{-1}. The advantage of porous graphene paper is more notable in terms of rate capability. Even at the current density of 2 A g^{-1}, the porous paper could still deliver a reversible capacity of 178 mAh g^{-1}, corresponding to a capacity retention rate of ~44%, which is much higher than ~20% of the nonporous one. Electrochemical impedance spectroscopy (EIS) further confirmed that both charge transfer and Li-ion diffusion kinetics were remarkably improved by making holes in the graphene sheets. It was also found that the optimal pore size for best electrochemical performance was 20–70 nm. Further increase of the pore size resulted in

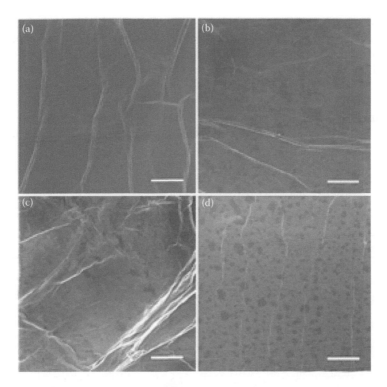

FIGURE 3.5 SEM images of holey GO sheets deposited on Si wafers with GO suspension/70% HNO_3 volume ratio of (a) 1:5, (b) 1:7.5, (c) 1:10, and (d) 1:12.5. Scale bar is 20 μm. (Reprinted with permission from Ref. 16, 8739–8749. Copyright 2011 American Chemical Society.)

lower capacity especially at high rates, but the mechanism is not quite clear at the moment. This research proved that generating pores in graphene paper is an effective way to enhance its electrochemical performance.

Later, several attempts to prepare graphene paper anode by different methods were carried out. For instance, Mukherjee et al. used a unique photothermal reduction method to produce graphene paper with an open-pore structure [17]. The preparation process involved irradiation of GO paper under laser of a camera flash. During this process, GO paper was instantly and extensively heated, which induced deoxygenation in the GO sheets. GO was thus reduced into graphene, while the rapid outgassing simultaneously generated microscale pores, cracks, and voids in the paper. The radiation caused large volume expansion of the paper in the vertical direction. The original thickness of the GO paper was 10–20 μm, while the thickness of the photoreduced graphene paper was ~100 μm. It was surprising to find that the porous graphene paper still maintained structural integrity and robustness even when it was highly expanded. SEM characterization showed that the expansion induced large gaps between adjacent graphene sheets and also generated microscale cracks on the surface of the paper. These voids could facilitate penetration of electrolytes deep into the interior of graphene paper, which would result in much faster Li-ion diffusion than in the case of nonporous graphene paper. Compared with hydrazine-reduced graphene paper, which has no such porous structures, porous graphene paper exhibited higher specific capacity and much better rate capability. Even at an ultrahigh rate of 150 C, the paper could still deliver a reversible capacity of 63 mAh g^{-1}, and a capacity retention of ~97% could be achieved after 6000 charge/discharge cycles. Such outstanding electrochemical performance endows porous graphene paper with attractive potential for various applications. However, the expanded structure induced low density of the paper electrode, which gave rise to relatively low volumetric energy density.

Porous graphene could also be produced through a CVD process that employed expanded vermiculite, an inorganic mineral, as the template [18]. In order to generate pores in the graphene sheets, vermiculite was pretreated by substoichiometric hydrochloric acid (HCl) to form iron (Fe) vacancies on the surface. As Fe-containing

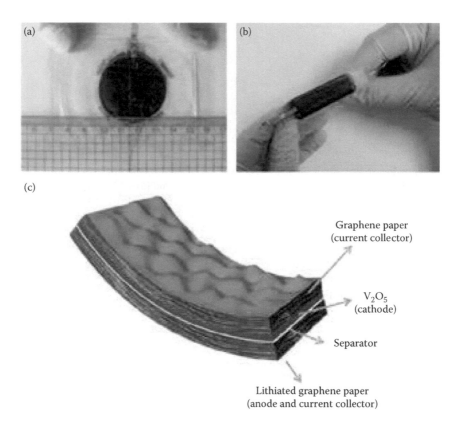

FIGURE 3.6 (a) The assembled flexible Li battery in which V_2O_5/graphene composite paper and electrochemically lithiated-graphene paper were used as cathode and anode, respectively. (b) The battery is thin, lightweight, and flexible enough to be rolled up or twisted. (c) Schematic drawing of the flexible Li battery. (From Gwon, H., Kim, H.-S., Lee, K. U. et al. Flexible energy storage devices based on graphene paper. *Energ. Environ. Sci.* 4:1277–1283, 2011. Reproduced by permission of The Royal Society of Chemistry.)

surfaces of expanded vermiculite are active sites for graphene formation, the growth of graphene was inhibited at Fe vacancies, which induced the formation of holes in the products. The as-prepared holey graphene sheets were then filtrated to form a continuous paper. The porous paper exhibited a very high reversible capacity of 1350 mAh g^{-1} at a current density of 50 mA g^{-1} and almost no capacity decay was observed after 100 charge/discharge cycles.

Flexibility is one of the most promising characteristics of graphene paper, which is especially suitable for the applications in flexible secondary batteries for bendable electronic devices. A prototype of graphene-paper-based flexible LIB was presented by Gwon et al. in 2011 [19]. As shown in Figure 3.6, a battery packed in polymer films has a diameter of several centimeters and can be easily rolled up. Figure 3.6c illustrates the composition and structure of the flexible battery. In the cathode, graphene paper was used as the current collector to support active vanadium oxide (V_2O_5) cathode materials. Compared with V_2O_5 coated on Al foils, the one with graphene paper presented a four-times higher volumetric capacity and significantly improved cycling stability. This was attributed to the composite-like gradient structure between two components formed during the pulsed laser deposition of V_2O_5 on graphene paper due to the surface roughness of the graphene paper. Therefore, the interface resistance and thus the polarization of the electrode could be effectively reduced. The opposite electrode (anode) was made from graphene paper that had a reversible capacity of ~300 mAh g^{-1} at a current density of 186 mA g^{-1}. In order to form a full cell, the graphene paper anode needed to be electrochemically lithiated since Li did not present in the cathode or the anode. The first charge and discharge capacity of the flexible cell was ~15 μAh cm^{-2} and negligible irreversible capacity was observed. The cell could still operate when it was rolled up or twisted. Though the capacity decay was relatively large during charge/discharge cycles, this primitive device shows attractive application potentials of graphene-based flexible batteries in next-generation bendable electronic equipment.

3.3 GRAPHENE-BASED COMPOSITE ELECTRODE MATERIALS

Graphene itself not only can be used as a high-performance anode material, but can also be applied as a functional component to improve the electrochemical performance of other cathode and anode materials. Significant progress has been achieved for graphene-based composite electrode materials in the last few years.

3.3.1 GRAPHENE-BASED COMPOSITE CATHODE MATERIALS

3.3.1.1 Li Metal Phosphate-Graphene Composite Cathode Materials

As a typical Li metal phosphate cathode material, lithium iron phosphate ($LiFePO_4$) has attracted much attention and has been widely used in commercial LIBs due to its excellent charge/discharge performance, especially outstanding cycling stability and good safety. However, the intrinsic electrical conductivity of $LiFePO_4$ is ultralow, which causes poor electrochemical performance if naked $LiFePO_4$ is directly employed as the cathode material. Thus, modification of $LiFePO_4$ to improve its conductivity is always required. In most cases, including commercial products, LiFePO$_4$ particles are coated by pyrolitic carbon to enhance electron transmission. Due to its amorphous nature, the electrical conductivity of pyrolitic carbon is much lower than that of graphitic carbon materials. Therefore, it can be reasonably expected that modification of LiFePO$_4$ with carbon materials with a high graphitic degree should give rise to better electrochemical performance. However, the structure of graphitic carbon needs be taken into account as it should be able to form intimate contact with individual LiFePO$_4$ nanoparticles. Graphene, which has high electrical conductivity, ultrathin sheetlike morphology, and flexible structure, is believed to be an ideal carbon material for LiFePO$_4$ modification. LiFePO$_4$/graphene composite cathode material has aroused much attention in recent years.

The first report on graphene-modified LiFePO$_4$ was published in 2010 [20]. Using a coprecipitation method, LiFePO$_4$ nanoparticles were deposited on graphene sheets to form a composite structure. Though the composite cathode material could deliver an obvious higher discharge capacity then pure LiFePO$_4$ and its rate capability was also improved, the structure of the composite needed further optimization to achieve better performance. Our group then proposed a novel composite structure of graphene-modified LiFePO$_4$, which gave rise to outstanding rate capability and cycling stability [21]. In our experiments, LiFePO$_4$ nanorods synthesized by solvothermal method were mixed with aqueous solution of GO. The mixture was then spray-dried and annealed to obtain a LiFePO$_4$/graphene composite cathode material. As shown in Figure 3.7, the composite exhibited quasi-spherical morphology that was characteristic of spray-dried samples. Magnification of an individual sphere showed that it was composed of randomly stacked LiFePO$_4$ nanoparticles. It was found that the surface of the particle was covered by a thin film that was almost transparent under SEM, which could be identified as graphene. TEM characterization further revealed the detailed structure of the composite in local areas. Graphene sheets were visible under TEM, which stretched within the secondary particle and wrapped primary LiFePO$_4$ nanoparticles. Elemental mapping (Figure 3.7d) further clearly displayed that each LiFePO$_4$ was covered by a thin layer of graphene. Due to the sheetlike structure of graphene, and its larger size than LiFePO$_4$ nanoparticles, the surface of LiFePO$_4$ was not fully and intimately covered by graphene sheets. Voids between graphene and LiFePO$_4$ could be found in some areas, which were beneficial for impregnation of electrolytes, and left a naked surface of LiFePO$_4$ for fast Li-ion diffusion. In order to find out the distribution of graphene in the composites, LiFePO$_4$ was then dissolved by HCl. It was surprising to find that the remaining graphene maintained similar spherical morphology as that of the LiFePO$_4$/graphene composite. Though the removal of LiFePO$_4$ nanoparticles left plenty of vacancies, graphene sheets still interconnected. This undoubtedly indicated that graphene homogenously distributed in the composite to form a continuous 3-D conductive network. High-resolution TEM image (inset of Figure 3.7e) displayed that the graphene film covering LiFePO$_4$ has a relatively ordered multilayer structure, apparently different from the amorphous structure of pyrolitic carbon. The average thickness of graphene film was ~2 nm, correspongding to 3–5 layers. The interlayer distance was ~0.4 nm, slightly larger than that of graphite. It should be pointed out that the GO sheets used in our experiments were mostly single-layered. Therefore, the multilayer structure of

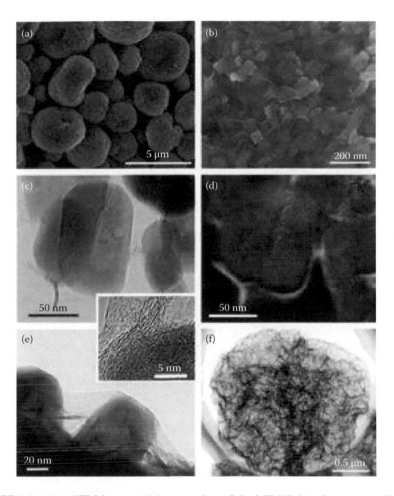

FIGURE 3.7 (a,b) SEM images of the overview of the LiFePO$_4$/graphene composite particles, (c) TEM image, and (d) corresponding elemental map showing graphene-sheets wrapping on LiFePO$_4$ nanoparticles, where the red areas represent the LiFePO$_4$ nanoparticles obtained from the P L-edge and the green areas represent graphene sheets obtained from C K-edge. (e) TEM image on the edge of individual microspheres. The inset is a high-resolution TEM image illustrating the 3–5 monolayer thickness of the graphene sheets on the surface of an individual LiFePO$_4$ nanoparticle. (f) TEM image showing a 3-D graphene network obtained by removing LiFePO$_4$ nanoparticles with an HCl solution. (From Zhou, X., Wang, F., Zhu, Y. et al., Graphene modified LiFePO$_4$ cathode materials for high power lithium ion batteries. *J. Mater. Chem.* 21:3353–3358, 2011. Reproduced by permission of The Royal Society of Chemistry.)

the graphene film wrapping on LiFePO$_4$ was caused by restacking or folding of GO sheets during the spray-drying process. Raman spectroscopy indicated that the graphene-wrapped sample had narrower D band and G band compared with pyrolitic carbon. XPS analysis showed that the former sample also had much lower content of O and higher content of sp^2 hybridized C than the latter. Both results suggested higher graphitization degree of graphene-based composite than conventional pyrolytic-carbon-modified sample.

Electrochemical performance was then evaluated for both samples. At the same content of carbon, $LiFePO_4$/graphene composite cathode material exhibited much better rate capability and cycling stability than pyrolitic-carbon-modified $LiFePO_4$ mainly due to the much higher electrical conductivity of graphene than pyrolitic carbon. It was interesting to note that the $LiFePO_4$ cathode materials modified by a mixture of graphene and pyrolitic carbon showed even better electrochemical performance. By replacing half the contents of graphene with pyrolitic carbon, the composite cathode material could deliver a discharge capacity of 70 mAh g^{-1} at an ultrahigh discharge rate of 60 C, which is ~47% of its initial capacity at 0.1 C. When charged at 10 C and discharged at 20 C for 1000 cycles, the cell only had a small capacity decay of ~5%. The sample using mixed carbon source had almost identical Raman spectrum and XPS data compared with $LiFePO_4$ modified with pure graphene, which indicated that the addition of pyrolitic carbon did not reduce the graphitic degree of the carbon layer. However, a high-resolution TEM image revealed that the stacking of graphene sheets in the mixed carbon sample was less ordered than that in graphene-modified one, which was probably due to the disturbance from amorphous pyrolitic carbon. Nevertheless, such disturbance might facilitate Li-ion diffusion on the surface of $LiFePO_4$, as it should be easier for Li ion to penetrate disordered carbon layers than orderly stacked graphene sheets. Consequently, $LiFePO_4$ modified by both graphene and pyrolitic carbon possessed best electrochemical performance. Our results shed some light on the application of graphene as a functional component to obtain high-performance electrode materials for LIBs. The effect of graphene is not only ascribed to its high electrical conductivity, but also to its flexible 2-D nanostructure which allows formation of a 3-D conductive network that is in intimate contact with active materials.

Later, reports on graphene-modified $LiFePO_4$ cathode material with a different structure and prepared by different methods were successively published. For instance, through a sol-gel process, $LiFePO_4$/graphene composites with a hierarchical porous structure could be produced [22]. Graphene sheets were inserted in the porous $LiFePO_4$ matrices, which enhanced the conductivity of $LiFePO_4$ and induced higher specific capacity and better rate capability than pure $LiFePO_4$. A microwave-assisted hydrothermal method was also developed to synthesize the composite, which is less time consuming than traditional hydrothermal route, since the whole reaction could be finished in 15 min [23]. The as-prepared $LiFePO_4$/graphene composite had a specific capacity approaching the theoretic value, long cycle life, and exceptional rate capability. Besides chemically reduced graphene, CVD-derived graphene sheets were also used to modify $LiFePO_4$ cathode materials as reported by Tang et al. [24]. Using Ni foam as the template, a 3-D graphene network was first prepared, which was then immersed in the suspension of $LiFePO_4$ nanoparticles. After evaporation of the solvent, 3-D graphene-supported $LiFePO_4$ cathode materials were finally obtained. Because of the excellent electrical conductivity of graphene prepared by the CVD method, the rate performance of $LiFePO_4$ could be significantly improved.

Though graphene has a very simple atomic structure, the structural parameters and the distribution state of graphene sheets in the composite cathode materials varies significantly depending on the experimental conditions. In order to understand the mechanism of graphene in improving the electrochemical performance of $LiFePO_4$ and to optimize the synthesis and performance of $LiFePO_4$/graphene composites, some deep investigations have been carried out recently. Yang et al. studied the impact of stacked graphene and unfolded graphene on $LiFePO_4$ cathode [25]. The stacked graphene, which consisted of multiple flacks packed in a perpendicular direction to the basal plane of graphene, was prepared by thermal exfoliation of GO, while the unfolded graphene was obtained by solution-based chemical reduction of GO. As expected, unfolded graphene induced much better charge/discharge performance than the stacked one, as the former enables better distribution of $LiFePO_4$. Therefore, restacking of graphene should be prevented to the largest extent in the synthesis of $LiFePO_4$/graphene composites, to realize the full potential of graphene as a conducting agent. The effect of graphene wrapping on the performance of $LiFePO_4$ cathode materials was also investigated as reported by Wei et al. [26]. Two different wrapping modes were compared in this work. The full wrapping of $LiFePO_4$ was realized by electrostatic interaction induced self-assembly of negatively charged graphene sheets and surface-modified $LiFePO_4$ nanoparticles that carried positive charges, while partial wrapping was achieved in the case where $LiFePO_4$ was not modified (Figure 3.8). Electrochemical measurement revealed that fully wrapped $LiFePO_4$ cathode materials presented poor electrochemical performance, even worse than pure $LiFePO_4$ with no carbon modification. In contrast, a partially wrapped sample displayed the best charge/discharge performance. EIS profiles indicated that the charge transfer resistance of a partially wrapped sample, which is related to the Li-ion insertion and extraction reactions, is much lower than that of the fully wrapped one. The above results suggested that partial wrapping of $LiFePO_4$ by graphene is beneficial to the performance of $LiFePO_4$, because full and tight graphene wrapping significantly blocked up the diffusion of Li-ions.

In another study, Bi et al. systematically compared the performance of $LiFePO_4$ cathode materials modified by three kinds of graphene samples obtained from different preparation methods, including CVD, chemical exfoliation, and Wurtz-type reductive coupling reaction [27]. At the same content of graphene, the sample prepared using CVD-derived graphene demonstrated better rate capability than the other two samples, which is mainly due to the higher electrical conductivity of graphene obtained by the CVD method than the other two kinds of graphene samples.

Lithium manganese phospate ($LiMnPO_4$) has an almost identical structure and specific capacity as that of $LiFePO_4$; however, its discharge voltage is ~0.6V higher than that of $LiFePO_4$. Accordingly, the energy density of $LiMnPO_4$ is ~20% higher than that of $LiFePO_4$, which makes $LiMnPO_4$ attractive as a next-generation cathode material. However, the electrical conductivity of $LiMnPO_4$ is even lower than that of $LiFePO_4$. The practical performance of $LiMnPO_4$ is still far from that of the demands for applications. Considering the analogy between $LiMnPO_4$ and $LiFePO_4$, the successful experience of modification of $LiFePO_4$ by graphene may also be applied in the case of $LiMnPO_4$. Based on the inspiring results of the $LiFePO_4$/graphene composite cathode material, our group successively conducted the synthesis of $LiMnPO_4$/graphene nanocomposites as a new cathode material [28]. Nano-sized $LiMnPO_4$ particles with controlled morphology were first synthesized. Then, modification of

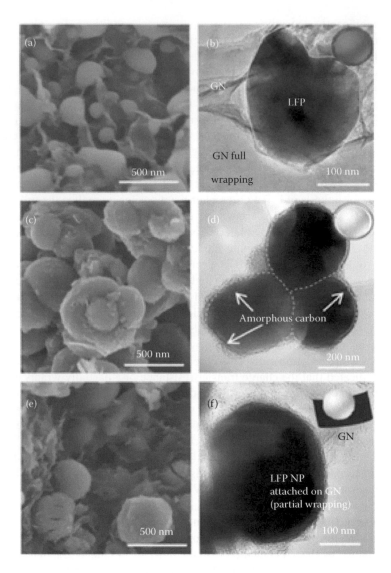

FIGURE 3.8 (a) SEM and (b) TEM images of LiFePO$_4$ wrapped by graphene sheets; (c) SEM and (d) TEM images of LiFePO$_4$ coated by pyrolytic carbon; (e) SEM and (f) TEM images of LiFePO$_4$ attached on graphene sheets. (Reprinted from *Carbon*, 57, Wei, W., Lv, W., Wu, M.-B. et al., The effect of graphene wrapping on the performance of LiFePO$_4$ for a lithium ion battery, 530–533, Copyright 2013, with permission from Elsevier.)

these nanoparticles by graphene was realized using a similar spray-drying technique. Homogenous wrapping of LiMnPO$_4$ nanoparticles with graphene sheets was achieved (Figure 3.9). Compared with pyrolitic-carbon-coated sample, graphene wrapping demonstrated higher specific capacity as well as a much-improved rate capability and cycling stability. The effectively enhanced electrical conductivity by the formation of

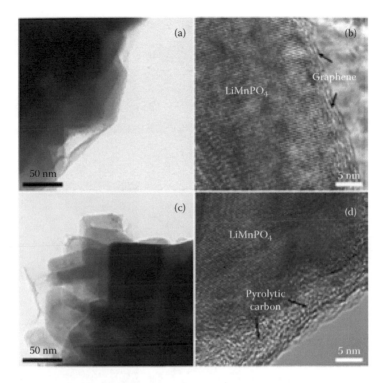

FIGURE 3.9 TEM images of (a and b) LiMnPO$_4$/graphene composite and (c and d) LiMnPO$_4$ coated by pyrolytic carbon. (From Qin, Z. H., Zhou, X. F., Xia, Y. G. et al., Morphology controlled synthesis and modification of high-performance LiMnPO$_4$ cathode materials for Li-ion batteries. *J. Mater. Chem.* 22:21144–21153, 2012. Reproduced by permission of The Royal Society of Chemistry.)

a 3-D conductive network composed of graphene was believed to be the key factor that gave rise to improved electrochemical performance of LiMnPO$_4$.

It has been discovered that doping of LiMnPO$_4$ with Fe could enhance the conductivity and stability of this cathode material. Therefore, synthesis of LiMn$_x$Fe$_{1-x}$PO$_4$ cathode material has been an alternate way to conquer the intrinsic defects of LiMnPO$_4$. In this case, graphene can also serve as a highly effective conducting agent to further enhance its electrochemical performance. Wang et al. first reported the synthesis of graphene-modified LiMn$_x$Fe$_{1-x}$PO$_4$ cathode material with ultrahigh rate performance [29]. As illustrated in Figure 3.10, in the experiment, Fe-doped manganese (II,III) oxide (Mn$_3$O$_4$) nanoparticles were initially anchored on GO sheets, which were then transformed to Fe-doped LiMnPO$_4$ nanorods by reacting with LiOH and phosphoric acid (H$_3$PO$_4$) solvothermally. Meanwhile, ascorbic acid was added to reduce Fe^{III} to Fe^{II} and also to reduce GO to graphene. At the manganese (Mn)/Fe ratio of 3:1, the obtained LiMn$_{0.75}$Fe$_{0.25}$PO$_4$/graphene composite had an electrical conductivity of 0.1–1 S cm^{-1}, which is 10^{13}–10^{14} times higher than that of pure LiMnPO$_4$. SEM and TEM images showed that uniform LiMn$_{0.75}$Fe$_{0.25}$PO$_4$ nanorods with lengths of 50–100 nm and widths of 20–30 nm were homogenously grown

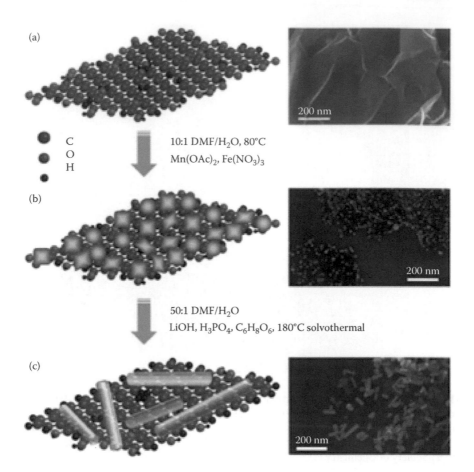

FIGURE 3.10   The schematic illustration of the growing process of $LiMn_{0.75}Fe_{0.25}PO_4$ nanorods on graphene and the corresponding SEM images of materials obtained at different stages of the reactions. (a) Structural model (left) and SEM image (right) of GO. (b) Structural model (left) and SEM image (right) of Fe doped Mn_3O_4 precursor nanoparticles grown on GO. (c) Structure model (left) and SEM image (right) of $LiMn_{0.75}Fe_{0.25}PO_4$ nanorods grown on reduced GO.

on large graphene sheets. Charge/discharge measurement showed that the composite cathode material had an outstanding rate capability that a discharge capacity of 107 mAh g^{-1} and 65 mAh g^{-1} could be delivered at 50 C and 100 C, respectively, which was comparable to some of the best reported cathode materials such as $LiFePO_4$. The material also demonstrated a good cycling stability in that only a 1.9% decay of the capacity occurred from the 11th to the 100th cycle. It was also discovered that graphene not only improved the electrical conductivity of cathode materials, but also played an important role in controlling the morphology of $LiMn_{0.75}Fe_{0.25}PO_4$. Without adding graphene, $LiMn_{0.75}Fe_{0.25}PO_4$ particles with irregular morphology instead of nanorod morphology were obtained. The functional groups and conjugated graphitic regions on GO sheets were believed to induce homogenous distribution of the precursors and controlled growth of nanocrystals.

Later, the application of x-ray absorption near-edge structure (XANES) spectroscopy to study the chemical bonding in the $LiMn_{0.75}Fe_{0.25}PO_4$/graphene composite cathode material was carried out, trying to get deep understanding of its ultrahigh rate performance [30]. The C K-edge XANES spectra indicated that graphene sheets in the composite cathode material had a better recovery of the sp^2 carbon crystallinity than freestanding reduced graphene sheets. This was ascribed to the intimate interaction between graphene and $LiMn_{0.75}Fe_{0.25}PO_4$, which boosted the reduction of GO during the reaction. Further combining with Mn, Fe L-edge, and O K-edge XANES spectra, it could be confirmed that $LiMn_{0.75}Fe_{0.25}PO_4$ bonded to graphene through the PO_4 unit, which led to the weakening of the P-O bond while strengthening the Fe-O bond via an induction effect. Such charge redistribution decreased the Fe 3d density of states and increased the O 2p density of states, which afforded a fast electron transportation channel. In addition, P L-edge and Li K-edge spectra suggested the existence of $Li_4P_2O_7$ on the surface of $LiMn_{0.75}Fe_{0.25}PO_4$, allowing fast Li-ion transportation. As a result, the $LiMn_{0.75}Fe_{0.25}PO_4$/graphene composite cathode material showed excellent rate capability.

Lithium vanadium phosphate ($Li_3V_2(PO_4)_3$) is also an attractive phosphate cathode material for LIBs, as it has a high theoretical capacity of 197 mAh g^{-1} and an average discharge voltage of ~3.8 V. Similar to $LiFePO_4$ and $LiMnPO_4$, $Li_3V_2(PO_4)_3$ also suffers from low electronic conductivity, which limits its rate capability. Thus, graphene modification is expected to be effective to improve the electrochemical performance of $Li_3V_2(PO_4)_3$ as well. Liu and coworkers applied a sol-gel process to prepare $Li_3V_2(PO_4)_3$/graphene nanocomposite, in which $Li_3V_2(PO_4)_3$ nanoparticles were adhered to and enwrapped by chemically reduced graphene sheets [31]. At a cutoff voltage of 3.0–4.3 V, the composite cathode material presented excellent rate performance and cycling stability. Even at a high rate of 20 C, it could still deliver a discharge capacity of 108 mAh g^{-1}, and no obvious capacity decay was observed when cycled at 20 C for 100 times. In contrast, pure $Li_3V_2(PO_4)_3$ without graphene modification could only reach a discharge capacity of 80 mAh g^{-1} at 5 C and apparent capacity decay occurred even at the low rate of 0.1 C. The significantly improved electrochemical performance of the $Li_3V_2(PO_4)_3$/graphene composite could be mainly attributed to the effective conducting network formed by graphene sheets. The same research group then compared graphene-modified $Li_3V_2(PO_4)_3$ cathode material with conventional carbon coated ones [32]. Using the CV data, it could be calculated that graphene modified ones had higher Li-ion diffusion coefficient in both the solid state cathode electrode and the electrolyte. The Nyquist plots indicated that the graphene-modified sample also had higher conductivity than the latter. Therefore, $Li_3V_2(PO_4)_3$/graphene composite presented better electrode reaction kinetics, which resulted in better electrochemical performance as reflected by the charge/discharge test. At the same discharge rate of 10 C, $Li_3V_2(PO_4)_3$/graphene composite could deliver a capacity of ~110 mAh g^{-1} while the carbon-coated sample could only reach ~85 mAh g^{-1}. Furthermore, the discharge capacity of the graphene-modified sample kept increasing from 17 mAh g^{-1} at the initial cycle to 46 mAh g^{-1} at the 1000th cycle. Such a phenomenon was not observed for the carbon-coated one.

Therefore, the addition of graphene could enormously improve the rate capability and cycling stability of $Li_3V_2(PO_4)_3$ cathode material. Besides higher electrical conductivity of graphene than amorphous carbon, graphene was also more efficient in forming a conductive network. In the carbon-coated sample, some carbon nanoparticles and some $Li_3V_2(PO_4)_3$ nanoparticles aggregated separately without effectively compounding each other. In the graphene-modified sample, $Li_3V_2(PO_4)_3$ nanoparticles were homogenously grown on graphene sheets, ensuring efficient contact of each active nanoparticle to the conductive network. Consequently, $Li_3V_2(PO_4)_3$ nanoparticles in the $Li_3V_2(PO_4)_3$/graphene composite were more available and electrochemically efficient than those in the carbon-coated sample.

Graphene-modified $Li_3V_2(PO_4)_3$ cathode material was also prepared by a spray-drying method similar to the previous report of graphene-modified $LiFePO_4$ cathode materials [33]. Spherical microsized particles composed of $Li_3V_2(PO_4)_3$ primary nanoparticles, interconnected graphene network, and citric-derived pyrolitic carbon were obtained. Graphene significantly enhanced the conductivity of $Li_3V_2(PO_4)_3$, while amorphous carbon interrupted the stacking of graphene sheets and minimized in-plane anisotropy of electron migration within the graphene layers. Accordingly, the composite cathode material could deliver a capacity almost reaching the theoretic value, and displayed good rate capability and cycle life.

A more complicated structure of graphene-modified $Li_3V_2(PO_4)_3$ was designed and synthesized by Rui et al. [34]. In this composite, ultrasmall $Li_3V_2(PO_4)_3$ nanoparticles with diameters of 5–8 nm were embedded in a nanoporous carbon matrix with a spherical morphology. The $Li_3V_2(PO_4)_3$/carbon composites were then attached to graphene sheets. Nanoporous carbon acted as a nanocontainer to enhance the interaction between the active materials and the electrolyte, while graphene enhanced the charge transfer and also facilitated the growth of $Li_3V_2(PO_4)_3$ nanograins. The sample could reach a discharge capacity of 90 mAh g^{-1} at an ultrahigh rate of 50 C, and still remained at 88 mAh g^{-1} after 1000 charge/discharge cycles.

3.3.1.2   Li-Metal-Oxide Graphene Composite Cathode Materials

Li metal oxide is a large family of cathode materials for LIBs, including some important members that have been widely used in commercial products, such as $LiCoO_2$ and $LiMn_2O_4$. Spinel-structured $LiMn_2O_4$ cathode material with the advantages of 3-D Li-ion diffusion paths, low cost, and environmental benignity has attracted much attention in recent years. Though $LiMn_2O_4$ has a higher electrical conductivity than phosphate cathode materials, it is still four orders of magnitude lower than that of $LiCoO_2$. Therefore, modification of $LiMn_2O_4$ with conductive agents is an important pathway to achieve better electrochemical performance. The strategy of employing graphene to enhance the conductivity of active materials can also be applied for $LiMn_2O_4$ cathode materials.

The first report on graphene-modified $LiMn_2O_4$ cathode material was published in 2011 by Bak and coworkers [35]. In the experiment, MnO_2/graphene hybrid was first prepared by *in situ* redox deposition of MnO_2 on chemically reduced GO. Then the hybrid was mixed with LiOH aqueous solution and underwent microwave-assisted hydrothermal treatment to obtain $LiMn_2O_4$/graphene composite. TEM characterization (Figure 3.11) showed that $LiMn_2O_4$ nanoparticles with diameters of 10–40 nm

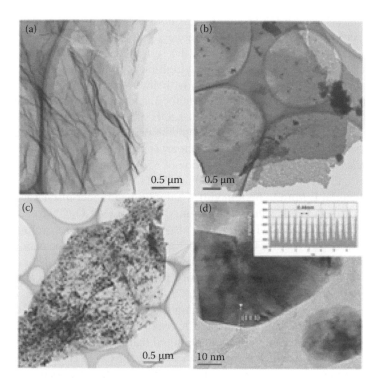

FIGURE 3.11 TEM images of (a) graphene nanosheets, (b) MnO_2/graphene hybridprecursor, and (c) spinel $LiMn_2O_4$/graphene hybrid. (d) High-resolution TEM image of an individual spinel $LiMn_2O_4$ nanoparticle on graphene RGO (the inset is the direct line scanning analysis profile of the $LiMn_2O_4$ nanoparticle). (From Bak, S.-M., Nam, K. W., Lee, C.-W, et al., Spinel $LiMn_2O_4$/reduced GO hybrid for high rate lithium ion batteries. *J. Mater. Chem.* 21:17309–17315, 2011. Reproduced by permission of The Royal Society of Chemistry.)

were homogenously dispersed on graphene sheets. Fourier-transformed x-ray absorption fine structure spectra indicated that the Debye-Waller factor of the Mn-O bond in the $LiMn_2O_4$/graphene composite was slightly smaller than that in pure $LiMn_2O_4$, which may suggest reduced Jahn-Teller distortion in the hybrid due to its unique hybrid structure. The graphene-modified $LiMn_2O_4$ cathode material exhibited excellent rate capability; 85% and 74% capacity with respect to the discharge capacity of 137 mAh g^{-1} at 1 C rate could be reached at 50 C and 100 C, respectively. The sample also displayed good cycling stability and a capacity retention of 90% at 1 C rate and 96% at 10 C rate was achieved after 100 cycles. Using the methodology suggested by Tolbert et al. 2010 [36], the authors carried out a deep investigation to understand the excellent rate performance of the composite. The analysis of the CV curves demonstrated that the surface reaction contributed 59% and 71% of the total charge storage for the $LiMn_2O_4$/graphene composite at the scan rate of 1 mV s^{-1} and 10 mV s^{-1}, respectively, which was unaffected by the increase in potential scan rate and was speculated to arise from the relatively large surface area of nano-sized $LiMn_2O_4$

particles. Such large capacitive storage was considered to greatly contribute to the high rate performance of the $LiMn_2O_4$/graphene cathode materials.

In another work, Zhao et al. reported the self-assembly of ultrasmall $LiMn_2O_4$ nanoparticles on CNT or graphene to be a high-performance cathode material [37]. In their experiments, Mn_3O_4 nanoparticles with diameters of ~5 nm were first synthesized using a catanionic reverse-micelle method, and then mixed with a suspension of CNT or graphene. After reaction with LiOH at 380°C in air, Mn_3O_4 was converted to $LiMn_2O_4$, and the $LiMn_2O_4$/graphene or $LiMn_2O_4$/CNT composite cathode materials were obtained. $LiMn_2O_4$ nanocrystals derived from ultrasmall Mn_3O_4 nanoparticles had a narrow size distribution of ~7 nm, and were deposited homogeneously as a thin layer on either individual CNTs or graphene sheets. Comparing with commercial $LiMn_2O_4$, both modified samples had higher discharge capacities approaching the theoretic value, (i.e., 144 mAh g^{-1} for $LiMn_2O_4$/CNT and 146 mAh g^{-1} for $LiMn_2O_4$/graphene). The samples also demonstrated good cycling stability. The average capacity loss for $LiMn_2O_4$/CNT and $LiMn_2O_4$/graphene per cycle for 80 cycles were 0.16% and 0.195%, respectively. The modification with CNT or graphene could also enhance the rate performance of $LiMn_2O_4$. Upon returning to 0.3 C after 5 C test, the capacity losses of $LiMn_2O_4$/CNT and $LiMn_2O_4$/graphene were 7.3% and 9.0%, respectively, in contrast to 30% for the commercial $LiMn_2O_4$. Meanwhile, the graphene-modified sample exhibited less polarization than the CNT-modified one during the entire cycling. It should be noted that the content of graphene in the composite was 6 wt%, which is lower than 10 wt% for CNT. Therefore, it could be deduced that graphene was more effective in improving the electrochemical performance of $LiMn_2O_4$ than CNT. As graphene prepared from chemical reduction of GO had more oxygen-functional groups than pristine CNT, graphene sheets might have stronger interaction with $LiMn_2O_4$ nanoparticles to minimize phase segregation during the preparation process and electrochemical cycling.

$LiMn_2O_4$/graphene composite cathode materials were also prepared by other methods, such as solvothermal growth of $LiMn_2O_4$ on graphene sheets [38] or mechanical mixing of $LiMn_2O_4$ with graphene using ultrasonic agitation [39]. Electrochemical performance, especially the rate capability, was enormously improved compared to pure $LiMn_2O_4$, which shed some light on the potential application of graphene in high-performance $LiMn_2O_4$ cathode materials.

$LiNi_{0.5}Mn_{1.5}O_4$ is another type of important cathode material mainly due to its high energy density. With a theoretical specific capacity of 147 mAh g^{-1} and a high working voltage of 4.7 V, the energy density of $LiNi_{0.5}Mn_{1.5}O_4$ is 20% and 30% higher than that of $LiCoO_2$ and $LiFePO_4$, respectively. However, the electrical conductivity of $LiNi_{0.5}Mn_{1.5}O_4$ is also low. Meanwhile, at the high working voltage, the side reaction on the interface between $LiNi_{0.5}Mn_{1.5}O_4$ and the electrolyte usually causes serious capacity fading during cycling. In addition, Mn^{2+} ions tend to dissolve in the electrolyte and then deposit on the anode, which also induces capacity fading. Therefore, surface protection becomes important to enhance the cycling stability of $LiNi_{0.5}Mn_{1.5}O_4$. Fang et al. employed mildly oxidized GO sheets as a protection layer to wrap $LiNi_{0.5}Mn_{1.5}O_4$ particles [40]. It could be observed by TEM that the GO layer coating on the surface of $LiNi_{0.5}Mn_{1.5}O_4$ had a thickness of ~5 nm. Compared to pristine $LiNi_{0.5}Mn_{1.5}O_4$, the sample coated with GO had a narrower voltage gap between charge and discharge curves, indicating that GO could reduce polarization of the cell. The GO-coated sample also presented better rate capability. It could deliver 56% of its 1 C capacity at the discharge rate of 10 C, while pristine $LiNi_{0.5}Mn_{1.5}O_4$ could deliver nearly no capacity at the same rate. Another advantage of GO modification is the improvement of the cycling stability. The battery only showed 0.039% capacity decay per cycle in 1000 cycles. The excellent electrochemical performance of $LiNi_{0.5}Mn_{1.5}O_4$/graphene composite was ascribed to three main reasons: (1) GO was able to enhance the conductivity of the cathode material and thus reduce the impedance of the cell as proved by EIS, although it was mildly oxidized. The oxygenous groups in GO sheets also increased the affinity between GO and $LiNi_{0.5}Mn_{1.5}O_4$, giving rise to more efficient wrapping; (2) GO coating on the surface of $LiNi_{0.5}Mn_{1.5}O_4$ particles could suppress dissolution of Mn^{3+}; and (3) the side reaction between Ni^{4+} and electrolyte could also be inhibited by GO. However, it should be pointed out here that oxygen functional groups on GO sheets are much more active than conjugated carbon atoms. The side reaction between oxygenous groups and electrolytes may also occur especially at high voltage, although the authors did not mention it in the paper. Deeper investigation and long-term evaluation need to be conducted to comprehensively understand the practical performance of GO in the electrode.

As a member of the multicomponent solid solution cathode material, $LiNi_xMn_yCo_{1-x-y}O_2$ has attracted much attention in recent years because of its structural stability and improved electronic conductivity compared with lithium nickel manganese oxide. Among the $LiNi_xMn_yCo_{1-x-y}O_2$ series, $LiNi_{1/3}Mn_{1/3}Co_{1/3}O_2$ is one of the most promising candidates as a high-performance cathode material because of its high specific capacity and long cycle life. However, $LiNi_{1/3}Mn_{1/3}Co_{1/3}O_2$ also suffers from low electrical conductivity, which limits its practical application. Many attempts including doping and surface coating by conductive materials have been carried out to increase the conductivity of $LiNi_{1/3}Mn_{1/3}Co_{1/3}O_2$. Graphene, being a highly conductive carbon material with low density, is expected to be more effective and has been successfully applied in modification of $LiNi_{1/3}Mn_{1/3}Co_{1/3}O_2$ cathode materials. Rao and coworkers used ball milling to mix preprepared $LiNi_{1/3}Mn_{1/3}Co_{1/3}O_2$ particles to form a composite with a graphene weight ratio of 10% [41]. $LiNi_{1/3}Mn_{1/3}Co_{1/3}O_2$ was homogeneously covered with graphene sheets in the sample. CV analysis showed that graphene-modified $LiNi_{1/3}Mn_{1/3}Co_{1/3}O_2$ had a higher current at the same scan rate than pristine $LiNi_{1/3}Mn_{1/3}Co_{1/3}O_2$, implying improved electrochemical performance by the addition of graphene. The graphene modification demonstrated apparent advantages in enhancing the rate capability of the cathode material. Discharge capacities of 172 and 153 mAh g^{-1} was reached for the graphene/$LiNi_{1/3}Mn_{1/3}Co_{1/3}O_2$ composite at the rate of 2 C and 5 C, respectively, while the pristine counterparts could only deliver 162 and 138 mAh g^{-1} at the same rate. It is important to note that graphene modification could also effectively decrease the irreversible capacity of $LiNi_{1/3}Mn_{1/3}Co_{1/3}O_2$, which is always a big problem in the application of $LiNi_xMn_yCo_{1-x-y}O_2$ cathode materials. The initial irreversible capacity could be reduced from 5.4% for pristine $LiNi_{1/3}Mn_{1/3}Co_{1/3}O_2$ to 1.6% after graphene addition. EIS results showed that the charge-transfer resistance of the $LiNi_xMn_yCo_{1-x-y}O_2$/graphene composite was 32 Ω, which was much lower than 181 Ω for the pristine $LiNi_{1/3}Mn_{1/3}Co_{1/3}O_2$, which implied that the enhanced electrochemical performance of graphene-modified $LiNi_xMn_yCo_{1-x-y}O_2$ was mainly attributed to the improvement of the electrical conductivity by graphene.

3.3.1.3 Metal-Oxide-Graphene Composite Cathode Materials

Metal oxide is a large family that has electrochemical activity and can be used as electrode materials in LIBs. Although most metal oxides have relatively low electrode potential and are suitable as anode materials, some members have the potential high enough to be used as cathode material. Among them, vanadium oxide is one of the most important and has been widely studied. There are mainly two types of vanadium oxide cathode material, VO_2 and V_2O_5, and they can be changed into each other by redox reactions.

VO_2 is a promising cathode material for both organic and aqueous LIBs owing to its high capacity and unique structure. Nevertheless, VO_2 has relatively high charge-transfer resistance, which causes poor cycling stability. Graphene was thus employed as a conductive agent to conquer the intrinsic defects of VO_2 and improve its electrochemical performance. Yang and coworkers reported the synthesis of VO_2/graphene composites with a unique nanoribbon structure for ultrafast Li storage [42]. The material was prepared by *in situ* reduction of VO_2 by GO under hydrothermal conditions. As shown in Figure 3.12, VO_2 had ultrathin nanoribbon morphology with 200–600 nm in width, ~10 nm in thickness, and several tens of micrometers in length. Some irregular-shaped thin sheets observed in the sample were graphene that attached to the surface of VO_2 nanoribbons. The graphene layers were not continuous. Some nano-sized pores were discerned in a high-resolution TEM image, which was believed to be favorable for diffusion of electrolytes and access of Li-ions to the active VO_2 material. The porous structure was also proved by nitrogen sorption isotherms in which a typical type II hysteresis loop could be observed. The calculated pore size distribution indicated that the composite material had mesopores with the pore diameter in the range of 3–20 nm. The VO_2 nanoribbons and graphene sheets formed a 3-D interpenetrating network with plenty of open macropores. With a VO_2 content of 78%, the composite cathode material could deliver a high reversible capacity of 415 mAh g^{-1} at 1 C and could still demonstrate a reversible capacity as high as 204 mAh g^{-1} at the extremely high rate of 190 C, corresponding to a power density of 110 kW kg^{-1}. After cycling at 190 C for 1000 times, the capacity retention of 93% was maintained. The electrochemical performance of the VO_2/graphene composite was also evaluated at elevated temperatures. The reversible capacity increased along with the rise of the temperature. The highest capacity of 410 mAh g^{-1} at 5 C was acquired at a temperature of 75°C. The excellent cycling stability was also maintained at high temperatures. A capacity retention of ~90% could be achieved after 200 cycles at 28 C. The extraordinary charge/discharge performance of graphene-modified VO_2 cathode material could be ascribed to its unique structure. On one hand, graphene intimately attached to VO_2 nanoribbons effectively increased the electrical conductivity of the active material. Meanwhile, nanopores in the graphene sheets facilitated diffusion of Li-ions. On the other hand, the ultrasmall dimension of the thickness of the nanoribbons gave rise to very short solid-state diffusion time

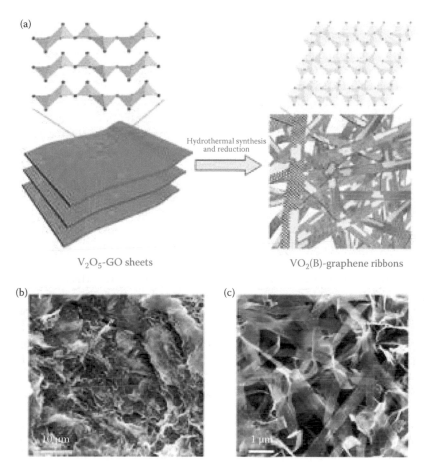

FIGURE 3.12 (a) Structural model of VO$_2$/graphene composite nanoribbons. (b) SEM image of the V$_2$O$_5$/graphite oxide composite and (c) SEM image of VO$_2$/graphene composite nanoribbons. (Reprinted with permission from Ref. 41, 1596–1601. Copyright 2013 American Chemical Society.)

of Li-ions (estimated to be less than 0.01 s). In addition, VO$_2$ with an edge-sharing structure could preserve the structural stability during long-term charge/discharge cycles. The VO$_2$/graphene composite cathode materials reported in this paper presented outstanding electrochemical performance, especially its ultrafast charge and discharge capability, which has attractive potential for applications in high-power LIBs.

In another report, VO$_2$ with a nanotube structure was modified by graphene using a hydrothermal method [43]. The simultaneous reduction of ammonium vanadate to VO$_2$ and reduction of GO to graphene under hydrothermal conditions resulted in the formation of VO$_2$ nanotube/graphene hybrid material. VO$_2$ nanotubes had a mean diameter of ~100 nm and length of several micrometers, and they were wrapped by graphene sheets. Since no similar tubular VO$_2$ had been reported before, the authors speculated that GO might play an important role in controlling the growth of VO_2 to form nanotube morphology. The V 2p peak in the XPS spectrum of the hybrid shifted to higher binding energy comparing with pristine VO_2, implying a weak interaction between VO_2 and graphene. Compared with previously reported pure nanostructured VO_2 cathode material, the VO_2/graphene composite showed better electrochemical performances in terms of reversible capacity and cycle life. The hybrid presented an initial reversible capacity of 450 mAh g^{-1} at the current density of 40 mA g^{-1}, and the capacity retention of ~70% could be achieved after 20 charge/discharge cycles. However, the capacity dropped rapidly when the current density was raised. Consequently, the researchers from the same group then tried to improve the performance of VO_2/graphene hybrid by a new structural design [44]. By simply doubling the content of GO in the reaction suspension and prolonging the hydrothermal time from 48 to 72 h while keeping other conditions identical to those in their earlier research, a flowerlike nanostructure composed of VO_2 nanosheets and N-doped graphene was obtained. In this hybrid, the ultrathin VO_2 nanosheets (~10 nm in thickness) enormously reduced the diffusion length of Li-ion and increased the contact area between the active material and the electrolyte, which gave rise to improved kinetics of Li insertion/extraction. N-doped graphene, which was derived from the reaction between GO and ammonium ions presented in the reaction mixture, could further enhance the electrical conductivity of the hybrid than pristine graphene. As a result, the flowerlike nanocomposite of VO_2/graphene showed much improved electrochemical performance than the VO_2 nanotube/graphene hybrid. At a current density of 200 mA g^{-1}, VO_2 nanosheet/N-doped graphene composite could deliver a capacity of 300 mAh g^{-1}, while only 200 mAh g^{-1} was achieved at an even lower current density of 100 mA g^{-1} for the VO_2 nanotube/graphene composite.

V_2O_5 has also attracted much attention as a cathode material due to its high energy density and low cost. V_2O_5 cathode material has a layered structure and possesses a theoretical specific capacity of 294 mAh g^{-1}. However, similar to VO_2, V_2O_5 also suffers from low electrical conductivity and sluggish kinetics of Li-ion transport, which causes poor rate capability and cycle stability. The hybridization of V_2O_5 with highly conductive materials, such as carbon materials, is useful to improve its charge transfer efficiency. Graphene, as a new member of the carbon materials group with outstanding physicochemical properties, has been proved to be a good candidate to enhance the electrochemical performance of V_2O_5 cathode materials.

Rui and coworkers designed and synthesized graphene-supported porous V_2O_5 spheres as a high-performance cathode material by a solvothermal method [45]. In the experiment, vanadium isopropoxide was mixed with the GO suspension in N-methylpyrrolidone. The following addition of ammonia resulted in the nucleation of $V_2O_5 \cdot nH_2O$ on GO sheets. The mixture was then sealed and heated at 180°C for 12 h. During this solvothermal process, the V^{5+} species was likely reduced to from amorphous vanadium oxide with particle size of 10–50 nm, while GO was also partially reduced. The obtained precipitate was finally calcined to form the V_2O_5/graphene composite cathode. As shown in Figure 3.13, V_2O_5 with diameters of 200–800 nm were attached to a graphene sheet. Large amounts nanopores with a diameter of ~50 nm could be observed on V_2O_5 spheres. Some broken spheres suggested hollow interior. TEM images further confirmed the hollow structure of the V_2O_5

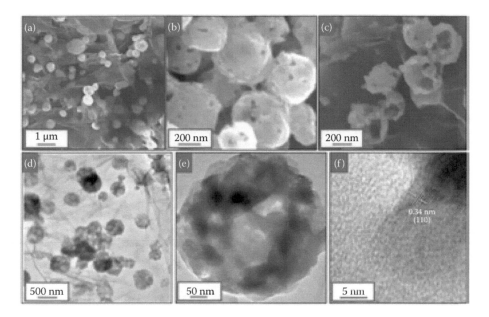

FIGURE 3.13 (a–c) SEM and (d–e) TEM images of the V_2O_5/graphene composite. (f) High-resolution TEM image of V_2O_5 in the sample. (From Rui, X., Zhu, J., Sim, D. et al., Reduced graphene oxide supported highly porous V_2O_5 spheres as a high-power cathode material for lithium ion batteries. *Nanoscale* 3:4752–4758, 2011. Reproduced by permission of The Royal Society of Chemistry.)

spheres, and the shell was composed of V_2O_5 nanocrystals having a size of ~50 nm. The V_2O_5 spheres were supposed to be formed by the oxidation and assembly of amorphous vanadium oxide nanoparticles during the annealing process as a result of lowering the surface energy. These spheres were strongly anchored on the graphene sheets as no detachment was observed even after the sample was ultrasonicated for 1 h. The strong interaction between V_2O_5 and graphene may facilitate fast electron transport in the hybrid, whereas the hollow structure of V_2O_5 itself was believed to be beneficial for fast Li diffusion. Consequently, the V_2O_5/graphene hybrid cathode material exhibited much improved electrochemical performance. It could deliver an initial discharge capacity of 238 mAh g^{-1} in the voltage range of 4.0–2.0 V, and no irreversible capacity was observed in the subsequent charge process. When cycled at a current density of 90 mA g^{-1}, capacity retention of 85% could be achieved after 50 cycles, corresponding to a capacity fading rate of 0.3% per cycle, and the Coulombic efficiency of each cycle within 50 cycles consistently maintained at ~100%. The cell also showed good rate capability. Even when the discharge rate was raised to 5700 mA g^{-1} (19 C), a capacity of 102 mAh g^{-1} was still retained, and it could still deliver a capacity of 93 mAh g^{-1} after 200 cycles at such a high rate. The high specific capacity and excellent rate performance provide this V_2O_5/graphene hybrid cathode with attractive potential in the application of high-performance LIBs.

In another report, $V_2O_5 \cdot nH_2O$ xerogel/graphene composite cathode material was presented. Hydrated vanadium pentoxide has a higher intercalation capacity than crystalline V_2O_5 [46]. A high specific capacity of 560 mAh g^{-1} is expected for the $V_2O_5 \cdot nH_2O$ xerogel. However, $V_2O_5 \cdot nH_2O$ also has low conductivity and Li-ion diffusion rate. Du and coworkers proposed the modification of $V_2O_5 \cdot nH_2O$ with graphene by simply filtrating the mixed suspension of $V_2O_5 \cdot nH_2O$ and graphene in water to form a thin film. $V_2O_5 \cdot nH_2O$ prepared by a hydrothermal method had a ribbon-like nanostructure with width of ~100 nm, and were homogeneously attached to graphene sheets. The electrochemical performance of the $V_2O_5 \cdot nH_2O$ hybrid was significantly affected by the graphene content. The hybrid with a graphene content of 39.6 wt% showed the best performance. It delivered an initial discharge capacity of 212 mAh g^{-1} in the voltage range of 1.5–4 V, and a discharge capacity of 190 mAh g^{-1} was maintained after 50 cycles. The improved electrochemical performance was ascribed to following factors. First, graphene could effectively enhance the electrical conductivity of $V_2O_5 \cdot nH_2O$. Second, graphene was able to buffer the volume change of $V_2O_5 \cdot nH_2O$ during cycling. Third, graphene could also help to maintain the layered structure of $V_2O_5 \cdot nH_2O$ with less amount of crystal water than in the case of pure $V_2O_5 \cdot nH_2O$, resulting in less side reactions and improved cycle stability.

3.3.2   GRAPHENE-BASED COMPOSITE ANODE MATERIALS

3.3.2.1   Metal-Graphene Composite Anode Materials

3.3.2.1.1   Si/Graphene Composite Anode Materials

Silicon is a promising anode material for LIBs because of its ultrahigh capacity. The theoretic capacity of Si anode is up to 4200 mAh g^{-1}, which is more than 10 times higher than that of commercial graphite anode. Consequently, Si has been anticipated to be the next-generation anode material for high-energy-density LIBs. Nevertheless, the large volume change (up to 300%) of Si during charge and discharge processes causes severe pulverization of Si particles, which generates loss of electrical contact and gives rise to fast capacity fading. Therefore, the short cycle life hinders the practical application of Si. Several strategies have been developed to conquer this problem. One way is to reduce the particle size of Si so as to ease the internal strains during volume expansion. Another way is to load Si on conductive matrices, which can buffer the volume change and also improve electrical conductivity. Carbon materials are suitable choices as the matrices because of their diverse structures and high conductivity. As a new member of the carbon materials group, graphene is expected to be advantageous over other carbon materials to improve the electrochemical performance of Si due to its excellent electrical conductivity, large surface area, and flexible 2-D nanostructure.

The first attempt to synthesize graphene-modified Si anode materials was conducted by Chou and coworkers by simply mixing graphene and commercial nano-sized Si particles in a mortar [47]. The Si/graphene composite could maintain a reversible capacity of 1186 mAh g^{-1} up to 30 cycles, while the capacity of pure Si dropped rapidly to 346 mAh g^{-1} after 30 cycles. The enormously enhanced cycling stability was attributed to an interconnected graphene framework that not only provided high electrical conductivity, but also accommodated large strains of Si nanoparticles during cycling. However, the simple mechanical mixing was not able to achieve high homogeneity in the composite, and the effect of graphene could not be fully utilized. Therefore, Si/graphene composites with diverse and more controllable structures were successively designed and prepared by different methods in order to optimize the electrochemical performance.

Zhou et al. used a freeze-drying technique to insert Si nanoparticles into graphene sheets [48]. In the as-prepared composite, wrinkled graphene sheets formed a network and well-covered highly dispersed Si nanoparticles (Figure 3.14). With a graphene content of 17.9 wt%, the composite anode material presented a high initial capacity of 3070 mAh g^{-1} at a current density of 200 mA g^{-1}, although the initial irreversible capacity was high (~1200 mAh g^{-1}). After the first cycle, the Coulombic efficiency increased and stabilized at 98%–100% in the following cycles. After 100 cycles, the reversible capacity of graphene-modified Si anode was still as high as 1153 mAh g^{-1}, while that for the pure Si nanoparticles was only 13 mAh g^{-1}. The rate capability of Si could also be effectively improved by the addition of graphene. The reversible capacity of 803 mAh g^{-1} could be reached at a high current density of 4 A g^{-1}. Compared with mechanical mixing of Si with graphene, the two

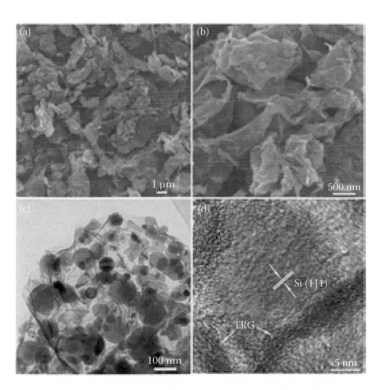

FIGURE 3.14 SEM images (a and b) and TEM images (c and d) of the Si/graphene composite anode material. (From Zhou, X., Yin, Y.-X., Wan, L.-J. et al., Facile synthesis of silicon nanoparticles inserted into graphene sheets as improved anode materials for lithium-ion batteries. *Chem. Commun.* 48:2198–2200, 2012. Reproduced by permission of The Royal Society of Chemistry.)

components reported in this paper were blended in a much more homogeneous way, which resulted in improved electrochemical performance.

Si/graphene composite with other unique structures were also reported. For instance, Si nanoparticles encapsulated by crumpled graphene sheets were prepared by a capillary-driven assembly route in aerosol droplets [49]. All Si nanoparticles were wrapped by a crumpled graphene shell that could accommodate expansion and contraction of Si during continuous charge/discharge processes without fracture. Therefore, the cycling performance of Si was remarkably improved. The sample could retain 83% of the initial capacity after 250 cycles, and after 15 cycles, the capacity loss within each cycle was as low as 0.05% in the subsequent cycles. The crumpled capsule structure was well maintained after the cycling test. A graphene/carbon-coated Si nanoparticle hybrid was reported by Zhou et al. [50], in which Si nanoparticles were wrapped between graphene sheets and amorphous carbon layers. The double protection by both graphene and amorphous carbon gave rise to outstanding electrochemical performance. In another report, Si/graphene nanocomposite was synthesized by electrostatic assembly between positively charged aminopropyltriethoxysilane-modified Si nanoparticles and negatively charged GO sheets [51]. Si nanoparticles were well embedded and homogeneously dispersed on graphene sheets without aggregation. Meanwhile, the ample nanospace that existed around individual Si nanoparticles provided enough buffering space for the volume change of Si. Si/graphene composite with a multilayer structure was also produced by repeated filtering liquid phase exfoliated graphene sheets and subsequent coating of amorphous Si film by plasma-enhanced CVD method. The graphene film sandwiched by Si layers in the composite could effectively circumvent aggregation of Si during charge/discharge processes and was also able to buffer the volume change at the same time. Consequently, the structural integrity could be well maintained after long-term cycles.

In order to enhance the stability of the Si/graphene composite, attempts to reinforce the interaction between two components were carried out. Yang and coworkers reported the synthesis of Si/graphene nanocomposite with covalent bonds formed between Si nanoparticles and graphene sheets [52]. In their experiment, graphene was first grafted by aminophenyl groups, which was then mixed with Si nanoparticles in the solvent of acetonitrile (CH_3CN). After the addition of iso-amylnitrite, the *in situ* acryl radicals grasped Si nanoparticles by forming covalent bonds that could be confirmed by Raman, thermogravimetric analysis (TGA), and XPS measurement. Such strong interactions between Si and graphene resulted in enhanced cycling stability. At a current density of 300 mA g^{-1}, the composite anode material could deliver a reversible capacity of 828 mAh g^{-1} after 50 cycles. Even at a high current density of 4 A g^{-1}, it could still reach 350 mAh g^{-1} after 40 cycles.

Another approach to form covalent bonds between Si and graphene could be realized by the reaction of amino-functionalized Si nanoparticles with GO [53]. It was interesting to note that at a weight ratio of Si/graphene = 15/1, the reversible capacity increased along with cycle times. The possible reason for this phenomenon proposed by the authors was that some Si nanoparticles were sandwiched by graphene sheets in the sample, which suppressed the infiltration of electrolytes. During long-term cycling, the volume expansion of Si gradually enlarged the space between adjacent graphene sheets, which allowed the penetration of electrolyte. A similar strategy using amino-functionalized Si to covalently bind to graphene was conducted by Wen et al. [54]. In addition, the authors combined the aerosol spray process to eventually obtain graphene-bonded and encapsulated Si nanoparticles as an anode material. Although Si was wrapped by graphene to form a capsulelike structure, some open ends of the graphene shell could be clearly identified, which allowed penetration of electrolyte into the capsules to react with the Si.

In all of the above research, Si nanoparticles were either purchased or synthesized separately, which were then mixed with graphene sheets to form the composites. *In situ* growth of Si on graphene, which may have better homogeneity and stronger interaction between two components, is another route to obtain Si/graphene composite anode materials. However, it is not easy to obtain nano-sized Si by a chemical process that is adapted to graphene. Fortunately, magnesiothermal reduction of silica to Si provides a suitable way to obtain Si nanostructures grafted on graphene sheets. On one hand, amorphous silica with abundant surface silanol groups has an excellent affinity to chemically derived graphene or GO, which allows formation of silica on the surface of graphene and can be then *in situ* converted to Si. On the other hand, the morphology and structure of amorphous silica can be easily tuned, which brings about the possibility to obtain Si with various nanostructures. Therefore, our group carried out the synthesis of Si/graphene using a magnesiothermal reduction method [55]. In the first step, homogenous amorphous silica thin layers were deposited onto both sides of GO sheets by hydrolysis of tetraethyl orthosilicate in an aqueous suspension of GO. The thickness of the silica layer was measured to be 3 nm on each side. The silica/GO composite nanosheets were then sealed in a container with magnesium (Mg) powder and heated at $700°C$ for 3 h. During this process, silica was reduced by Mg vapor to form Si nanoparticles, while GO was thermally reduced to graphene. The main by-product magnesium oxide (MgO) was removed by washing in HCl. Si nanoparticles had diameters in the range of 10–30 nm, which was smaller than those reported previously. No detachment of Si nanoparticles from graphene sheets were observed even after the ultrasonication treatment of the composite, implying a strong interaction between two components. XPS data revealed the existence of silicon oxide (SiO_x) ($0 < x < 2$) in the composite, which was derived from incomplete reduction of SiO_2 by Mg or surface oxidation of freshly prepared Si nanoparticles in air or in the solution. The as-prepared Si/graphene composite sheets were then further mixed with additional GO in aqueous solution and spray-dried to form spherical micro-sized particles with a 3-D porous structure as shown in Figure 3.15. High-magnification SEM images showed that the microspheres were composed of interconnected Si/graphene composite sheets and pure graphene sheets and enclosed plenty of cavities with sub-micron dimensions. STEM characterization showed that Si nanocrystals loaded on graphene sheets formed a nanoporous structure that was expected to be beneficial for the diffusion of electrolytes. The nanopores among Si nanocrystals could also act as buffering areas for the volume expansion during Li insertion. Elemental mapping using electron energy-loss spectroscopy indicated that the composite contained three elements of Si, C, and O, and their spatial distribution was demonstrated in Figure 3.15f. It was found that O was enriched on the outer surface of Si nanoparticles, suggesting a thin SiO_x layer that covered the Si nanocrystals. The oxidized layer also

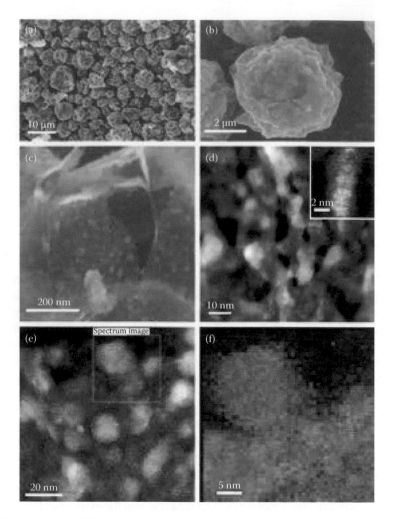

FIGURE 3.15   (a–c) SEM images of the Si/graphene nanocomposite. (d) STEM image of a Si/graphene composite microsphere, and Si nanocrystals (inset). (e) STEM image of the Si/graphene nanocomposite and (f) the corresponding elemental map in the selected area in (e), with green for C, red for Si, and blue for O. (From Xin, X., Zhou, X., Wang, F. et al. A 3D porous architecture of Si/graphene nanocomposite as high-performance anode materials for Li-ion batteries. *J. Mater. Chem.* 22:7724–7730, 2012. Reproduced by permission of The Royal Society of Chemistry.)

linked Si nanocrystals with graphene sheets. The porous structure was further confirmed by N_2 sorption characterization. The BET surface area was calculated to be 289.2 m^2 g^{-1} for the composite, and a major pore size of ~10 nm was obtained by the Barrett–Joyner–Halenda (BJH) method, which probably corresponded to the cavities among Si nanoparticles. The Si/graphene composite anode material exhibited excellent electrochemical performance. It could constantly deliver a reversible capacity as high as 1000 mAh g^{-1} within 30 cycles at the current density of 100 mA g^{-1}. When increasing the charge/discharge rate, the Si/graphene composite materials could still maintain relatively high capacity. Even at a high current density of 10 A g^{-1}, a reversible capacity of 440 mAh g^{-1} could still be achieved. The advantageous performance of this Si/graphene anode material could be mainly ascribed to three factors. First, Si nanoparticles were strongly anchored on graphene sheets, which effectively enhanced the electrical conductivity and structural stability. Second, the porous structure provides the composite material with abundant cavities for electrolyte penetration and fast Li-ion diffusion. Third, flexible graphene sheets and amorphous SiO_x shells could simultaneously buffer the volume change of Si.

Besides Si with an isotropic particulate morphology, anisotropic Si nanowires were also prepared and modified with graphene to become anode materials for LIBs. Lu et al. reported the *in situ* growth of Si nanowires on graphene sheets by a supercritical fluid–liquid–solid process using gold nanoparticles attached to graphene sheets as the catalysts [56]. The nanowires had diameters of 20–50 nm and lengths of several micrometers. Though the nanocomposite has a low initial Coulombic efficiency of 57.3%, a high reversible capacity of 2227 mAh g^{-1} was attained in the second cycle and a capacity of 1386 mAh g^{-1} could still be maintained after 30 cycles. In another report, Si nanowire/graphene composite was produced by electrostatic self-assembly. By mixing positively charged Si nanowires using primary amine functionalization with negatively charged GO in an aqueous solution, wrapping of Si nanowires with GO sheets to form a core-shell structure occurred spontaneously because of electrostatic interaction between two components. Compared with pristine Si nanowires, graphene wrapping could significantly enhance the rate capability and cycling stability.

As discussed previously in Section 3.2.3, the unique 2-D sheet-like nanostructure of graphene endows this novel carbon material with significant advantages in the formation of a paperlike electrode for flexible LIBs. This idea is also adopted in the preparation of Si/graphene composite anode with a flexible paperlike morphology. The paper anode could be easily prepared by filtering a mixed suspension of Si nanoparticles and graphene or GO sheets, as reported by several research groups [57–59]. Due to the sheet-like morphology of graphene, a continuous thin film could be formed in which graphene sheets and Si nanoparticles were homogenously distributed. Graphene sheets provided high electrical conductivity for the paper anode by forming 3-D conducting network and also served as a mechanically strong framework to anchor and stabilize Si nanoparticles. Consequently, the cycling stability and the rate performance of Si could be effectively enhanced. In addition, the composite paper could be directly used as the anode without any binders, which gave rise to higher energy density than traditional powdery counterparts. However, the relatively large graphene sheets with an extremely high aspect ratio may suppress the Li diffusion in a paper electrode. Therefore, Zhao et al. employed graphene sheets with vacancies to tackle this problem [60], similar to what they have done in the preparation of holey graphene paper anode. As illustrated in Figure 3.16, Li-ions could easily penetrate into the interior of the Si/graphene composite paper through holes in the graphene sheets. Accordingly, the rate capability was significantly improved. Even at a high current density of 8 A g^{-1}, the paper anode could still maintain 34% of the theoretical capacity of Si.

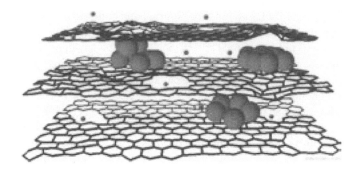

FIGURE 3.16 Schematic drawing of a section of a Si/porous graphene composite electrode material. Li-ions scan diffuses easily across graphene sheets by passing through the in-plane vacancies. (Zhao, X., Hayner, C. M., Kung, M. C. et al.: *Adv. Energ. Mater.* 2011. 1. 1079–1084. Copyright Wiley-VCH Verlag GmbH & Co. KGaA, Weinheim. Reproduced with permission.)

3.3.2.1.2 Germanium/Graphene Composite Anode Materials

Since both germanium (Ge) and graphene are group IV materials, Ge has a similar structure to that of Si and is also a potential anode material with a high theoretical capacity of 1600 mAh g^{-1}. Although Ge is less cost-effective than Si, the Li-ion diffusion coefficient of Ge is 400 times higher than that of Si and its electrical conductivity is 10^4 higher. However, as with Si, Ge also suffers from poor cycling stability due to the large volume change during Li uptake and release. The strategies employed in buffering the volume change of a Si anode can also be applied for a Ge anode. Therefore, graphene is expected to be a significant candidate to improve the electrochemical performance of Ge considering its successful application in the modification of Si anode materials.

The preparation of Ge/graphene composite anode material could be realized by loading of Ge nanoparticles on the surface of graphene sheets. The Ge nanoparticles could be produced either by chemical reduction of Ge (IV) ions [61] or chemical vapor deposition [62]. Graphene sheets not only significantly increased the electrical conductivity, but also served as a flexible support that stabilized nano-sized Ge particles. As a result, the composite anode materials exhibited excellent electrochemical performance, especially much improved cycling stability.

In another case, a double protection strategy was applied to improve the electrochemical performance of Ge by deposition of core-shell structure Ge@C nanoparticles on graphene sheets [63]. As displayed in Figure 3.17, Ge nanoparticles capped with oleylamine were first prepared and then heat-treated at 500°C under reductive atmosphere. The carbonization of oleylamine formed a carbon layer that encapsulated Ge nanoparticles. The as-prepared Ge@C core-shell nanoparticles were mixed with chemically reduced graphene in ethanol and the final product was obtained after the solvent was evaporated. In this composite anode material, Ge nanoparticles had a mean diameter of ~10 nm and the carbon shell had thicknesses of 2–3 nm. The core-shell nanoparticles were homogeneously deposited on graphene sheets, while severe aggregation occurred in the case of pure Ge@C nanoparticles,

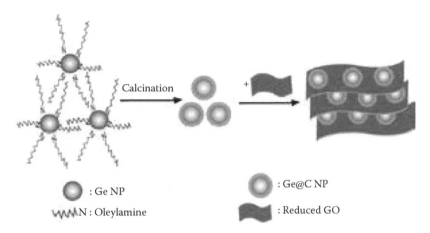

Calcination   +

⬤ : Ge NP

〰〰N : Oleylamine

⬤ : Ge@C NP

▰ : Reduced GO

FIGURE 3.17 Schematic illustration of the synthesis route for the Ge@C/graphene nano-composite. (Reprinted with permission from Ref. 62, 2512–2515. Copyright 2012 American Chemical Society.)

indicating that graphene plays an important role in achieving good dispersion of Ge@C nanoparticles. Galvanostatic charge/discharge measurement revealed that Ge@C/graphene composite anode material possessed excellent electrochemical performance. Although a large irreversible capacity, which was mainly ascribed to the formation of solid electrolyte interphase, was observed in the initial cycle, the reversibility of the cell was dramatically improved since the second cycle, with an average Coulombic efficiency of >99%. After 50 cycles at the current density of 50 mA g^{-1}, the Ge@C/graphene nanocomposite still maintained a high reversible capacity of 940 mAh g^{-1}. In contrast, Ge@C nanoparticles with no graphene modification only delivered a specific capacity of ~490 mAh g^{-1} under the same conditions. The rate capability was also significantly enhanced by the addition of graphene. Even at a high current density of 3600 mA g^{-1}, the Ge@C/graphene composite still exhibited a reversible capacity of 380 mAh g^{-1}, whereas the Ge@C nanoparticles only showed a low specific capacity of 100 mAh g^{-1}. After high-rate measurement, Ge@C/graphene was able to recover to the initial capacities, suggesting excellent reversibility of this composite anode material. EIS data further revealed that the charge-transfer resistance of the Ge@C/graphene composite was more than 40% lower than that of Ge@C nanoparticles, which proved the enhanced rate capability of the former.

In another report, Ge nanowires were employed instead of Ge nanoparticles and modified by graphene [64]. By a vapor–liquid–solid process, employing germanium tetrachloride ($GeCl_4$) as the precursor, Ge nanowires with diameters of ~100 nm were obtained. Then, graphene was grown on the surface of Ge nanowires by a metal-catalyst-free CVD process. High-resolution TEM characterization clearly revealed that the sample was dominated by single- or few-layer graphene, which could also be confirmed by Raman spectroscopy. In some areas, disconnection points along the growing axis of graphene were observed. Such defects might be beneficial for the access of Li-ions to the Ge nanowires through graphene coating layers. The successful growth of graphene on Ge nanowires could be probably ascribed to the following two reasons. First, Ge nanowires were flat enough to accommodate carbonaceous species to be nucleated. Second, the binary phase diagram of Ge and C was similar to that of Cu and C, which implied that the formation mechanism of graphene on Ge was analogous to that in the case of Cu. Graphene was formed on Ge nanowires preferably by adsorption of C on Ge followed by nucleation and propagation of graphene sheets. The Ge nanowire/graphene composite was then assembled into a half-cell to measure its electrochemical performance. The composite anode material presented an ultrahigh initial capacity approaching 1900 mAh g^{-1}, which was even larger than the theoretical value of Ge (1600 mAh g^{-1}), and a Coulombic efficiency of 69% in the first cycle. It was most likely due to the formation of solid electrolyte interphase (SEI) during the first charging step. Since the second cycle, the Ge nanowire/graphene composite showed outstanding cycling stability. The capacity retention was well retained for various rates up to 4 C without significant capacity drop. Capacities of 1210 and 1059 mAh g^{-1} could be delivered after 200 cycles at the rates of 0.5 C and 4 C, respectively, corresponding to capacity retention of 95% and 92%, respectively. In contrast, Ge nanowires without graphene modification only exhibited capacity retention of 41% after 200 cycles at 4 C. The composite anode material also presented excellent rate capability. At a high rate of 20 C, a reversible capacity of 363 mAh g^{-1} could still be reached. The capacity of bare Ge nanowires, however, dropped rapidly to 20 mAh g^{-1} at 20 C. The enormously improved electrochemical performance of Ge by modification of CVD-derived graphene could be ascribed to the following reasons. First, graphene produced by the CVD method had an excellent electrical conductivity, which was beneficial for fast electron transport in the electrode. Second, the intimate contact between graphene and Ge nanowires hold Ge tightly to effectively accommodate the mechanical strain during charge/discharge cycles. It was observed that most Ge wires kept intact and were still coated by a thin layer of graphene after 200 cycles. Third, the defects in the graphene coating layers provided pathways for fast Li-ion diffusion.

Graphene-encapsulated Ge nanowires could also be produced by arc-discharge [65]. Graphite and germanium dioxide (GeO$_2$) were employed as the precursors to graphene and Ge nanowires, respectively. Ge nanowires have a mean diameter of 20–30 nm, and were uniformly covered by ultrathin graphene sheets. The shell had 2–4 layers of graphene with an interlayer distance of 0.35 nm. Besides graphene-encapsulating Ge nanowires, large graphene sheets that attached to Ge nanowires were also observed, which connected all nanowires together to form a highly conductive network. As expected, the cycling stability and rate capability were effectively improved by graphene encapsulation.

3.3.2.1.3   *Tin/Graphene Composite Anode Materials*

Tin (Sn) is another attractive metal-based anode material for LIBs due to its high specific capacity (992 mAh g^{-1}). However, as with Si, Sn also suffers from severe volume expansion/contraction during the charge/discharge process, which dramatically reduces the cycle life of the battery. The successful experience of using graphene to improve the electrochemical performance of a Si anode thus can be applied in the case of Sn. Sn/graphene composite anode material with various structures has been prepared in recent years.

Sn/graphene nanocomposites could be simply prepared by deposition of Sn nanoparticles on graphene sheets as reported by several groups [66,67]. For instance, using $NaBH_4$ as the reducing agent, Sn^{2+} was reduced to Sn^0, resulting in growth of Sn nanoparticles on chemically reduced graphene sheets [68]. The large surface area of graphene and oxygenous groups on graphene sheets facilitated homogeneous distribution of Sn nanoparticles, while Sn nanoparticles also prevented restacking of graphene sheets. Because of the enhanced electrical conductivity and the buffering effect brought by graphene, the electrochemical performance, especially the cycling stability of the nanocomposite, was much improved compared with pure Sn anode.

In order to further improve the electrochemical performance, Sn/graphene composites with more complicated structures were also prepared. Ji and coworkers designed a multilayer structure by sandwiching Sn nanopillar arrays between graphene layers [69]. As shown in Figure 3.18, graphene film was first prepared by vacuum filtration. Then, a thin film of Sn was deposited on the graphene film using a thermal evaporator. By repeating this process, two layers of Sn film embedded between three layers of graphene films were obtained. Finally, the Sn/graphene composite was heat-treated at 300°C, and a unique nanopillar-like morphology of Sn was obtained after annealing. The transformation from Sn thin film to Sn nanopillar was mainly ascribed to wettability and surface energy minimization as suggested

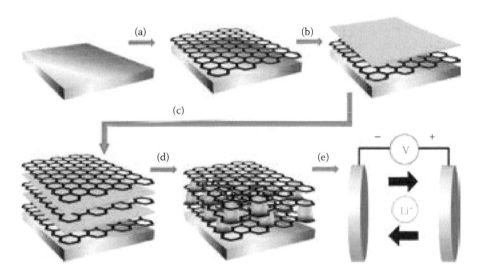

FIGURE 3.18 Schematic illustration of the graphene/Sn-nanopillar nanostructure preparation procedures. (a) Transferring graphene film to a copper foil current collector; (b) depositing the Sn film on the transferred graphene surface via thermal evaporation; (c) repeating steps a and b followed by coating another layer of graphene; (d) thermal annealing of the as-formed nanostructure to obtain the multilayer graphene/Sn-nanopillar nanostructure; and (e) assembling Li-ion cells to evaluate the electrochemical performance of the graphene/Sn-nanopillar nanostructured electrodes. (From Ji, L., Tan, Z., Kuykendall, T. et al., Multilayer nanoassembly of Sn-nanopillar arrays sandwiched between graphene layers for high-capacity lithium storage. *Energ. Environ. Sci.* 4:3611–3616, 2011. Reproduced by permission of The Royal Society of Chemistry.)

by the authors. Poor wettability of Sn on graphene sheets induced the formation of Sn nanoparticles during thermal evaporation. The following annealing process at a temperature higher than the melting point of Sn caused the merging of Sn nanoparticles and the formation of nanopillars to reduce the surface energy. The nanopillars had diameters of 150–200 nm and most of the Sn nanopillars had both ends directly connected to graphene. The Sn/graphene composite could be directly used as the anode without adding binders. At a current density of 0.05 A g^{-1}, the multilayered Sn/graphene composite anode could deliver reversible capacities of 723 and 679 mAh g^{-1} after the 15th and 30th cycles, respectively, corresponding to the capacity retention of 98.4% and 92.5% from the first cycle. At a high current density of 5 A g^{-1}, a reversible capacity of 408 mAh g^{-1} could still be attained. In this composite anode, graphene sheets not only electrically bridged active Sn nanopillars, but also provided effective mechanical support to bear volume expansion of Sn. In addition, the multilayered Sn/graphene composite could be directly employed as the anode without binders and the integrity of the anode could be well preserved during long-term charge/discharge cycles.

In another report, Luo et al. designed a unique structure that Sn nanosheets were confined between two graphene layers [70]. In their experiment, SnO$_2$ nanoparticles were first deposited on the surface of graphene sheets via a hydrolysis process. The graphene-supported SnO$_2$ nanoparticles were then coated with glucose through a hydrothermal method. Subsequent annealing at a high temperature resulted in the final product. During the thermal treatment, SnO$_2$ was reduced into metallic Sn by glucose-derived carbon, whereas glucose-derived carbon was transformed to graphene under the catalysis of Sn. The thickness of Sn nanosheets was as small as 10 nm, which was mainly ascribed to the confined voids between graphene and glucose-derived carbon. The graphene sheets attached at both sides of Sn nanosheets had a mean thickness of 5 nm. Sn/graphene nanocomposites with such a sandwich-like structure exhibited excellent electrochemical performance. After 60 cycles at a current rate of 50 mA g^{-1}, a reversible capacity of 590 mAh g^{-1} could still be reached and at a high current density of 1600 mA g^{-1}, a reversible capacity of 265 mAh g^{-1} was still retained. The surface-to-surface contact between Sn nanosheets and graphene afforded fast electron transport channels, and the 2-D feature of the composite facilitated the diffusion and transport of Li-ions. However, it was discovered that after long-term charge/discharge cycling, Sn nanosheet sandwiched between graphene layers partially fragmented into nanoparticles, which implied that the confined structure reported in this paper might not be strong enough to prevent Sn from pulverization.

Employing an extra component in the Sn/graphene composite anode material to more effectively confine nano-sized Sn had also been carried out in several publications. Core-shell Sn@C nanoparticles embedded in graphene sheets was reported by Wang and coworkers [71]. Through a one-step CVD procedure, SnO$_2$ nanoparticles deposited on graphene sheets were reduced into Sn by ethylene, while carbon shell derived from the decomposition of ethylene was uniformly coated on Sn nanoparticles to form the core-shell structure. The core-shell Sn@C particles were homogeneously distributed on graphene sheets. Sn particles exhibited a size distribution from 50 nm to 2 μm. Large Sn particles were formed due to the fusion of adjacent Sn nanoparticles. The carbon shell had a mean thickness of ~10 nm and was composed of staggered and shortened graphene sheets with a lattice fringe of 0.355 nm. As the reaction took place at a high temperature when Sn was in the liquid state, the volume shrinkage from Sn droplets to solid Sn nanoparticles during the cooling process gave rise to void space between the Sn core and the carbon shell, which was able to relieve the stress from volume expansion/contraction of Sn nanoparticles. The composite anode material could deliver an initial capacity of 1069 mAh g^{-1} and a reversible capacity of 566 mAh g^{-1} could be still reached after 100 cycles at a current density of 75 mA g^{-1}. When the current density was increased to 3750 mA g^{-1}, it could still deliver a reversible capacity of 286 mAh g^{-1}. In contrast, the capacity of pure core-shell Sn@C nanoparticles without graphene modification faded rapidly from the initial 1036 mAh g^{-1} to 188 mAh g^{-1} at the 100th cycle. Therefore, graphene sheets played an important role in enhancing the electrochemical performance of the Sn anode in conjunction with the carbon shell. Similar carbon shell could also be produced via a hydrothermal process using glucose as the precursor.

Another case of Sn@CNT nanostructures deposited on graphene was reported by Zou et al. [72]. In their experiment, SnS$_2$/graphene composite was first prepared in a microwave reactor. SnS$_2$ was then reduced by C$_2$H$_2$ to form Sn and carbon disulfide (CS$_2$). CS$_2$ evaporated with the carrier gas, while melted Sn droplets catalyzed the formation of CNTs. During the growth of CNTs, liquid Sn was infiltrated into the nanotube cavities by capillary force. Finally, Sn nanorods encapsulated by CNTs and grown on graphene sheets were obtained. Each Sn nanorod was single-crystal and grown along the [200] direction. The nanocomposite with a hierarchical structure exhibited high reversible capacity, excellent rate capability, and cycling stability, which was mainly ascribed to the increased electrochemical activities of Sn by CNT protection and graphene support, and prevention of graphene agglomeration by core-shell Sn@CNT decoration.

3.3.2.2 Metal Oxide-Graphene Composite Anode Materials

3.3.2.2.1 SnO₂-Graphene Composite Anode Materials

Tin dioxide (SnO$_2$) is expected to be a good substitute for the graphite anode in LIBs due to its high theoretical capacity of 782 mAh g^{-1}. There are two electrochemical processes evolved in the SnO$_2$-based electrodes:

$$SnO_2 + 4Li^+ + 4e^- \rightarrow Sn + 2Li_2O$$

$$Sn + xLi^+ + xe^- \leftrightarrow Li_xSn \; (0 \leq x \leq 4.4)$$

The first process is usually reported to be irreversible, whereas the second process is well known to be reversible. It can be observed that the second process is identical to the alloying and dealloying reaction of metallic Sn anode. Therefore, the large volume change of Sn anode (up to 300%) also occurs in SnO$_2$-based anode materials, which causes poor cycle life. In order to improve its cyclability, modification of SnO$_2$ with graphene has been extensively studied in recent years.

The first research on SnO$_2$/graphene composite anode material was carried out by Paek and coworkers, which was also the first publication of graphene-based composite electrode material for LIBs [73]. The idea was to generate nanostructured SnO$_2$/graphene composite electrode in which SnO$_2$ nanoparticles were surrounded by graphene sheets to limit the volume expansion and the nanopores existing between two components could be used as buffered spaces, as illustrated in Figure 3.19. Such a structure could be easily realized by mixing preprepared SnO$_2$ nanoparticles and graphene sheets in ethanol. TEM images revealed that graphene sheets were randomly distributed between loosely packed SnO$_2$ nanocrystals to form a 3-D hybridized nanostructure with plenty of nanospaces existing between two components. Such a nanostructure enormously improved the electrochemical performance, especially the cycling stability in comparison with pure SnO$_2$ anode material. Although a relatively large irreversible capacity in the first charge/discharge cycle was observed, the reversible capacity started to be stabilized at the second cycle. At a current density of 50 mA g^{-1}, a reversible capacity of 810 mAh g^{-1} was delivered in the second cycle and a capacity of 570 mAh g^{-1} (70% retention) could still be remained after 30 cycles. In contrast, pure SnO$_2$ nanoparticles experienced rapid capacity fading from 570 mAh g^{-1} to 60 mAh g^{-1} within only 15 cycles. Consequently, it was strongly

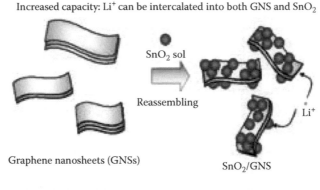

FIGURE 3.19 Schematic illustration of the synthesis and the structure of SnO$_2$/graphene composite anode material. (Reprinted with permission from Ref. 72, 72–75. Copyright 2009 American Chemical Society.)

proved that the addition of graphene could effectively tackle the problem caused by volume expansion and pulverization of SnO_2 during repeated Li alloying and dealloying processes. The successful improvement of the electrochemical performance of SnO_2 by graphene modification reported in this paper soon attracted much interest in the preparation of SnO_2/graphene composite anode materials.

Compared with mechanical mixing of SnO_2 nanoparticles with graphene nanosheets, *in situ* growth of SnO_2 nanocrystals on graphene sheets is an alternate way that may be advantageous in achieving homogeneous distribution of two components in the nanoscale, and have been attempted by different research groups. Our group proposed a redox-reaction-based strategy to grow ultrasmall SnO_2 nanocrystals on graphene sheets [74]. $SnCl_2$ and GO were used as the precursors. In acidic aqueous solutions, the simultaneous hydrolysis of Sn^{2+} and oxidation of Sn(II) to Sn(IV) by GO gave rise to the formation of SnO_2 nanoparticles. Meanwhile, GO was partially reduced to graphene. The abundant oxygenous groups on GO had a strong affinity with Sn^{2+}, which induced the growth of SnO_2 nanoparticles on the surface of graphene sheets. Calcination of the as-prepared sample was carried out to further increase the reduction degree of graphene. As revealed by TEM, highly crystalized SnO_2 nanoparticles with diameters of 3–5 nm were uniformly distributed on both sides of graphene sheets. The molar ratio of SnO_2/graphene was varied in our experiments to investigate the influence of the graphene content to the electrochemical performance of the composite anode material. It was found that the sample with a SnO_2/graphene molar ratio of 3:2, corresponding to a graphene content of 2.4 wt%, exhibited the best performance. It delivered an initial discharge capacity of 1700 mAh g^{-1} and charge capacity of 978 mAh g^{-1} at a current density of 67 mA g^{-1}. The relatively low Coulombic efficiency was mainly due to the formation of SEI on SnO_2 nanoparticles with large surface area. The capacity retention kept well in the following cycles. The Coulombic efficiency was above 98%, especially after five cycles. After 30 cycles, the cell could still maintain a charge capacity of 840 mAh g^{-1}, corresponding to a capacity retention of 86%. The SnO_2/graphene nanocomposite also presented good rate capability. At a current density of 400 mA g^{-1}, the charge capacity remained at 590 mAh g^{-1} for up to 50 cycles. When the rate was further increased to 1000 mAh g^{-1}, the cell could still deliver a charge capacity of 270 mAh g^{-1} after 50 cycles. The excellent cycling stability as well as the rate performance could be mainly ascribed to the special roles of graphene sheets. Graphene acted as a buffer to prevent harmful effects derived from large volume change of SnO_2 during charge/discharge processes. At the same time, graphene could also help to prevent the aggregation of Sn nanoparticles. However, the cycle life of this composite anode material needs further improvement for practical applications.

In order to achieve optimal electrochemical performance of the SnO_2/graphene composite anode materials, the structure of the nanocomposite was tuned by using various synthetic methods. For instance, Ding et al. reported the growth of SnO_2 nanosheets on graphene through a hydrothermal process [75]. SnO_2 nanosheets had thicknesses of 5–10 nm and lengths of ~100 nm and were mostly grown upright with a random orientation on the graphene sheets. Such anisotropic sheetlike nanostructure was expected to be beneficial for the electrochemical reactions due to the large exposed surface. After 50 charge/discharge cycles at a current density of 160 mA g^{-1}, the nanocomposite could still deliver a capacity of 518 mAh g^{-1} with only a 0.51% capacity loss per cycle. The capacity loss for the particulate counterpart, however, was almost two times larger (1%), which indicated the advantage of nanosheet over traditional particulate morphology in terms of electrochemical properties.

In another case, researchers employed the atomic layer deposition (ALD) technique to control the morphology and crystallinity of SnO$_2$ on graphene sheets [76]. As a surface-controlled and layer-by-layer process, ALD could provide precise control over the SnO$_2$ deposition. In this paper, by varying the deposition temperature either amorphous or crystalline SnO$_2$ could be obtained. Both types of SnO$_2$/graphene composite exhibited promising electrochemical performance because of the addition of graphene; amorphous SnO$_2$, however, displayed better cycling stability compared with the crystalline counterpart. At a current density of 400 mA g^{-1}, the amorphous SnO$_2$/graphene composite could deliver a reversible capacity of 793 mAh g^{-1} after 150 charge/discharge cycles, whereas the corresponding capacity for the crystalline SnO$_2$/graphene sample was only 499 mAh g^{-1}. The amorphous sample also showed higher Coulombic efficiency than the crystalline one during the charge/discharge cycles. Morphological characterization of the composite anode materials after long-term cycling revealed that crystalline SnO$_2$ nanoparticles with sizes of 30–40 nm were pulverized into Sn nanoparticles with diameters of 4–7 nm after 150 cycles, which caused electric-contact loss of some Sn nanocrystals. For the amorphous sample, the pristine SnO$_2$ existed as thin films covering both sides of graphene. Though the film were broken up into small Sn nanoparticles during cycling, the amorphous nature could accommodate substantial volume change and only a small fraction of Sn lost the electrical contact to graphene sheets. Therefore, the amorphous SnO$_2$/graphene nanocomposite is a promising anode material for LIBs because of its excellent cycling stability.

Despite the improvement of the cycle life of SnO$_2$ anode material by graphene modification, the integrity of SnO$_2$ could still not be well maintained during charge/discharge cycles, which caused capacity fading. In order to overcome this drawback, Wang and coworkers designed and synthesized SnO$_2$/graphene composites coated with a buffer layer that was further cross-linked to the binder to form a robust network, as shown in Figure 3.20 [77]. Using a hydrothermal method, SnO$_2$/graphene aerogel with a porous structure was first prepared. The aerogel was then dispersed

FIGURE 3.20 Structure model of SnO$_2$/graphene composite anode materials. (Reprinted with permission from Ref. 76, 1711–1716. Copyright 2013 American Chemical Society.)

in water by ultrasonication to form homogenous suspension of SnO_2/graphene composite nanosheets. The nanosheets were coated afterward with polydopamine (PD) via self-polymerization of dopamine in weak base conditions. Finally, the PD-coated sample was mixed with poly(acrylic acid) (PAA) as the binder and carbon black as the conductive additive, and then pasted on Cu foil and cured at 150°C in vacuum, which resulted in the cross-linking between PD and PAA. In this composite, SnO_2 nanoparticles with a mean diameter of ~5 nm were densely distributed on the surface of graphene sheets. The PD coating could endure large volume change of SnO_2 nanoparticles during charge/discharge cycles by adjusting its own elastic deformation. Meanwhile, the existence of PD avoided direct contact between SnO_2 nanoparticles with electrolytes, which could diminish the side reactions at the electrode–electrolyte interface. Further cross-link between PD and PAA through forming amide bonds produced a robust network that could stabilize the whole anode during cycling. Aided by PD coating and cross-linking, the composite anode material could deliver a relatively high initial reversible capacity of 931 mAh g^{-1} with a Coulombic efficiency of 68%. The electrode without cross-link, however, only presented a reversible capacity of 800 mAh g^{-1}, corresponding to a Coulombic efficiency of 57%. Bare SnO_2/graphene-composite-based anode without PD coating displayed an even lower capacity of 520 mAh g^{-1} (Coulombic efficiency of 40%). The electrode with cross-link also exhibited significantly enhanced cycling stability and rate capability compared with the one without cross-link. A capacity up to 718 mAh g^{-1} could be reached after 200 cycles at a current density of 100 mA g^{-1}. EIS measurement further indicated that PD coating and cross-linking had less effect on ohmic resistance of the electrode and was advantageous to stabilize the charge-transfer resistance of the electrode during cycling. The carefully designed structure reported in this paper provided some new concepts to the design and synthesis of high-performance electrode materials for LIBs.

Besides the control of the structure and morphology of SnO_2, the modification of graphene itself is another way to enhance the performance of SnO_2/graphene composite anode materials. Nitrogen doping has been proved to be effective to improve the electrochemical performance of graphene anode materials for LIBs; therefore, N-doped graphene is employed in the case of SnO_2/graphene hybrids. For example, Wang et al. prepared paper-like N-doped graphene/SnO_2 composite as a high-performance anode for LIBs. In the experiment, 7,7,8,8-tetracyanoquinodimethane (TCNQ) ions were mixed with chemically reduced graphene in organic solvents, which formed negatively charged graphene suspension [78]. After the addition of Sn^{II} salts, Sn ions and TCNQ decorated graphene sheets quickly self-assembled into a sandwich-like structure because of the strong electrostatic interaction between Sn^{2+} and $TCNQ^-$. The precipitate was then calcined in inert atmosphere to obtain sandwich-type SnO_2/graphene paper. A certain amount of carbon atoms in the graphene sheets were substituted by N atoms from TCNQ during the annealing process to form N-doped graphene. In the paperlike composite, ultrasmall SnO_2 nanoparticles were sandwiched by graphene layers. Energy-dispersive x-ray mapping revealed that both Sn and N were homogeneously distributed in the sample. The content of N was measured to be ~8% by XPS. The XPS data also confirmed the existence of pyridinic, pyrrolic, and graphitic type of N atoms in the sample, strongly indicating the successful doping of graphene by N. The N-doped graphene/SnO_2 composite displayed outstanding rate performance. Even at a high current density of 5 A g^{-1}, the cell could still deliver a reversible capacity of 504 mAh g^{-1}. Meanwhile, the cycling stability was also improved. At a current density of 50 mA g^{-1}, an ultralow capacity loss of 0.02% per cycle was achieved from the second to the 50th cycle. In contrast, the capacity loss for pure SnO_2 was 2.9%. The enormously improved electrochemical performance of this sample could be attributed to the following reasons. First, N-doping introduced a large amount of surface defect on the graphene sheets, which might favor a larger reversible capacity than pristine graphene. Second, the sandwichlike structure dramatically increased the electrical conductivity as the electronic transport length was effectively shortened. Third, the flexible paperlike structure provided elastic buffer to endure volume change of SnO_2.

In another report, SnO_2 nanocrystals strongly anchored on N-doped graphene sheets through the formation of covalent Sn-N bonds was realized by *in situ* hydrazine monohydrate vapor reduction of the SnO_2/GO composite [79]. XPS analysis revealed 3% atomic content of N and the existence of pyrrolic and pyridinic N species in the hybrid. SEM, TEM, and Auger electron spectroscopy characterization suggested that SnO_2 nanoparticles with a mean diameter of ~5 nm were well encapsulated by graphene sheets. A combination of XANES, STEM, and energy dispersive x-ray (EDX) further confirmed the formation of Sn-N bonds in the hybrid, which was expected to enhance the stability of this anode material. The initial charge and discharge capacity of the composite anode material were 1144 and 1865 mAh g^{-1}, respectively, corresponding to a Coulombic efficiency of 61.3%. The Coulombic efficiency dramatically increased to more than 97% after the initial cycle, and a stable reversible capacity of 1021 mAh g^{-1} was obtained, which was the highest ever reported for the SnO_2-based anode materials. After 500 cycles, the capacity of the hybrid anode even increased to 1346 mAh g^{-1}, which was probably due to the improved Li-ion accessibility during the cycling process. The rate performance of the composite anode material was also excellent. Even at an ultrahigh current density of 20 A g^{-1}, the cell could still deliver a reversible capacity of 417 mAh g^{-1}. Such extraordinary electrochemical performance was attributed to the *in situ* hydrazine monohydrate vapor reduction. If the SnO_2/GO composite was simply reduced by thermal treatment, the product could only deliver a capacity of 551 mAh g^{-1} after 200 cycles at a current density of 0.5 A g^{-1}, which was less than half the value for the N-doped sample.

3.3.2.2.2 *Titanium-Dioxide–Graphene Composite Anode Materials*

Titanium dioxide (TiO_2) is another important type of metal oxide anode material. Although the theoretical capacity of TiO_2 is lower than most metal oxides and even lower than graphite, it is extremely stable during charge/discharge cycles. Almost no volume expansion occurs even when TiO_2 is fully lithiated. Therefore, TiO_2 anode material exhibits outstanding cycling stability. Moreover, TiO_2 has a relatively higher charge voltage (~1.7 V) compared with that of graphite anode, which can avoid the formation of SEI layers and electroplating of lithium, resulting in better safety and less capacity loss. However, the electrical conductivity of TiO_2 is extremely low, which causes poor rate capability. As a result, modification of TiO_2 by conducting agents is required to improve its electrochemical performance. Graphene, which has superior electrical conductivity, large surface area, and excellent mechanical flexibility, is considered to be an ideal additive to circumvent the intrinsic drawbacks of TiO_2 anode materials.

Wang and coworkers reported the first case of TiO_2/graphene composite anode materials for LIBs [80]. A low-temperature crystallization process was applied in their experiments to grow TiO_2 nanocrystals on graphene sheets using $TiCl_3$ as the precursor, as shown in Figure 3.21. Either a rutile or anatase phase could be formed by adjusting the reaction conditions. The rutile nanocrystals have rod-like morphology, whereas anatase nanocrystals have particulate nanostructure. Aided by the surfactant of sodium dodecyl sulfate, hydrophilic TiO_2 nanocrystals could homogeneously grow on hydrophobic graphene sheets. Both rutile and anatase TiO_2/graphene hybrids showed remarkably enhanced electrochemical performance compared with bare TiO_2. At a rate of 30 C, rutile/gaphene composite and anatase/graphene composite could deliver a reversible capacity of 87 and 96 mAh g^{-1}, respectively, while pure rutile and anatase could only reach 35 and 25 mAh g^{-1}, respectively. Both composite anode materials also exhibited good cycle stability. No obvious capacity decay occurred after 100 cycles at 1 C. The enhanced electrochemical performance could be mainly ascribed to the increased conductivity of the hybrid by the incorporation of graphene. EIS analysis revealed that with the addition of only 0.5 wt% graphene, the resistivity of the cell could be reduced from 93 to 73 Ω. Graphene was also more effective than CNTs in improving the charge/discharge performance of TiO_2 anode. With the same content of 0.17 wt%, the cell with graphene could deliver ~70% higher capacity than the one using CNTs.

A hard-template-assisted method was also employed to prepare a sandwich-like porous TiO_2/graphene hybrid with a high surface area as a high-performance anode material [81]. In the first step, a thin layer of mesoporous silica was grown on both sides of GO sheets, which was then applied as the hard template to the deposition of TiO_2 within the mesopores. TiO_2 was obtained by hydrolysis and dehydration of $(NH_4)_2TiF_6$ at a controlled concentration and pH value. After crystallization of TiO_2 at 500°C and subsequent etching of silica in NaOH, the TiO_2/graphene hybrid anode was finally obtained. The hybrid had nanosheet morphology with a lateral size from 200 nm to several micrometers and a thickness of ~50 nm. A high-resolution TEM image revealed that TiO_2 nanoparticles with crystalline sizes of ~5 nm constructed the

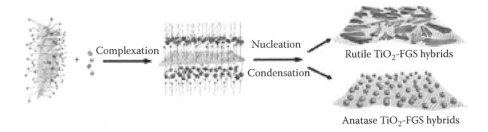

FIGURE 3.21 Schematic illustration of the synthesis process and the structure of TiO_2/graphene composite anode material. (Reprinted with permission from Ref. 79, 907–914. Copyright 2009 American Chemical Society.)

porous architecture on graphene. The mesoporous nature TiO_2 was further confirmed by N_2 sorption measurements, which showed type-IV isotherms with a H3 type hysteresis loop typical of mesoporous materials. The BET surface was calculated to be 202 m^2 g^{-1}, higher than most reported porous TiO_2. The TiO_2/graphene hybrid could reach a high initial discharge capacity of 269 mAh g^{-1}, which was even higher than the theoretical value of titania, which implied that additional lithium storage existed in the hybrid anode material. The detailed charge/discharge curves revealed that addition capacity was mainly derived from surface lithium storage similar to supercapacitive behavior because of the large surface area of nanostructured TiO_2. The sheetlike TiO_2/graphene hybrid also showed excellent cycling stability and rate performance. A reversible capacity of ~180 mAh g^{-1} could be retained after 30 cycles at 0.2 C, and the cell could still deliver a reversible capacity of 123 mAh g^{-1} at a high rate of 10 C. In contrast, the graphene-removed sample prepared by calcination of TiO_2/graphene composite in air only reached a reversible capacity of 36 mAh g^{-1} at the same rate of 10 C, less than one-third of the value for the hybrid. Using EIS, the ohmic resistance and charge-transfer resistance were measured to be 4.8 and 97.2 Ω, respectively, for the TiO_2/graphene hybrid, both of which were significantly lower than the values for pure TiO_2 (8.1 and 199.1 Ω). These results indicated that graphene did play a crucial role in improving the electrochemical performance of TiO_2 anode materials.

The structure and morphology of TiO_2/graphene hybrid anode materials were also tuned to optimize the electrochemical performance and expand their applications. For example, by a seed-assisted hydrothermal method, rutile TiO_2 with nanorod morphology could grow on graphene sheets to form vertically aligned arrays [82]. The density and length of the nanorods could be finely tuned. Such a unique structure provided a large contact area between the active materials and the electrolytes, facilitated the electron transfer within the electrode by graphene sheets, and also prevented the aggregation of active materials during charge/discharge cycles. As a consequence, the TiO_2 nanorod/graphene composite displayed excellent rate capability and cycling performance. Chen et al. reported the synthesis of graphene-wrapped TiO_2 with a hollow structure [83]. Via electrostatic interaction, negatively charged GO sheets can be uniformly linked to functionalized TiO_2 hollow particles with a positively charged surface. Compared with pure TiO_2, the graphene-modified TiO_2 sample exhibited significantly enhanced Li storage properties, including higher specific capacitance and better cycling stability. TiO_2 nanotube/graphene nanocomposite was also proposed [84], which was synthesized by hydrothermal treatment of the mixture of TiO_2 powder and GO in a high-concentration NaOH solution. The tubular TiO_2 with diameters of ~10 nm and lengths of hundreds of nanometers to a few micrometers was uniformly dispersed on graphene sheets to form a 3-D porous network with two components interpenetrated with each other. Such a unique structure could effectively facilitate the charge transfer during charge/discharge processes. As a result, even at a high current density of 8 A g^{-1}, a stable reversible capacity of ~80 mAh g^{-1} could still be retained within 1000 cycles for the composite anode material. In another case, Zhang and coworkers presented TiO_2/graphene composite nanofibers as high-performance anode materials [85]. Using the mixed suspension of TiO_2 and graphene as the precursors, composite nanofibers with diameters of 100–200 nm could be obtained by electrospun and subsequently heat treatment. Anatase TiO_2 nanocrystals with an average size of ~6 nm and graphene nanosheets were homogeneously distributed in the fibers. In addition, a large amount of mesopores with a mean diameter of 8.6 nm also existed in the fibers, giving rise to a surface area of ~200 $m^2\ g^{-1}$, which is about three times higher than that for bare TiO_2 nanofibers. Cells using the composite nanofiber as the anode material could retain 84% of the reversible capacity after 300 cycles as the current density of 150 mAh g^{-1}, which is 25% higher than that of bare TiO_2. TiO_2/graphene hybrid with a paper-like morphology was also proposed by Hu et al. as a flexible anode for LIBs [86]. In their experiment, wet graphene paper prepared by filtration of graphene suspension containing surfactant was immersed in the solution containing TiO_2 precursors, which was then heat-treated at 120°C. After the intercalation of TiO_2 nanoparticles into the interlayer space in the graphene paper, the thickness of the paper could be expanded from ~2.6 to ~9.3 μm. The existence of TiO_2 nanocrystals also helped to form 3-D open structure inside the paper electrode, which is beneficial for fast Li-ion diffusion. Meanwhile, graphene sheets in the paper constructed a highly conductive network. As a result, excellent rate performance and cycling stability was achieved for the flexible paperlike electrode and a similar preparation strategy might be applied to the synthesis of other graphene/metal oxide composite paper for flexible energy storage devices.

In most cases of TiO_2/graphene nanocomposite, TiO_2 is obtained by hydrolysis of Ti-containing precursors. As it is known that the hydrolysis process of Ti species is rapid, it is not easy to control the structure and morphology of as-prepared titania. In some cases, harsh experimental conditions are required to control the formation of TiO_2, such as using strong acid or toxic reagents. Thus, it is important to develop modified approaches to obtain TiO_2 in a more controllable way and under mild experimental conditions. Recently, our group proposed the synthesis of TiO_2/graphene composite anode materials by a hydroxyl titanium oxalate (HTO) mediated synthesis method instead of the direct hydrolysis process, which was able to obtain flowerlike nanostructures and excellent charge/discharge performances [87]. As shown in Figure 3.22, tetrabutyl titanium as the precursor first reacted with oxalate

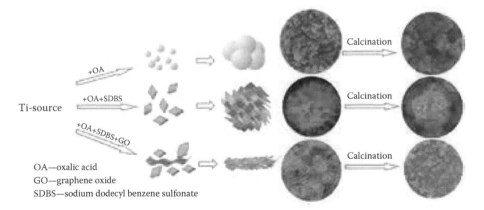

FIGURE 3.22 Illustration of the preparation process and growth mechanism of TiO_2-based nanocomposites. (Reprinted with permission from Ref. 86, 11035–11043. Copyright 2012 American Chemical Society.)

acid to form amorphous HTO. This reaction proceeds much more slowly than that of the hydrolysis process. The formation of HTO lasted for tens of minutes, and thus it was easy to control the whole reaction process and the structure of the HTO intermediate. By adding GO and the surfactant of sodium dodecylbenzenesulfonate (SDBS) in the reaction solution, HTO with a unique flowerlike structure formed and could be uniformly distributed on GO sheets. The existence of SDBS is important to form the unique flowerlike structure. Otherwise, only aggregated, nonuniform HTO nanoparticles were obtained. GO in the solution could serve as the substrate for the growth of HTO. The following calcination process transformed amorphous HTO into anatase phase. At the same time, GO was thermally reduced into graphene. TEM characterization revealed that the anatase TiO_2 still retained the flowerlike morphology and homogenously dispersed on graphene sheets. A high-resolution TEM image showed that TiO_2 nananocrystals had a mean diameter of 5–10 nm. The specific surface area of the TiO_2/graphene nanocomposite is measured to be 211.6 m^2 g^{-1}, which is much larger than the pure TiO_2 without graphene modification (38.8 m^2 g^{-1}). The enormously increased surface area is ascribed to the large surface area of graphene, as well as smaller particle size of TiO_2 nanocrystals in the presence of graphene. Pore size distribution analysis showed that the nanocomposite has uniform mesopores with a mean pore diameter of ~10 nm, also indicating the small particle size of TiO_2 nanocrystals on the graphene sheets. Such a unique nanostructure gave rise to outstanding electrochemical performance. An initial discharge and charge capacity of 280 and 226 mAh g^{-1} was achieved, respectively, for the flower-like TiO_2/graphene nanocomposite. It also showed excellent rate capability. Even at a high rate of 50 C, a reversible capacity of 80 mAh g^{-1} could still be delivered, which is 35% of its initial capacity and is 30% higher than that for the pure TiO_2 anode materials without graphene modification. Furthermore, approximately 97%, 98%, and 99% of its initial charge capacity could be retained after 100 charge/discharge cycles at 1 C, 10 C, and 20 C respectively, indicating good cycling stability. In the nanocomposite, the ultrasmall TiO_2 nanoparticles guarantees short diffusion length of Li-ions, whereas graphene sheets serve as conducting agents to improve the electrical conductivity of TiO_2. In addition, 2-D graphene sheets with large surface area also effectively inhibit the aggregation of TiO_2 nanoparticles, raising the homogeneity of the Li-ion insertion/extraction. It should be pointed out that compared with the synthesis of similar TiO_2/graphene composite anode materials, the synthetic route reported in this paper is nontoxic and easier to operate, which can be readily scaled up. A scale 1 kg per batch was realized in the lab, and a prototype of 18650-type cylindrical cell using flowerlike TiO_2/graphene nanocomposite as the anode material and $LiMn_2O_4$ as the cathode material was manufactured in order to evaluate the practical performance of this novel anode material in commercial batteries. The cell discharged at an average voltage of 2.1 V and an initial discharge capacity of 700 mAh was reached. When the discharge rate was increased to 12 C, the cell could still retain a capacity of 430 mAh, showing good rate capability. The cell had a capacity retention rate of ~90% after 200 cycles at a rate of 1 C, and ~80% after 300 cycles at 5 C, which is better than commercial $LiMn_2O_4$ batteries using graphite anode. However, the cell exhibited relatively large irreversible capacity, which needs further improvement to meet the qualifications for practical use.

Anatase and rutile are the two most common polymorphs of TiO_2 and their electrochemical properties as anode materials for LIBs have been widely investigated. However, they both have relatively low specific capacities. In comparison, the theoretical specific capacity (335 mAh g^{-1}) of another polymorph, $TiO_2(B)$ is almost two times higher than those of the previous two counterparts. Therefore, the electrochemical application of $TiO_2(B)$ has attracted much attention in recent years. Similar to anatase and rutile, $TiO_2(B)$ also suffers from poor electrical conductivity. Thus, it can be anticipated that graphene can enhance the electrochemical performance of $TiO_2(B)$ due to its extremely high conductivity, which has recently been confirmed. Huang and coworkers reported the synthesis of graphene-modified $TiO_2(B)$ nanobelts as high-performance anode materials [88]. The hybrid was prepared by hydrothermal treatment of commercial TiO_2 nanoparticles in 10 M NaOH solutions with the presence of GO. In the composite, $TiO_2(B)$ nanobelts with widths of 50–100 nm and lengths of several micrometers lie horizontally on graphene sheets, which forms a tight contacting sheet-belt nanostructure. The nanobelt morphology can shorten the Li-ion diffusion length, whereas the intimate contact between two components can effectively enhance the electron transport of $TiO_2(B)$. In addition, it was observed that abundant mesopores with a pore diameter of 10–20 nm existed in the nanobelts, which generate large electrode–electrolyte contact area. Such a unique structure of the $TiO_2(B)$/graphene hybrid gave rise to outstanding electrochemical performance. The initial discharge and charge capacity of the hybrid at a current density of 0.05 A g^{-1} was 909 and 746 mAh g^{-1}, respectively, which is much higher than the corresponding values for the pure $TiO_2(B)$ nanobelts (332 and 277 mAh g^{-1}) and also exceeds most of the previously reported anatase or rutile-based TiO_2/graphene composite anode materials. The ultrahigh capacity is partially ascribed to the contribution from graphene. The hybrid also exhibited excellent cycling stability. At a current density of 0.15 A g^{-1}, 93% capacity retention could be achieved after 100 cycles. By comparison, the reversible capacity dropped by 28% for the pure $TiO_2(B)$ anode material under the same conditions. Graphene can also enhance the rate capability of $TiO_2(B)$. Even at a high current density of 3 A g^{-1}, the hybrid could still reach a reversible capacity of 210 mAh g^{-1}, which is remarkably higher than that of pure $TiO_2(B)$ (20 mAh g^{-1}). The EIS results further proved that $TiO_2(B)$/graphene hybrid possessed lower contact and charge-transfer resistances than pure $TiO_2(B)$.

3.3.2.2.3 *Iron-Oxide–Graphene Composite Anode Materials*

Iron oxide is another important type of metal oxide anode material with great application potential in LIBs because of its high specific capacity, low cost, and eco-friendliness. Similar to SnO_2, iron oxides also suffer from poor cycle stability, mainly due to the large volume variation during charge/discharge processes that leads to pulverization of the particles and hence breakdown of the electrical connection between anode materials and current collectors. The strategy applied in modifying SnO_2 anode materials to achieve better electrochemical performance can also be introduced to improve the performance of iron oxide anode materials. Integration of iron oxide with graphene has recently been broadly investigated as an effective approach.

Iron (II, III) oxide (Fe_3O_4) is one potential iron oxide anode material. It has a high theoretical specific capacity of 927 mAh g^{-1} and relatively high voltage plateau. In 2010, Zhou and coworkers published the first report of Fe_3O_4/graphene composite anode material with a well-organized flexible interleaved structure, as illustrated in Figure 3.23, aiming to improve the reversible capacity and cycle stability of Fe_3O_4 [89]. In the first step of their experiment, graphene nanosheets were dispersed in the aqueous solution of iron (III) chloride ($FeCl_3$). The hydrolysis of $FeCl_3$ then formed spindle-shaped ferric oxide hydrate (FeOOH) intermediates embedded in graphene sheets. The following heat treatment under Ar transformed FeOOH into Fe_3O_4 nanoparticles that were attached to graphene sheets. Structural characterization revealed that the Fe_3O_4/graphene composites had a layer-by-layer assembled structure in which Fe_3O_4 particles were laterally wrapped by graphene sheets. The composite also has a relatively large surface area (53 m^2 g^{-1}) and pore volume (0.23 cm^3 g^{-1}), both of which are much higher than that of commercial Fe_3O_4 with similar particle sizes, indicating that the composite has an open porous structure. The authors then systematically compared the electrochemical performance of the Fe_3O_4/graphene composite with commercial Fe_3O_4 samples. At a current density of 35 mA g^{-1}, an initial charge capacity of 900 mAh g^{-1} could be delivered by the Fe_3O_4/graphene composite anode material, which is adjacent to the theoretical value. After cycling at the same current density 30 times, the reversible capacity did not drop, and even gradually increased to 1026 mAh g^{-1}, exhibiting outstanding cycle stability of the composite anode material. The increasing capacity was probably ascribed to the kinetically activated electrolyte degradation. In contrast, commercial Fe_3O_4 anode materials only delivered an initial reversible capacity of 770 mAh g^{-1}, and it rapidly dropped to 475 mAh g^{-1} after 30 cycles, which apparently reflected the effect of graphene in improving the cyclability of Fe_3O_4. The Fe_3O_4/graphene composite anode material also possesses excellent rate capability. Even at a high current density of 1750 mAh g^{-1}, it could still reach a reversible capacity of 520 mAh g^{-1}, ~53% of its

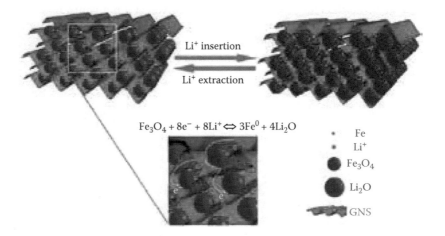

FIGURE 3.23 The structure model of Fe_3O_4/graphene composite anode material. (Reprinted with permission from Ref. 88, 5306–5313. Copyright 2010 American Chemical Society.)

initial capacity. However, only 10% of the initial capacity could be retained for the commercial Fe_3O_4 at the same current density. To better understand the dramatic difference of the electrochemical performance between two anode materials, the morphology and microstructure of both materials after 30 charge/discharge cycles were analyzed. It was observed by SEM that for commercial Fe_3O_4, the average particle size decreased from 735 to 428 nm after continuous Li-ion insertion/extraction, indicating that severe pulverization occurred during charge/discharge cycles. In contrast, Fe_3O_4 particles were still strongly wrapped by graphene sheets after 30 cycles, and their morphology and size remained almost the same, suggesting the crucial role of graphene in maintaining the integrity of the Fe_3O_4 particles. Overall, the significantly enhanced charge/discharge performance of the hybrid anode material is mainly attributed to the function of graphene as well as its unique microstructure. First, graphene sheets serve as an excellent conducting agent to enhance the electron transport within the anode. Second, flexible graphene sheets wrapping Fe_3O_4 particles can also prevent the detachment and agglomeration of pulverized Fe_3O_4, thus extending the cycle life. Third, a certain amount of pores exist in the composite anode material that facilitate Li-ion diffusion. Consequently, the Fe_3O_4/graphene composite anode material exhibits excellent cycle stability and rate capability.

Fe_3O_4/graphene composites with similar morphology and excellent electrochemical performance were also prepared by different research groups using different methods. For example, Fe_3O_4 nanoparticles could be loaded onto graphene sheets by a gas/liquid interface reaction, and the nanocomposite could remain 99% of its initial reversible capacity after 90 charge/discharge cycles [90]. In another case, Fe_3O_4/graphene nanocomposites were prepared by mixing of monodispersed Fe_3O_4 nanoparticles having diameters of ~10 nm with GO and subsequent thermal reduction [91]. Exceptional cycle stability over 800 charge/discharge cycles was observed for the composite anode material. Using a hydrothermal method, Sathish et al. were able to anchor Fe_3O_4 nanoparticles on graphene sheets, and the electrochemical performance of the nanocomposites with different weight ratio between Fe_3O_4 and graphene was investigated [92]. It was found that the nanocomposite with a Fe_3O_4/graphene ratio of 40:60 exhibited the best electrochemical performance. Lower Fe_3O_4 content lead to lower specific capacity, whereas higher Fe_3O_4 content caused agglomeration of Fe_3O_4 nanoparticles, giving rise to fast capacity fading during cycling. A unique synthetic route using CO_2-expanded ethanol as the solvent was proposed by Zhuo and coworkers to prepare Fe_3O_4/graphene nanocomposite with uniform loading of Fe_3O_4 nanoparticles on graphene sheets [93]. In the experiment, the vessel containing an ethanol solution of GO and Fe^{3+} was charged with CO_2 to 10 MPa. It was then heated at 120°C for 6 h. The intermediate obtained in the sealed vessel was then calcined to obtain the Fe_3O_4/graphene nanocomposite. During the liquid phase reaction, the compressed CO_2 in the vessel dissolved in ethanol, which altered the polarity of the solvent and thus is beneficial for the precipitation of Fe^{3+}. Meanwhile, CO_2 in the supercritical state behaved as an antisolvent to reduce the solvent strength, which could reduce the aggregation of nanoparticles. In addition, the low viscosity, high mass transfer rate, and zero surface tension of supercritical CO_2 helped the homogeneous deposition of nanoparticles on graphene sheets. As a result, uniform distribution of Fe_3O_4 nanoparticles on the graphene sheets and strong adhesion between two components were achieved. If the synthesis was carried out in pure ethanol with no CO_2, most Fe_3O_4 particles were deviated from graphene and aggregated to form large units, which further confirmed the significant effect of supercritical CO_2 in controlling the nucleation and growth of Fe_3O_4 nanoparticles. The initial charge capacity of the composite anode material could reach 941 mAh g^{-1}, and a reversible capacity of 838 mAh g^{-1} could still be retained after 100 cycles at a current density of 1 A g^{-1}. As a comparison, the reversible capacity of the sample synthesized without CO_2 rapidly dropped to 370 mAh g^{-1} after 100 cycles. These results evidently verified the importance of the microstructure of the composite anode materials to their electrochemical performance.

Besides particulate morphology, Fe_3O_4 with other morphologies were also reported. For instance, Hu et al. published a paper reporting the Fe_3O_4 nanorods/graphene composite anode material that was synthesized by chemical reduction of GO and ammonium ferrous sulfate $((NH_4)_2Fe(SO_4)_2)$ in aqueous solution under ambient pressure [94]. Fe_3O_4 nanorods had a mean diameter of ~10 nm and lengths of hundreds of nanometers, and were uniformly anchored on graphene sheets. The composite anode materials delivered a relatively high initial reversible capacity of 925 mAh g^{-1}, and only 5% capacity loss occurred after 100 cycles at the rate of 1 C. In another case, hollow Fe_3O_4 nanospheres encapsulated by graphene sheets were reported as high-performance anode materials [95]. Using a hydrothermal method, monodispersed hollow Fe_3O_4 particles (~220 nm in diameter) whose shells were composed of tiny Fe_3O_4 nanocrystals were first obtained. The hollow particles were then modified with amino groups on the surface and subsequently mixed with the aqueous suspension of GO. Because of the electrostatic interaction between negatively charged GO and positively charged Fe_3O_4 hollow particles, the self-assembly occurred between two components to form GO-encapsulated hollow Fe_3O_4 particles. Finally, GO was chemically reduced to graphene by hydrazine. In the nanocomposite, individual hollow Fe_3O_4 particles were caged by graphene sheets, while graphene itself formed an interconnected conducting web. The composite had a relatively high surface area of 132 m^2 g^{-1} and porosity of 0.39 cm^3 g^{-1}, which could be attributed to the existence of abundant voids between aggregated Fe_3O_4 nanoparticles in the shells and the formation of secondary pores between Fe_3O_4 particles and graphene sheets. Electrochemical measurement revealed that a high reversible capacity of ~900 mAh g^{-1} could be steadily maintained within 50 charge/discharge cycles at the current density of 100 mA g^{-1}. When the current density was raised to 800 mA g^{-1}, 58% of the capacity at 50 mA g^{-1} could still be retained.

In order to further enhance the stability of Fe_3O_4/graphene composite anode materials, Fe_3O_4 particles anchoring on graphene sheets were wrapped by thin carbon shells as proposed by Li and coworkers [96]. Using glucose and urotropine as the carbon source, amorphous carbon shells with a thickness of 2–3 nm could be successfully coated on the surface of Fe_3O_4 nanoparticles through a hydrothermal process. The core/shell Fc_3O_4@carbon particles were also homogenously loaded on graphene sheets. The special limitation of Fe_3O_4 within closed carbon shells could avoid the morphological changes of the active materials and prevent aggregation of adjacent Fe_3O_4 nanoparticles during charge/discharge processes. Meanwhile, carbon shells could also improve the electrical conductivity of the electrode along with graphene. The Fe_3O_4-carbon-graphene composite anode material could deliver an initial reversible capacity of 952 mAh g^{-1} at 200 mA g^{-1}, and after 100 cycles at the same current, a reversible capacity of 843 mAh g^{-1} could still be retained. The Coulombic efficiency was always above 95% after the fourth cycle and kept increasing during the cycling process. The composite anode material also exhibited excellent rate capability at high current densities up to 5 A g^{-1}.

Fe_2O_3 is another important anode material for LIBs, which has a theoretical specific capacity of 1005 mAh g^{-1}, even higher than that of Fe_3O_4. There have been a series of reports on the modification of Fe_2O_3 by graphene so far. Zhu et al. used a two-step synthesis method to prepare Fe_2O_3/graphene composite anode material by homogeneous precipitation of $FeCl_3$ in the GO suspension with urea and subsequent chemical reduction of GO into graphene under microwave irradiation [97]. In this nanocomposite, Fe_2O_3 nanoparticles with mean diameters of ~60 nm were uniformly loaded on the surface of graphene sheets. At a current density of 100 mA g^{-1}, the initial discharge and charge capacity for the composite anode material was 1355 and 982 mAh g^{-1}, respectively. If the weight of graphene in the composite was eliminated, the corresponding capacity values based only on the mass of Fe_2O_3 were 1693 and 1227 mAh g^{-1}, respectively, which were even higher than the theoretical value. The excessive capacity was derived from by the formation of organic layer on the surface of Fe_2O_3, as well as the Li-ion storage of graphene sheets. After cycling at 100 mA g^{-1} for 50 cycles, a reversible capacity of 1027 mAh g^{-1} (based on the mass of Fe_2O_3) could be retained, suggesting good cycle stability of the composite anode material. When the current density was increased to 800 mA g^{-1}, the reversible capacity was still as high as ~800 mAh g^{-1}. For comparison, the electrochemical performance of Fe_2O_3 nanoparticles physically mixed with graphene was also measured. The charge capacity dramatically dropped from 930 mAh g^{-1} to 129 mAh g^{-1} after 30 cycles at a current density of 100 mA g^{-1}, clearly indicating the significance of synergetic effect between Fe_2O_3 and graphene in the composite anode material.

The morphology of Fe_2O_3 in the Fe_2O_3/graphene composite anode materials can be tuned by using different synthetic methods. Bai and coworkers reported the synthesis of Fe_2O_3 nanospindles/graphene nanocomposites by a hydrothermal method [98]. During the reaction, Fe^{3+} ions are attracted by negatively charged GO sheets and served as nucleation sites. After the addition of $NH_3 \cdot H_2O$, $Fe(OH)_3$ nucleated and grew on the surface of graphene, while the existence of ethylene glycol prevented the agglomeration of $Fe(OH)_3$ nanoparticles. The following hydrothermal treatment transformed $Fe(OH)_3$ into β-FeOOH intermediates and finally Fe_2O_3. Meanwhile GO was reduced to graphene by ethylene glycol under hydrothermal conditions. Electron microscopy images (Figure 3.24) revealed that Fe_2O_3 particles had spindle-like morphology and an average size of ~120 nm along the long axis and ~60 nm along the minor axis. Graphene in the nanocomposites were naturally crumpled and curved with a petal-like shape. Fe_2O_3 nanospindles were densely dispersed on graphene sheets. The nanocomposite showed a large initial reversible capacity of 1556 mAh g^{-1} and the current density of 100 mA g^{-1} and a capacity of 969 mAh g^{-1} could be retained after 100 cycles. At a high current density of 5 A g^{-1}, a relatively high reversible capacity of 336 mAh g^{-1} was still reached after 100 cycles. The authors also examined the morphology of the electrode after charge/discharge processes.

FIGURE 3.24 (a, b) SEM and (c, d) TEM images of the Fe_2O_3/graphene nanocomposite. (e) High-resolution TEM image of part of an area of a selected Fe_2O_3 nanospindle in (d). (From Bai, S., Chen, S., Shen, X. et al., Nanocomposites of hematite (α-Fe_2O_3) nanospindles with crumpled reduced GO nanosheets as high-performance anode material for lithium-ion batteries. *RSC Adv.* 2:10977–10984, 2012. Reproduced by permission of The Royal Society of Chemistry.)

It was found that the spindle-like morphology and particle size of Fe_2O_3 remained almost the same as in their initial state. In contrast, bare Fe_2O_3 nanospindles with no graphene modification pulverized into small and irregular particles. The results evidently verified the important role of graphene in accommodating the stress and strain of volume change of Fe_2O_3 during Li-ion insertion/extraction, which gave rise to remarkably improved cycle stability.

In another report, disklike Fe_2O_3 nanoparticles incorporated with graphene was presented, which was prepared by a silicate-anion-assisted hydrothermal method [99]. The Fe_2O_3/graphene nanocomposite possessed disk-like morphology with a mean diameter of ~200 nm. The TEM image revealed that the nanodisks had layered structure and were composed of Fe_2O_3 nanosheets ~5 nm in thickness (Figure 3.25). It was difficult to distinguish graphene in the nanocomposite. Only edges of graphene could be observed in TEM images. Elemental analysis verified that graphene was homogeneously distributed in the nanocomposite. Considering the anisotropic nanosheet-like morphology of both Fe_2O_3 and graphene, it could be speculated that graphene might not only be wrapped by Fe_2O_3 but was also sandwiched between Fe_2O_3 nanosheets. Such tight and face-to-face contact between two components is different from the point-to-point contact mode in the case of particulate Fe_2O_3 and

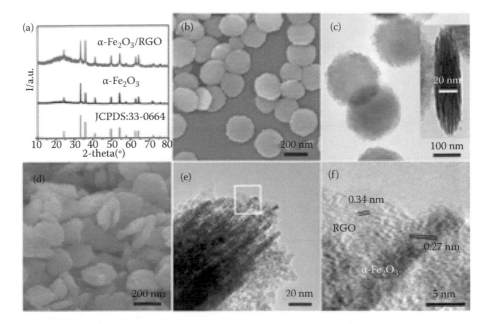

FIGURE 3.25 (a) XRD pattern of the Fe$_2$O$_3$/graphene composite and Fe$_2$O$_3$. (b) SEM and (c) TEM images of pure Fe$_2$O$_3$. (d) SEM and (e) TEM images of Fe$_2$O$_3$/graphene composite. (f) High-resolution TEM image of the white square in (e). (Reprinted with permission from Ref. 98, 3932–3936. Copyright 2013 American Chemical Society.)

graphene, and could be verified by a series of structural characterization results. The G peak of graphene in the Raman spectrum for the Fe$_2$O$_3$/graphene nanocomposite blue-shifted by ~11 cm^{-1} compared with that of pure graphene, indicating charge transfer between two components. TGA analysis showed an apparent decrease of oxidation temperature of the nanocomposite than pure graphene, suggesting tight contact between Fe$_2$O$_3$ and graphene. XPS results also indicated the formation of Fe-O-C bonds in the nanocomposite. The oxygen bridge between Fe$_2$O$_3$ and graphene is believed to be beneficial to Li-ion storage, as it could facilitate electron hopping from graphene to Fe$_2$O$_3$. At a current density of 200 mA g^{-1}, the Fe$_2$O$_3$/graphene composite anode material delivered an initial charge capacity of 1088 mAh g^{-1} and a capacity of 931 mAh g^{-1} could still be reached after 100 cycles. As a comparison, the physical mixture of Fe$_2$O$_3$ and graphene only had a charge capacity of 244 mAh g^{-1} after 50 cycles, showing much worse cycle stability. The Fe$_2$O$_3$/graphene nanocomposite also exhibited much improved rate capability. Even at a high current density of 10 A g^{-1}, a reversible capacity of 337 mAh g^{-1} could still be delivered, while there was nearly no capacity observed for the physical mixture of Fe$_2$O$_3$ and graphene. The Nyquist plots also indicated that the Fe$_2$O$_3$/graphene nanocomposite possessed much lower contact and charge transfer resistances than those of the mixture, which further proved that the tight contact between two components could effectively reduce the contact resistance and enhance charge transfer, resulting in remarkably improved cycle stability and rate performance.

Besides tuning the morphology and microstructure of Fe_2O_3, the modification of graphene to achieve better electrochemical performance was also carried out in recent studies. Nitrogen doping of graphene can provide extra lone pair electrons, resulting in improved electrical conductivity, which is beneficial for the enhancement of charge/discharge performance of graphene-based composite electrode materials. Du and coworkers reported the synthesis of Fe_2O_3/nitrogen-doped graphene composite as a high-performance anode material [100]. In the experiment, $Fe(NH_4)_2(SO_4)_2$ served as both precursors to Fe_2O_3 and nitrogen source was dissolved in the aqueous solution of GO. The following hydrothermal treatment led to the formation of Fe_2O_3 nanoparticles, as well as nitrogen doping and reduction of GO into N-doped graphene. Fe_2O_3 nanoparticles had diameters of 100–200 nm and were uniformly distributed on graphene sheets. The successful doping of nitrogen was evidently proved by XPS spectra, in which graphitic, pyrrolic, and pyridinic nitrogen species in the graphene sheets could be clearly identified. The atomic ratio of N/C in the nanocomposite measured by XPS data was 0.0325. For comparison, Fe_2O_3/graphene nanocomposite with no nitrogen-doping was also prepared, using $FeSO_4$ as the precursor instead of N-containing iron salt. The electrical conductivity of the Fe_2O_3/N-doped graphene composite measured by the four probe method was 19.3 S m^{-1}, which was almost two times higher than that of the Fe_2O_3/graphene composite (10.3 S m^{-1}) and more than three orders of magnitude higher than that of pure Fe_2O_3 (0.00298 S m^{-1}). The advantage of nitrogen doping in increasing the conductivity of Fe_2O_3 resulted in improved electrochemical performance, especially the rate capability and cycle stability. At a current density of 800 mA g^{-1}, the Fe_2O_3/N-doped graphene composite anode material could deliver a reversible capacity of ~800 mAh g^{-1}, which was almost four times higher than that of the nondoped sample. When the cell was cycled at 100 mA g^{-1} for 100 cycles, a reversible capacity of 1012 mAh g^{-1} could still be retained for the Fe_2O_3/N-doped graphene composite and the Coulombic efficiency remained above 97%. In contrast, the Fe_2O_3/graphene composite anode material could only deliver a reversible capacity of 430 mAh g^{-1} under the same condition.

In some recent publications, the 3-D microstructure of the Fe_2O_3/graphene composite anode materials was rationally designed and fabricated to facilitate Li-ion diffusion and electron transportation, thus to promote the electrochemical performance. For instance, Wang et al. reported Fe_2O_3/N-doped graphene anode material with a unique structure in which hexagonal Fe_2O_3 platelets were sandwiched by N-doped graphene sheets [101]. The key to obtain such a novel structure is mainly ascribed to the effect of 7,7,8,8-tetracyanoquinodimethane anion (TCNQ$^-$) existed in the reaction solution. During the reaction, TCNQ$^-$ anions were absorbed on graphene sheets because of the π–π interaction, resulting in homogeneous dispersion of graphene due to the electrostatic repellence between negatively charged TCNQ$^-$. Such electrostatic repellence also promoted the formation of sandwichlike reaction chambers between adjacent graphene sheets. The following added Fe^{3+} cations were absorbed by TCNQ$^-$ anions, and stayed in the above-mentioned reaction chambers. Then, Fe_2O_3 nucleated and grew along the <110> orientation into hexagonal platelets being sandwiched by graphene sheets. Nitrogen atoms in TCNQ$^-$ acted as N source to the formation of N-doped graphene during the final calcination process. The atomic content of N was measured to be 6.9% for the nanocomposite. If no TCNQ$^-$ was added in the synthesis, only spherical Fe_2O_3 nanoparticles were formed and anchored on graphene sheets. Fe_2O_3 nanoplatelet/N-doped graphene composite anode material has great advantage over Fe_2O_3 nanoparticle/graphene composite in terms of electrochemical performance. The capacity loss per cycle for the Fe_2O_3 nanoplatelet/N-doped graphene composite at 0.2 C was only 0.4%, which was one-seventh of that for the Fe_2O_3 nanoparticle/graphene composite. The former also exhibited excellent rate capability. A reversible capacity of 531 mAh g^{-1} could be reached at 5 C, whereas the latter only delivered 88 mAh g^{-1} at 2 C. EIS analysis revealed that the charge transfer resistance of the former samples was 12.8 Ω, much smaller than that for the latter one (79.4 Ω). The significantly improved electrochemical performance of the Fe_2O_3 nanoplatelet/N-doped graphene composite anode material could be ascribed to the synergetic effect of small Li-ion diffusion lengths in ultrathin nanoplateles, the sandwiched space to buffer volume expansion, as well as the enhanced electrical conductivity of graphene by nitrogen doping.

In another report, Xiao et al. proposed the self-assembly of Fe_2O_3/graphene aerogels with large porosity and excellent charge/discharge performance [102]. The composite material was prepared by hydrothermal thermal treatment of aqueous suspension containing GO and iron source in a sealed vessel. During this process, hydrolysis of Fe^{3+} resulted in the formation of FeOOH on the surface of GO sheets. Subsequent heating at higher temperatures transformed FeOOH into Fe_2O_3. Meanwhile, Fe_2O_3-loaded GO sheets self-assembled into 3-D monolithic networks and hydrothermally reduced to graphene. The aerogel was finally obtained by freeze-drying of the as-prepared Fe_2O_3/graphene hydrogel. In the aerogel, Fe_2O_3 particles had diameters of 50–200 nm and were uniformly distributed on both sides of the graphene sheets. Graphene sheets were interconnected to form 3-D network microstructure with uniformly dispersed pores of several micrometers in diameter, and all the pores were interconnected as well. The specific surface area and pore volume of the aerogel was measured to be 77 m^2 g^{-1} and 0.29 cm^3 g^{-1}, respectively, indicating high porosity. The continuous graphene framework in the aerogel provided efficient pathways for fast electron migration, while the open porous structure facilitated diffusion of electrolytes in the electrode and effectively reduced the diffusion lengths of Li-ions. Consequently, the Fe_2O_3/graphene aerogel exhibited better electrochemical performance than traditional Fe_2O_3/graphene composite with sheet-like morphology. A high reversible capacity of 995 mAh g^{-1} could be retained after 50 cycles at a current density of 100 mA g^{-1} for the aerogel anode and even at a high rate of 5 A g^{-1}, it could still deliver a reversible capacity of 372 mAh g^{-1}. For comparison, the reversible capacity of sheetlike Fe_2O_3/graphene composite anode materials quickly dropped to 263 mAh g^{-1} after 50 cycles.

Zhou and coworkers reported the preparation of graphene-sheet-wrapped Fe_2O_3 composite with a 3-D hierarchical structure by a spray-drying technique [103]. The aqueous suspension containing both Fe_2O_3 nanoparticles and GO sheets was spray-dried at the inlet temperature of 200°C. During the evaporation process, GO sheets spontaneously assembled on the surface of the droplets and subsequently shrunk and encapsulated Fe_2O_3 nanoparticles. GO sheets were finally reduced by calcination. The as-prepared composite had a spherical morphology with diameters of 1–3 μm and was composed of randomly aggregated Fe_2O_3 nanoparticles whose diameters were 30–100 nm. Each Fe_2O_3 nanoparticle was intimately wrapped by crumpled graphene sheets. The composite anode material delivered an initial charge capacity of ~950 mAh g^{-1} and the reversible capacity of ~900 mAh g^{-1} could be maintained after 200 cycles. When the current density was increased to 1.6 A g^{-1}, it could still deliver a reversible capacity of ~660 mAh g^{-1}. The excellent electrochemical performance was ascribed to the synergetic effect contributed by both components. On one hand, 3-D graphene network in the composite not only facilitated fast electron transport but also provided elastic voids to buffer the volume expansion of Fe_2O_3 nanoparticles and prevented pulverization and aggregation of Fe_2O_3 nanoparticles during cycling. On the other hand, Fe_2O_3 particles homogeneously distributed in the composite avoided the restaking of graphene sheets during the spray-drying process, giving rise to the full utilization of graphene.

3.3.2.2.4 Co_3O_4-Graphene Composite Anode Materials

With a similar structure as Fe_3O_4, cobalt (II, III) oxide (Co_3O_4) can also serve as a potential anode material for LIBs. It has attracted extensive attention recently, mainly due to its high theoretical capacity of 890 mAh g^{-1}. However, Co_3O_4 also suffers from large volume expansion/extraction during charge/discharge processes, which causes poor cycle stability of this anode material. The successful experience of using graphene to improve the electrochemical performance of metal oxide anode materials has also applied in the case of Co_3O_4 anode materials in recent years.

The synthesis of Co_3O_4/graphene composite anode material was first reported by Wu et al. [104]. As illustrated in Figure 3.26, the addition of $NH_3 \cdot H_2O$ in the solution containing Co^{2+} and graphene led to the formation of $Co(OH)_2$/graphene composite. Subsequent calcination at 450°C caused the decomposition of $Co(OH)_2$ and the formation of Co_3O_4/graphene composite. Electron microscopy images showed that Co_3O_4 nanoparticles had diameters of 10–30 nm and were homogeneously anchored on graphene sheets. There was strong interaction between two components, as Co_3O_4 nanocrystals did not fall off from graphene sheets even after a long-time sonication treatment. It was noted that no Co_3O_4 nanparticles formed if graphene was absent in the synthesis, which indicated the significance of graphene in inducing the nucleation and growth of Co_3O_4 nanoparticles. Electrochemical tests revealed that the composite anode material had an initial discharge and charge capacity of 1097 and 753 mAh g^{-1}, respectively, at a current density of 50 mA g^{-1}. A gradual increase of the reversible capacity was observed when cycled at the same rate. After 30 cycles, the reversible capacity increased to 935 mAh g^{-1},

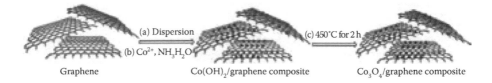

Graphene $Co(OH)_2$/graphene composite Co_3O_4/graphene composite

FIGURE 3.26 Schematic representation of the fabrication process of Co_3O_4/graphene composite. (a) Dispersion of chemically derived graphene in isopropyl alcohol/water solution. (b) Formation of $Co(OH)_2$/graphene composite in basic solution. (c) Phase transformation from $Co(OH)_2$/graphene composite to Co_3O_4/graphene composite by calcination.

showing outstanding cycle stability. As a comparison, the reversible capacity of pure Co_3O_4 dropped quickly from 817 mAh g^{-1} in the first cycle to 184 mAh g^{-1} after 30 cycles. The synergetic effect between Co_3O_4 and graphene was believed to be the main factor that induced excellent electrochemical performance.

Later, Co_3O_4/graphene composite anode materials with diverse morphologies and microstructures by using different methods were successively reported. For example, using hexamethylenetetramine instead of ammonia in the previous work by Wu et al., $Co(OH)_2$ nanosheet could form on the surface of graphene under microwave irradiation [105]. The following calcination then transformed $Co(OH)_2$ to porous Co_3O_4 nanosheet loaded on graphene sheets. The pore size of the Co_3O_4 nanosheet was 60–100 nm and its thickness was ~100 nm. The composite anode material with a sheet-on-sheet structure could deliver an initial reversible capacity of 1235 mAh g^{-1} and a high capacity of 1065 mAh g^{-1} could still be retained after 30 cycles. TEM observation revealed that the porous sheetlike structure of Co_3O_4 still remained on graphene sheets after 30 cycles, indicating the stability of the composite anode material. When the rate was increased to 5 C, the composite could still deliver a high capacity of 931 mAh g^{-1}, showing excellent rate capability.

By reducing the concentration of reactants, ultrasmall Co_3O_4 nanocrystals could be produced on graphene sheets as reported by Kim et al. The Co_3O_4 nanoparticles had a mean diameter of ~5 nm and were strongly anchored on the graphene sheets [106]. Such small size was helpful to further relieve the volume change of Co_3O_4 during charge/discharge cycles. As a result, the composite anode material exhibited excellent cycle stability. After cycling at a current density of 200 mA g^{-1} 42 times, a reversible capacity of 778 mAh g^{-1} could still be retained with Coulombic efficiency of 97%.

Yue et al. reported the preparation of graphene-encapsulated mesoporous Co_3O_4 particles as a novel anode material [107]. Mesoporous Co_3O_4 was prepared by a hard-template method. Using ordered mesoporous silica as the template, melted Co salt was infiltrated into the mesopores because of the capillary force and decomposed to form oxide inside the mesopores. Removal of the silica template resulted in mesoporous Co_3O_4 with ordered pore structures. The as-prepared mesoporous Co_3O_4 particles were then dispersed in an aqueous solution with a pH value of 5–6 in which the surface of Co_3O_4 was positively charged. By mixing the above mesoporous Co_3O_4 suspension with GO solution at a pH of 7–8, the electrostatic interaction between Co_3O_4 and GO gave rise to the encapsulation of mesoporous Co_3O_4 particles with GO sheets. The final chemical reduction of GO by ascorbic acid resulted in the formation of mesoporous Co_3O_4/graphene composite anode material. Mesoporous Co_3O_4 particles wrapped by graphene sheets had a cubic mesostructure and mean pore diameter of 3.8 nm (Figure 3.27). Such porous structure is beneficial for fast diffusion of Li-ions, and the pores can partially endure the volume expansion of the active materials. However, capacity fading still occurs because of the instability of the mesoporous structure. Graphene modification could effectively increase the stability as well as the electrical conductivity of mesoporous Co_3O_4, resulting in significantly improved cycle stability. A high reversible capacity of 1150 mAh g^{-1} could be reached for the mesoporous Co_3O_4/graphene composite anode material after 30 cycles at 1 C, whereas the capacity quickly faded to zero for bare mesoporous Co_3O_4 only after less than 10 cycles.

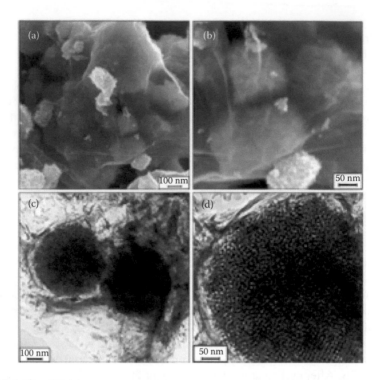

FIGURE 3.27 SEM images (a and b) and TEM images (c and d) of mesoporous Co_3O_4/ graphene nanocomposite. (From Yue, W., Lin, Z., Jiang, S. et al., Preparation of graphene-encapsulated mesoporous metal oxides and their application as anode materials for lithium-ion batteries. *J. Mater. Chem.* 22:16318–16323, 2012. Reproduced by permission of The Royal Society of Chemistry.)

Co_3O_4 nanoparticles anchored on graphene sheets could also be produced by the pulse-microwave-assisted method [108]. Compared with a conventional heating process, pulse-microwave-assisted synthesis has advantages in terms of efficiency, uniformity, and simplicity in operation, which shows commercial feasibility. In the experiment, a household microwave oven was used to heat a solution containing Co^{2+} and GO for 30 min at a power of 700 W and pulse frequency of 1.5. The as-prepared intermediates were then calcinated to obtain Co_3O_4/graphene composite material. The morphology of Co_3O_4 was strongly dependent on the solvent. If water was used, cubiclike Co_3O_4 particles with sizes of 300–500 nm were obtained, while small Co_3O_4 nanoparticles with diameters of 10–15 nm were produced when ethylene glycol was employed as the solvent. It is known that the heat under microwave irradiation comes from interaction of irradiation with the polar bond in the reactant. As water has a larger dipole than ethylene glycol, the nucleation and growth rate of Co_3O_4 nanocrystals in water is faster than that in ethylene glycol. Thus, much larger particles were observed in the case of water-based synthesis. In addition, a small amount of cobalt (II) oxide (CoO) impurities were observed in the water system, which was probably derived from the decomposition of $Co(OH)_2$. No CoO was found in the ethylene glycol system. The EIS analysis revealed that the diffusion coefficient of Li-ions for the ethylene-glycol-derived sample was 5.82×10^{-12} cm^2 s^{-1}, which was about 5.8 times higher than that of the water-derived sample (9.99×10^{-13} cm^2 s^{-1}). The internal resistance of the former sample (64.9 Ω) was also smaller than that of the latter one (116.5 Ω). This result indicated that Li-ions were more easily able to get into the framework of the former anode material than the latter one. Electrochemical measurement showed that anode material produced in ethylene glycol had an initial charge capacity of 934 mAh g^{-1} at a current density of 70 mA g^{-1}, which was higher than that of the water-derived sample (861 mAh g^{-1}). When the charge/discharge rate was increased to 3.5 A g^{-1}, a relatively high reversible capacity of 400 mAh g^{-1} could be reached for composite material obtained in ethylene glycol, which was more than 40% higher for the one prepared in water. The improved rate capability could be mainly ascribed to be lower internal resistance and higher Li-ion diffusion coefficient for the ethylene glycol-derived sample. It also exhibited excellent cycle stability. When cycled at a current density of 700 mA g^{-1} 50 times, a reversible capacity of 650 mAh g^{-1} could still be retained.

In another report, Kim and coworkers reported the synthesis Co_3O_4/graphene thin film as the anode for LIBs using an electrodeposition method [109]. In their experiment, GO was initially modified by poly(ehtyleneimine) (PEI) by stirring of GO in the aqueous solution of PEI. PEI-decorated GO was then added into the aqueous solution containing cobalt (II) nitrate ($Co(NO_3)_2$) and sodium nitrate ($NaNO_3$) for electrodeposition. The electrodeposition process was carried out potentiostatically at −1.0 V (vs. Ag/AgCl) for 5 min at room temperature and used stainless steel foil as the substrate. The resultant film was then calcined to obtain Co_3O_4/graphene composite film. PEI is a cationic polymeric electrolyte containing a large amount of amine groups that have strong interaction with negatively charged GO sheets in water. Meanwhile, amine groups in PEI molecules provide abundant ligand sites to capture Co^{2+} in the aqueous solution. Therefore, modification of GO by PEI resulted in the complexation between GO and Co^{2+}, which could facilitate the coelectrodeposition of two components. If no PEI was added, agglomeration of GO sheets occurred after the addition of Co^{2+}, resulting in precipitation before electrodeposition. EDS results revealed that Co and C elements were homogeneously distributed within the whole composite film, suggesting high uniformity of both graphene and Co_3O_4. Pure Co_3O_4 film in the absence of graphene was also prepared by similar electrodeposition method for comparison. Electrochemical measurement showed that the Co_3O_4/graphene composite film delivered initial discharge and charge capacities of 1342 and 1042 mAh g^{-1}, while the corresponding values for pure Co_3O_4 film were only 1209 and 899 mAh g^{-1}. The former anode also exhibited better cycle stability. When cycled at a current density of 700 mA g^{-1} 50 times, the Co_3O_4/graphene film reached a capacity retention of 113%, whereas the value for the pure Co_3O_4 anode was 103%. The increase of the capacity during cycling could be ascribed to the formation of a gel-like polymeric layer in the electrode. EIS analysis indicated that the charge transfer resistance for the composite film was apparently smaller than that for pure Co_3O_4, suggesting the remarkable effects of graphene in improving the electrical conductivity of Co_3O_4 anode materials.

3.3.2.3   Lithium Titanate-Graphene Composite Anode Materials

Lithium titanate ($Li_4Ti_5O_{12}$) with a spinel structure has drawn much attention as an appealing anode material in recent years. It has a fast Li-ion insertion/extraction capability and zero strain during charge/discharge processes, resulting in outstanding rate capability and cycling stability. Compared with conventional graphite anode materials, $Li_4Ti_5O_{12}$ (LTO) has a higher operating voltage at 1.55 V (vs. Li/Li^+), which will effectively prevent the formation of Li dendrite on the surface of anode, giving rise to improved safety of the batteries. Meanwhile, the equilibrium potential of the Ti^{3+}/Ti^{4+} redox couple is above the reduction potential of common electrolyte solvents, which can inhibit the formation of solid electrolyte interface film. Therefore, LTO is considered as good candidate anode material for high-power LIBs with good safety. Nevertheless, the electrical conductivity of LTO is extremely low, which causes serious polarization of the electrode when pristine LTO is used. Therefore, it is desirable to develop a suitable modification method of LTO to improve its conductivity. Hybridizing LTO with a conductive second phase is one of the most effective methods. The successful experience of employing high conductive graphene in the modification of electrode materials for LIBs has also been applied in the case of LTO in recent years.

The LTO/graphene hybrid anode materials can be easily prepared by homogeneous mixing of two components at the nanoscale. For instance, Shi et al. reported the preparation of the LTO/graphene composite anode material by simply mixing finely milled LTO nanoparticles and GO in organic solvent and subsequent reduction of GO at high temperatures [110]. Graphene, which had a weight content of 5% in the composite, was homogeneously distributed among LTO nanoparticles whose diameters were in the range of 100–400 nm. The intimate contact between two components at the nanoscale could effectively increase the electrical conductivity of LTO by graphene network. Compared with pristine LTO, the modification of LTO with graphene could reduce the charge transfer resistance of the cell by ~30%, suggesting an apparent increase of the electrical conductivity of insulating LTO by graphene. Meanwhile, the diffusion coefficient of Li-ions was also increased by graphene modification. At a 1-C rate, the LTO/graphene composite anode material could deliver an initial discharge capacity of 171.7 mAh g^{-1}, very close to the theoretical capacity of LTO (175 mAh g^{-1}). As a comparison, the discharge capacity for pure LTO was only 159.7 mAh g^{-1}. When the discharge rate was increased to 30 C, the hybrid cathode could still deliver a capacity of 122 mAh g^{-1}, which was 30 mAh g^{-1} higher than the corresponding value for pure LTO. It also showed excellent cycle stability. Capacity retention of 94.8% could be reached after the cell was cycled at a high rate of 20 C 300 times. The hybridization of LTO with graphene could also be realized with the spray-drying technique [111]. The primary LTO nanoparticles aggregated to form secondary microspheres after spray-drying, and graphene nanosheets formed 3-D conductive network that wrapped LTO nanoparticles in the secondary particles. Remarkably enhanced rate capability and cycle stability was also observed for the hybrid anode material.

Although it is easy to operate the mechanical mix of preprepared LTO with graphene to form hybrid anode materials, it is expected that *in situ* growth of LTO on graphene might be more effective to achieve more homogeneous distribution of two components and enhance the interaction between them. Kim et al. presented such an *in situ* synthesis process by growing LTO nanoplatelets on graphene sheets through a two-step microwave-assisted solvothermal reaction [112]. In their experiment, TiO_2/graphene composite was initially prepared by the solvothermal reaction of titanium ethoxide in diethylene glycol with the presence of GO. The as-prepared TiO_2/graphene hybrid was then mixed with LiOH solution and heated under microwave-assisted solvothermal conditions to form the Li-Ti-O/graphene intermediate. The subsequent calcination then produced the LTO/graphene composite anode material. The TiO_2 deposited on graphene sheets first had nanoparticulate morphology with particle size of 3–5 nm. After lithiation reaction, the morphology of the Li-Ti-O/graphene intermediate turned to be nanoplateletlike. Further calcination of the intermediate had little effect on the morphology but gave rise to the formation of phase pure and highly crystalline LTO nanoplatelets, as shown in Figure 3.28. It could be seen that uniform LTO nanoplatelets with a size of ~20 nm were homogeneously distributed on graphene sheets. Both XRD and high-resolution TEM characterization confirmed the high crystallinity of LTO. Galvanostatic charge and discharge tests showed that the LTO/graphene hybrid anode material could deliver an initial discharge capacity of 154 mAh g^{-1} at the rate

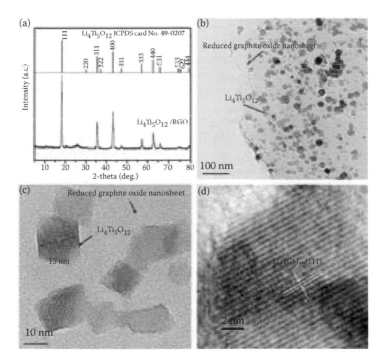

FIGURE 3.28 (a) XRD patterns of $Li_4Ti_5O_{12}$/graphene composite anode material. (b–d) TEM images of $Li_4Ti_5O_{12}$/graphene composite. (Reprinted with permission from *Electrochem. Commun.*, 12, Kim, H.-K., Bak, S.-M. and Kim, K.-B., 1768–1771, Copyright 2010, with permission from Elsevier.)

of 1 C. Relatively high capacity could still be maintained when further increasing the rate. The discharge capacity at 50 and 100 C were 128 and 101 mAh g^{-1}, respectively, corresponding to capacity retention of 83% and 60% compared with the value at 1 C. It was also noted that the voltage gap between charge and discharge curves did not significantly increase along with the increase of the rate, suggesting small polarization of the electrode at high rates. The excellent rate performance could be definitely ascribed to the high electrical conductivity of graphene and its intimate contact with LTO. Meanwhile, the nanoplatelet-like morphology of LTO could dramatically reduce the diffusion length of Li-ions, also contributing to the enhanced rate capability.

In another case, LTO nanosheets supported on graphene as a high-performance hybrid anode material was reported [113]. The authors used a similar synthetic strategy that employed TiO$_2$/graphene composite as the precursor, as has been published in a previous paper. By varying the reaction conditions, such as the solvent and concentration of reactants, LTO nanosheets with a lateral size of 100–200 nm and a thickness of several nanometers were formed on graphene sheets. It was also observed that most nanosheets stood upright on the graphene support. The charge transfer resistance of the cell using LTO/graphene as the anode material was measured to be 70 Ω, which was much lower than the value in the case of pure LTO nanosheets (95 Ω), suggesting the effect of graphene in improving the conductivity of LTO. As a result, the LTO/graphene hybrid anode material exhibited good cycle stability and a high discharge capacity of 140 mAh g^{-1} at a 20-C rate. Similar hydrothermal or solvothermal method had also been applied by other groups to the preparation of LTO/graphene composite anode materials with variable microstructures, and the effect of graphene in enhancing the electrochemical performance of LTO was confirmed [114,115].

In a special case, fiberlike LTO/graphene composite material was prepared using the electrospinning method [116]. The raw solution for the spinning was composed of titanium isopropoxide, lithium acetylacetonate, GO, polyvinylpyrrolidone, and ethanol as the solvent. After electrospinning, GO was reduced by hydrazine hydrate vapor. Finally, the composite nanofibers were calcined at 550°C in N$_2$ atmosphere. The as-prepared LTO/graphene composite nanofibers had a uniform diameter below 1 μm. The carbon content was measured to be 7.2% in the nanofiber, in which 1 wt% was contributed by graphene while others were derived from pyrolysis of the polymer after calcination. Small LTO nanocrystals were aggregated in the electrospun nanofibers, whereas graphene sheets were embedded among LTO nanoparticles. The nanofibers also had a high specific surface area of 170 m^2 g^{-1}, suggesting abundant nano-sized cavities among LTO nanocrystals inside these fibers. Compared with the LTO sample with no graphene, the addition of only 1 wt% of graphene could dramatically reduce the charge transfer resistance by almost six times. Consequently, the LTO/graphene composite nanofibers exhibited much improved electrochemical performance. At a high rate of 22 C, it could still deliver a discharge capacity of 110 mAh g^{-1}, which was more than twice the capacity of electrospun TiO$_2$ with no graphene. Moreover, after 1300 charge/discharge cycles at 22 C, an ultrahigh capacity retention of 91% was retained, showing excellent cycle stability of the electrospun LTO/graphene composite nanofibers.

REFERENCES

1. Yoo, E., Kim, J., Hosono, E. et al. 2008. Large reversible Li storage of graphene nanosheet families for use in rechargeable lithium ion batteries. *Nano Lett.* 8:2277–2282.
2. Wang, G., Shen, X., Yao, J. et al. 2009. Graphene nanosheets for enhanced lithium storage in lithium ion batteries. *Carbon* 47:2049–2053.
3. Lian, P., Zhu, X., Liang, S. et al. 2010. Large reversible capacity of high quality graphene sheets as an anode material for lithium-ion batteries. *Electrochim. Acta* 55:3909–3914.
4. Bhardwaj, T., Antic, A., Pavan, B. et al. 2010. Enhanced electrochemical lithium storage by graphene nanoribbons. *J. Am. Chem. Soc.* 132:12556–12558.
5. Pan, D., Wang, S., Zhao, B. et al. 2009. Li storage properties of disordered graphene nanosheets. *Chem. Mater.* 21:3136–3142.
6. Lee, W., Suzuki, S. and Miyayama, M. 2013. Lithium storage properties of graphene sheets derived from graphite oxides with different oxidation degree. *Ceram. Int.* 39:S753–S756.
7. Pollak, E., Geng, B., Jeon, K.-J. et al. 2010. The interaction of Li+ with single-layer and few-layer graphene. *Nano Lett.* 10:3386–3388.
8. Vargas, C., O. A., Caballero, Á. and Morales, J. 2012. Can the performance of graphene nanosheets for lithium storage in Li-ion batteries be predicted? *Nanoscale* 4:2083–2092.
9. Reddy, A. L. M., Srivastava, A., Gowda, S. R. et al. 2010. Synthesis of nitrogen-doped graphene films for lithium battery application. *ACS Nano* 4:6337–6342.
10. Wu, Z. S., Ren, W. C., Xu, L. et al. 2011. Doped graphene sheets as anode materials with superhigh rate and large capacity for lithium ion batteries. *ACS Nano* 5:5463–5471.
11. Li, X., Geng, D., Zhang, Y. et al. 2011. Superior cycle stability of nitrogen-doped graphene nanosheets as anodes for lithium ion batteries. *Electrochem. Commun.* 13:822–825.
12. Cai, D., Wang, S., Lian, P. et al. 2013. Superhigh capacity and rate capability of high-level nitrogen-doped graphene sheets as anode materials for lithium-ion batteries. *Electrochim. Acta* 90:492–497.
13. Wang, Z. L., Xu, D., Wang, H. G. et al. 2013. In situ fabrication of porous graphene electrodes for high-performance energy storage. *ACS Nano* 7:2422–2430
14. Abouimrane, A., Compton, O. C., Amine, K. et al. 2010. Non-annealed graphene paper as a binder-free anode for lithium-ion batteries. *J. Phys. Chem. C* 114:12800–12804.
15. Hu, Y., Li, X., Geng, D. et al. 2013. Influence of paper thickness on the electrochemical performances of graphene papers as an anode for lithium ion batteries. *Electrochim. Acta* 91:227–233.
16. Zhao, X., Hayner, C. M., Kung, M. C. et al. 2011. Flexible holey graphene paper electrodes with enhanced rate capability for energy storage applications. *ACS Nano* 5:8739–8749.
17. Mukherjee, R., Thomas, A. V., Krishnamurthy, A. et al. 2012. Photothermally reduced graphene as high-power anodes for lithium-ion batteries. *ACS Nano* 6:7867–7878.
18. Ning, G., Xu, C., Cao, Y. et al. 2013. Chemical vapor deposition derived flexible graphene paper and its application as high performance anodes for lithium rechargeable batteries. *J. Mater. Chem. A* 1:408–414.
19. Gwon, H., Kim, H.-S., Lee, K. U. et al. 2011. Flexible energy storage devices based on graphene paper. *Energy Environ. Sci.* 4:1277–1283.
20. Ding, Y., Jiang, Y., Xu, F. et al. 2010. Preparation of nano-structured LiFePO$_4$/graphene composites by co-precipitation method. *Electrochem. Commun.* 12:10–13.
21. Zhou, X., Wang, F., Zhu, Y. et al. 2011. Graphene modified LiFePO$_4$ cathode materials for high power lithium ion batteries. *J. Mater. Chem.* 21:3353–3358.
22. Yang, J., Wang, J., Wang, D. et al. 2012. 3D porous LiFePO$_4$/graphene hybrid cathodes with enhanced performance for Li-ion batteries. *J. Power Sources* 208:340–344.

23. Shi, Y., Chou, S.-L., Wang, J.-Z. et al. 2012. Graphene wrapped LiFePO$_4$/C composites as cathode materials for Li-ion batteries with enhanced rate capability. *J. Mater. Chem.* 22:16465–16470.
24. Tang, Y., Huang, F., Bi, H. et al. 2012. Highly conductive three-dimensional graphene for enhancing the rate performance of LiFePO$_4$ cathode. *J. Power Sources* 203:130–134.
25. Yang, J., Wang, J., Tang, Y. et al. 2013. LiFePO$_4$–graphene as a superior cathode material for rechargeable lithium batteries: Impact of stacked graphene and unfolded graphene. *Energy Environ. Sci.* 6:1521–1528.
26. Wei, W., Lv, W., Wu, M.-B. et al. 2013. The effect of graphene wrapping on the performance of LiFePO$_4$ for a lithium ion battery. *Carbon* 57:530–533.
27. Bi, H., Huang, F., Tang, Y. et al. 2013. Study of LiFePO$_4$ cathode modified by graphene sheets for high-performance lithium ion batteries. *Electrochim. Acta* 88:414–420.
28. Qin, Z. H., Zhou, X. F., Xia, Y. G. et al. 2012. Morphology controlled synthesis and modification of high-performance LiMnPO$_4$ cathode materials for Li-ion batteries. *J. Mater. Chem.* 22:21144–21153.
29. Wang, H. L., Yang, Y., Liang, Y. Y. et al. 2011. LiMn1-xFexPO4 nanorods grown on graphene sheets for ultrahigh-rate-performance lithium ion batteries. *Angew. Chem. Int. Ed.* 50:7364–7368.
30. Zhou, J. G., Wang, J., Zuin, L. et al. 2012. Spectroscopic understanding of ultra-high rate performance for LiMn$_{0.75}$Fe$_{0.25}$PO$_4$ nanorods-graphene hybrid in lithium ion battery. *Phys. Chem. Chem. Phys.* 14:9578–9581.
31. Liu, H. D., Gao, P., Fang, J. H. et al. 2011. Li$_3$V$_2$(PO$_4$)$_3$/graphene nanocomposites as cathode material for lithium ion batteries. *Chem. Commun.* 47:9110–9112.
32. Liu, H. D., Yang, G., Zhang, X. F. et al. 2012. Kinetics of conventional carbon coated-Li$_3$V$_2$(PO4)$_3$ and nanocomposite Li$_3$V$_2$(PO$_4$)$_3$/graphene as cathode materials for lithium ion batteries. *J. Mater. Chem.* 22:11039–11047.
33. Jiang, Y., Xu, W. W., Chen, D. D. et al. 2012. Graphene modified Li$_3$V$_2$(PO$_4$)$_3$ as a high-performance cathode material for lithium ion batteries. *Electrochim. Acta* 85:377–383.
34. Rui, X. H., Sim, D. H., Wong, K. M. et al. 2012. Li$_3$V$_2$(PO$_4$)$_3$ nanocrystals embedded in a nanoporous carbon matrix supported on reduced graphene oxide sheets: Binder-free and high rate cathode material for lithium-ion batteries. *J. Power Sources* 214:171–177.
35. Bak, S.-M., Nam, K.-W., Lee, C.-W. et al. 2011. Spinel LiMn$_2$O$_4$/reduced graphene oxide hybrid for high rate lithium ion batteries. *J. Mater. Chem.* 21:17309–17315.
36. Tolbert, S. H., Brezesinski, T., Wang J. et al. 2010. Ordered mesoporous alpha-MoO$_3$ with iso-oriented nanocrystalline walls for thin-film pseudocapacitors. *Nat. Mater.* 9:146–151.
37. Zhao, X., Hayner, C. M. and Kung, H. H. 2011. Self-assembled lithium manganese oxide nanoparticles on carbon nanotube or graphene as high-performance cathode material for lithium-ion batteries. *J. Mater. Chem.* 21:17297–17303.
38. Jo, K.-Y., Han, S.-Y., Lee, J. M. et al. 2013. Remarkable enhancement of the electrode performance of nanocrystalline LiMn$_2$O$_4$ via solvothermally-assisted immobilization on reduced graphene oxide nanosheets. *Electrochim. Acta* 92:188–196.
39. Xu, H. Y., Cheng, B., Wang, Y. P. et al. 2012. Improved electrochemical performance of LiMn$_2$O$_4$/graphene composite as cathode material for lithium ion battery. *Int. J. Electrochem. Sci.* 7:10627–10632.
40. Fang, X., Ge, M., Rong, J. et al. 2013. Graphene-oxide-coated LiNi$_{0.5}$Mn$_{1.5}$O$_4$ as high voltage cathode for lithium ion batteries with high energy density and long cycle life. *J. Mater. Chem. A* 1:4083–4088.
41. Rao, C. V., Reddy, A. L. M., Ishikawa, Y. et al. 2011. LiNi$_{1/3}$Co$_{1/3}$Mn$_{1/3}$O$_2$-graphene composite as a promising cathode for lithium-ion batteries. *ACS Appl. Mater. Interfaces* 3:2966–2972.

42. Yang, S., Gong, Y., Liu, Z. et al. 2013. Bottom-up approach toward single-crystalline VO_2-graphene ribbons as cathodes for ultrafast lithium storage. *Nano Lett.* 13:1596–1601.
43. Nethravathi, C., Viswanath, B., Michael, J. et al. 2012. Hydrothermal synthesis of a monoclinic VO_2 nanotube–graphene hybrid for use as cathode material in lithium ion batteries. *Carbon* 50:4839–4846.
44. Nethravathi, C., Rajamathi, C. R., Rajamathi, M. et al. 2013. N-doped graphene–VO_2(B) nanosheet-built 3D flower hybrid for lithium ion battery. *ACS Appl. Mater. Interfaces* 5:2708–2714.
45. Rui, X., Zhu, J., Sim, D. et al. 2011. Reduced graphene oxide supported highly porous V_2O_5 spheres as a high-power cathode material for lithium ion batteries. *Nanoscale* 3:4752–4758.
46. Du, G., Seng, K. H., Guo, Z. et al. 2011. Graphene–V_2O_5·nH_2O xerogel composite cathodes for lithium ion batteries. *RSC Adv.* 1:690–697.
47. Chou, S.-L., Wang, J.-Z., Choucair, M. et al. 2010. Enhanced reversible lithium storage in a nanosize silicon/graphene composite. *Electrochem. Commun.* 12:303–306.
48. Zhou, X., Yin, Y.-X., Wan, L.-J. et al. 2012. Facile synthesis of silicon nanoparticles inserted into graphene sheets as improved anode materials for lithium-ion batteries. *Chem. Commun.* 48:2198–2200.
49. Luo, J., Zhao, X., Wu, J. et al. 2012. Crumpled graphene-encapsulated Si nanoparticles for lithium ion battery anodes. *J. Phys. Chem. Lett.* 3:1824–1829.
50. Zhou, M., Cai, T., Pu, F. et al. 2013. Graphene/carbon-coated Si nanoparticle hybrids as high-performance anode materials for Li-ion batteries. *ACS Appl. Mater. Interfaces* 5:3449–3455.
51. Zhou, M., Pu, F., Wang, Z. et al. 2013. Facile synthesis of novel Si nanoparticles–graphene composites as high-performance anode materials for Li-ion batteries. *Phys. Chem. Chem. Phys.* 15:11394–11401.
52. Yang, S., Li, G., Zhu, Q. et al. 2012. Covalent binding of Si nanoparticles to graphene sheets and its influence on lithium storage properties of Si negative electrode. *J. Mater. Chem.* 22:3420–3425.
53. Zhao, G., Zhang, L., Meng, Y. et al. 2013. Decoration of graphene with silicon nanoparticles by covalent immobilization for use as anodes in high stability lithium ion batteries. *J. Power Sources* 240:212–218.
54. Wen, Y., Zhu, Y., Langrock, A. et al. 2013. Graphene-bonded and encapsulated Si nanoparticles for lithium ion battery anodes. *Small* 9:2810–2816.
55. Xin, X., Zhou, X., Wang, F. et al. 2012. A 3D porous architecture of Si/graphene nanocomposite as high-performance anode materials for Li-ion batteries. *J. Mater. Chem.* 22:7724–7730.
56. Lu, Z., Zhu, J., Sim, D. et al. 2012. In situ growth of Si nanowires on graphene sheets for Li-ion storage. *Electrochim. Acta* 74:176–181.
57. Lee, J. K., Smith, K. B., Hayner, C. M. et al. 2010. Silicon nanoparticles–graphene paper composites for Li ion battery anodes. *Chem. Commun.* 46:2025–2027.
58. Wang, J.-Z., Zhong, C., Chou, S.-L. et al. 2010. Flexible free-standing graphene-silicon composite film for lithium-ion batteries. *Electrochem. Commun.* 12:1467–1470.
59. Zhang, Y. Q., Xia, X. H., Wang, X. L. et al. 2012. Silicon/graphene-sheet hybrid film as anode for lithium ion batteries. *Electrochem. Commun.* 23:17–20.
60. Zhao, X., Hayner, C. M., Kung, M. C. et al. 2011. In-plane vacancy-enabled high-power Si-graphene composite electrode for lithium-ion batteries. *Adv. Energy Mater.* 1:1079–1084.
61. Cheng, J. and Du, J. 2012. Facile synthesis of germanium–graphene nanocomposites and their application as anode materials for lithium ion batteries. *Cryst. Eng. Comm.* 14:397–400.

62. Ren, J.-G., Wu, Q.-H., Tang, H. et al. 2013. Germanium–graphene composite anode for high-energy lithium batteries with long cycle life. *J. Mater. Chem. A* 1:1821–1826.
63. Xue, D.-J., Xin, S., Yan, Y. et al. 2012. Improving the electrode performance of Ge through Ge@C core–shell nanoparticles and graphene networks. *J. Am. Chem. Soc.* 134:2512–2515.
64. Kim, H., Son, Y., Park, C. et al. 2013. Catalyst-free direct growth of a single to a few layers of graphene on a germanium nanowire for the anode material of a lithium battery. *Angew. Chem. Int. Ed.* 52:5997–6001.
65. Wang, C., Ju, J., Yang, Y. et al. 2013. In situ grown graphene-encapsulated germanium nanowires for superior lithium-ion storage properties. *J. Mater. Chem. A* 1:8897–8902.
66. Wen, Z., Cui, S., Kim, H. et al. 2012. Binding Sn-based nanoparticles on graphene as the anode of rechargeable lithium-ion batteries. *J. Mater. Chem.* 22:3300–3306.
67. Yue, W., Yang, S., Ren, Y. et al. 2013. In situ growth of Sn, SnO on graphene nanosheets and their application as anode materials for lithium-ion batteries. *Electrochim. Acta* 92:412–420.
68. Wang, G., Wang, B., Wang, X. et al. 2009. Sn/graphene nanocomposite with 3D architecture for enhanced reversible lithium storage in lithium ion batteries. *J. Mater. Chem.* 19:8378–8384.
69. Ji, L., Tan, Z., Kuykendall, T. et al. 2011. Multilayer nanoassembly of Sn-nanopillar arrays sandwiched between graphene layers for high-capacity lithium storage. *Energy Environ. Sci.* 4:3611–3616.
70. Luo, B., Wang, B., Li, X. et al. 2012. Graphene-confined Sn nanosheets with enhanced lithium storage capability. *Adv. Mater.* 24:3538–3543.
71. Wang, D., Li, X., Yang, J. et al. 2013. Hierarchical nanostructured core–shell Sn@C nanoparticles embedded in graphene nanosheets: Spectroscopic view and their application in lithium ion batteries. *Phys. Chem. Chem. Phys.* 15:3535–3542.
72. Zou, Y. Q. and Wang, Y. 2011. Sn@CNT nanostructures rooted in graphene with high and fast Li-storage capacities. *ACS Nano* 5:8108–8114.
73. Paek, S. M., Yoo, E. and Honma, I. 2009. Enhanced cyclic performance and lithium storage capacity of SnO_2/graphene nanoporous electrodes with three-dimensional delaminated flexible structure. *Nano Lett.* 9:72–75.
74. Wang, X., Zhou, X., Yao, K. et al. 2011. A SnO_2/graphene composite as a high stability electrode for lithium ion batteries. *Carbon* 49:133–139.
75. Ding, S., Luan, D., Boey, F. Y. C. et al. 2011. SnO_2 nanosheets grown on graphene sheets with enhanced lithium storage properties. *Chem. Commun.* 47:7155–7157.
76. Li, X., Meng, X., Liu, J. et al. 2012. Tin oxide with controlled morphology and crystallinity by atomic layer deposition onto graphene nanosheets for enhanced lithium storage. *Adv. Func. Mater.* 22:1647–1654.
77. Wang, L., Wang, D., Dong, Z. et al. 2013. Interface chemistry engineering for stable cycling of reduced GO/SnO_2 nanocomposites for lithium ion battery. *Nano Lett.* 13:1711–1716.
78. Wang, X., Cao, X., Bourgeois, L. et al. 2012. N-doped graphene-SnO_2 sandwich paper for high-performance lithium-ion batteries. *Adv. Func. Mater.* 22:2682–2690.
79. Zhou, X., Wan, L.-J. and Guo, Y.-G. 2013. Binding SnO_2 nanocrystals in nitrogen-doped graphene sheets as anode materials for lithium-ion batteries. *Adv. Mater.* 25:2152–2157.
80. Wang, D. H., Choi, D. W., Li, J. et al. 2009. Self-assembled TiO_2-graphene hybrid nanostructures for enhanced Li-ion insertion. *ACS Nano* 3:907–914.
81. Yang, S., Feng, X. and Müllen, K. 2011. Sandwich-like, graphene-based titania nanosheets with high surface area for fast lithium storage. *Adv. Mater.* 23:3575–3579.
82. He, L., Ma, R., Du, N. et al. 2012. Growth of TiO_2 nanorod arrays on reduced graphene oxide with enhanced lithium-ion storage. *J. Mater. Chem.* 22:19061–19066.

83. Chen, J. S., Wang, Z., Dong, X. C. et al. 2011. Graphene-wrapped TiO₂ hollow structures with enhanced lithium storage capabilities. *Nanoscale* 3:2158–2161.

84. Wang, J., Zhou, Y., Xiong, B. et al. 2013. Fast lithium-ion insertion of TiO₂ nanotube and graphene composites. *Electrochim. Acta* 88:847–857.

85. Zhang, X., Suresh Kumar, P., Aravindan, V. et al. 2012. Electrospun TiO₂–graphene composite nanofibers as a highly durable insertion anode for lithium ion batteries. *J. Phys. Chem. C* 116:14780–14788.

86. Hu, T., Sun, X., Sun, H. et al. 2013. Flexible free-standing graphene–TiO₂ hybrid paper for use as lithium ion battery anode materials. *Carbon* 51:322–326.

87. Xin, X., Zhou, X. F., Wu, J. H. et al. 2012. Scalable synthesis of TiO₂/graphene nanostructured composite with high-rate performance for lithium ion batteries. *ACS Nano* 6:11035–11043.

88. Huang, H., Fang, J., Xia, Y. et al. 2013. Construction of sheet–belt hybrid nanostructures from one-dimensional mesoporous TiO₂(B) nanobelts and graphene sheets for advanced lithium-ion batteries. *J. Mater. Chem. A* 1:2495–2500.

89. Zhou, G., Wang, D.-W., Li, F. et al. 2010. Graphene-wrapped Fe₃O₄ anode material with improved reversible capacity and cyclic stability for lithium ion batteries. *Chem. Mater.* 22:5306–5313.

90. Lian, P., Zhu, X., Xiang, H. et al. 2010. Enhanced cycling performance of Fe₃O₄–graphene nanocomposite as an anode material for lithium-ion batteries. *Electrochim. Acta* 56:834–840.

91. Behera, S. K. 2011. Enhanced rate performance and cyclic stability of Fe₃O₄–graphene nanocomposites for Li ion battery anodes. *Chem. Commun.* 47:10371–10373.

92. Sathish, M., Tomai, T. and Honma, I. 2012. Graphene anchored with Fe₃O₄ nanoparticles as anode for enhanced Li-ion storage. *J. Power Sources* 217:85–91.

93. Zhuo, L., Wu, Y., Wang, L. et al. 2013. CO₂-expanded ethanol chemical synthesis of a Fe₃O₄@graphene composite and its good electrochemical properties as anode material for Li-ion batteries. *J. Mater. Chem. A* 1:3954–3960.

94. Hu, A., Chen, X., Tang, Y. et al. 2013. Self-assembly of Fe₃O₄ nanorods on graphene for lithium ion batteries with high rate capacity and cycle stability. *Electrochem. Commun.* 28:139–142.

95. Chen, D., Ji, G., Ma, Y. et al. 2011. Graphene-encapsulated hollow Fe₃O₄ nanoparticle aggregates as a high-performance anode material for lithium ion batteries. *ACS Appl. Mater. Interfaces* 3:3078–3083.

96. Li, B., Cao, H., Shao, J. et al. 2011. Enhanced anode performances of the Fe₃O₄–Carbon–rGO three dimensional composite in lithium ion batteries. *Chem. Commun.* 47:10374–10376.

97. Zhu, X. J., Zhu, Y. W., Murali, S. et al. 2011. Nanostructured reduced graphene oxide/Fe₂O₃ composite as a high-performance anode material for lithium ion batteries. *ACS Nano* 5:3333–3338.

98. Bai, S., Chen, S., Shen, X. et al. 2012. Nanocomposites of hematite (α-Fe₂O₃) nanospindles with crumpled reduced graphene oxide nanosheets as high-performance anode material for lithium-ion batteries. *RSC Adv.* 2:10977–10984.

99. Qu, J., Yin, Y.-X., Wang, Y.-Q. et al. 2013. Layer structured α-Fe₂O₃ nanodisk/reduced graphene oxide composites as high-performance anode materials for lithium-ion batteries. *ACS Appl. Mater. Interfaces* 5:3932–3936.

100. Du, M., Xu, C., Sun, J. et al. 2012. One step synthesis of Fe₂O₃/nitrogen-doped graphene composite as anode materials for lithium ion batteries. *Electrochim. Acta* 80:302–307.

101. Wang, X., Tian, W., Liu, D. et al. 2013. Unusual formation of α-Fe₂O₃ hexagonal nanoplatelets in N-doped sandwiched graphene chamber for high-performance lithium-ions batteries. *Nano Energy* 2:257–267.

102. Xiao, L., Wu, D., Han, S. et al. 2013. Self-assembled Fe_2O_3/graphene aerogel with high lithium storage performance. *ACS Appl. Mater. Interfaces* 5:3764–3769.

103. Zhou, G.-W., Wang, J., Gao, P. et al. 2013. Facile spray drying route for the three-dimensional graphene-encapsulated Fe_2O_3 nanoparticles for lithium ion battery anodes. *Ind. Eng. Chem. Res.* 52:1197–1204.

104. Wu, Z. S., Ren, W. C., Wen, L. et al. 2010. Graphene anchored with Co_3O_4 nanoparticles as anode of lithium ion batteries with enhanced reversible capacity and cyclic performance. *ACS Nano* 4:3187–3194.

105. Chen, S. Q. and Wang, Y. 2010. Microwave-assisted synthesis of a Co_3O_4–graphene sheet-on-sheet nanocomposite as a superior anode material for Li-ion batteries. *J. Mater. Chem.* 20:9735–9739.

106. Kim, H., Seo, D.-H., Kim, S.-W. et al. 2011. Highly reversible Co_3O_4/graphene hybrid anode for lithium rechargeable batteries. *Carbon* 49:326–332.

107. Yue, W., Lin, Z., Jiang, S. et al. 2012. Preparation of graphene-encapsulated mesoporous metal oxides and their application as anode materials for lithium-ion batteries. *J. Mater. Chem.* 22:16318–16323.

108. Hsieh, C.-T., Lin, J.-S., Chen, Y.-F. et al. 2012. Pulse microwave deposition of cobalt oxide nanoparticles on graphene nanosheets as anode materials for lithium ion batteries. *J. Phys. Chem. C* 116:15251–15258.

109. Kim, G.-P., Nam, I., Kim, N. D. et al. 2012. A synthesis of graphene/Co_3O_4 thin films for lithium ion battery anodes by coelectrodeposition. *Electrochem. Commun.* 22:93–96.

110. Shi, Y., Wen, L., Li, F. et al. 2011. Nanosized $Li_4Ti_5O_{12}$/graphene hybrid materials with low polarization for high rate lithium ion batteries. *J. Power Sources* 196:8610–8617.

111. Zhang, Q., Peng, W., Wang, Z. et al. 2013. $Li_4Ti_5O_{12}$/reduced graphene oxide composite as a high rate capability material for lithium ion batteries. *Solid State Ionics* 236:30–36.

112. Kim, H.-K., Bak, S.-M. and Kim, K.-B. 2010. $Li_4Ti_5O_{12}$/reduced graphite oxide nano-hybrid material for high rate lithium-ion batteries. *Electrochem. Commun.* 12:1768–1771.

113. Tang, Y., Huang, F., Zhao, W. et al. 2012. Synthesis of graphene-supported $Li_4Ti_5O_{12}$ nanosheets for high rate battery application. *J. Mater. Chem.* 22:11257–11260.

114. Shen, L., Yuan, C., Luo, H. et al. 2011. In situ synthesis of high-loading $Li_4Ti_5O_{12}$–graphene hybrid nanostructures for high rate lithium ion batteries. *Nanoscale* 3:572–574.

115. Ding, Y., Li, G. R., Xiao, C. W. et al. 2013. Insight into effects of graphene in $Li_4Ti_5O_{12}$/carbon composite with high rate capability as anode materials for lithium ion batteries. *Electrochim. Acta* 102:282–289.

116. Zhu, N., Liu, W., Xue, M. et al. 2010. Graphene as a conductive additive to enhance the high-rate capabilities of electrospun $Li_4Ti_5O_{12}$ for lithium-ion batteries. *Electrochim. Acta* 55:5813–5818.

4 Applications of Graphene in New-Concept Batteries

Xufeng Zhou and Zhaoping Liu

CONTENTS

4.1 Introduction 137
4.2 Application of Graphene in Li-S Batteries 138
    4.2.1 Introduction 138
    4.2.2 Progress of the Application of Graphene in Li-S Batteries 139
4.3 Application of Graphene in Li-Air Batteries 156
    4.3.1 Introduction 156
    4.3.2 Progress of the Application of Graphene in Li-Air Batteries 156
        4.3.2.1 Nonaqueous Li-Air Batteries 156
        4.3.2.2 Aqueous Li-Air Batteries 165
References 168

4.1 INTRODUCTION

In the last two decades, great progress has been achieved regarding the application of Li-ion batteries in various areas. They already dominate the electrochemical energy storage systems used for portable electronic devices and have exhibited potential in the application in electric vehicles and stationary electricity storage in recent years. However, the development of Li-ion batteries for large-scale energy storage, especially electric cars, has encountered a major obstacle, which is its relatively low energy density. Although the development of better electrode materials and the improvement of the manufacturing processes have continuously enhanced the energy density of Li-ions, their current level is still less than 200 Wh kg^{-1}, and a theoretical limitation of 300 Wh kg^{-1} is anticipated, which is still far less than the energy density of traditional fossil fuels. This means that Li-ion batteries can only sustain electric cars for a small amount of miles after fully charged unless extremely heavy batteries are installed. Therefore, it is urgent to develop new energy storage devices with much higher energy densities than Li-ion batteries.

Among various new energy storage techniques, lithium-sulfur (Li-S) batteries and Li-air batteries are two of the most attractive ones. Due to the different electrochemical energy storage mechanism with traditional Li-ion batteries, both Li-S and Li-air batteries have significantly enhanced energy density than Li-ion batteries. The theoretical energy density of Li-S batteries based on the total mass of the cathode and anode is estimated to be ~2600 Wh kg^{-1}, and the corresponding value for Li-air batteries exceeds 10,000 Wh kg^{-1}. As a result, these two new Li batteries have been given much attention as substitutions to Li-ion batteries for future applications in clean and sustainable energy storage.

Compared with commercial Li-ion batteries, Li-S and Li-air batteries are still not yet ready for practical application, as their performance at the current level cannot meet the requirements for commercialization. Much research has been carried out in recent years to improve electrochemical performance, such as capacity and cycling stability, of Li-S and Li-air batteries. Electrode materials, especially the S cathode in Li-S batteries and air cathode in Li-air batteries, have become the main focus of this research because battery performance is directly related to the performance of the electrode materials. Recently, application of graphene in the design and synthesis of electrodes for Li-S and Li-air batteries has drawn much attention. Effective improvement of the charge/discharge performance of Li-S and Li-air batteries have been achieved after the addition of graphene, and some important progress in this area will be presented in this chapter.

4.2   APPLICATION OF GRAPHENE IN LI-S BATTERIES

4.2.1   Introduction

The Li-S battery, which possesses a much higher energy density than a Li-ion battery, has attracted increasing worldwide attention in recent years. In a typical Li-S battery configuration, S and Li metal are applied as the cathode and anode, respectively, and the battery is operated in organic electrolytes. The Li-S couple can yield a theoretical specific capacity of 1675 mAh g^{-1} and a corresponding theoretical specific energy of ~2600 Wh kg^{-1} based on the complete reaction of Li with S to form Li_2S. Additionally, S is abundant in various minerals, is cheap, and is environmentally friendly. However, the Li-S battery has not reached mass commercialization, because some inherent problems in the cell chemistry still remain. First, S and its reduction products are insulating, which causes poor electrochemical activity. Second, the polysulfide intermediates (Li_2S_n, $3 \leq n \leq 6$) generated during the charge/discharge cycles are soluble in the electrolyte, which gives rise to a shuttle mechanism (moving back and forth between the cathode and the anode during cycling) and causes fast capacity fading. Third, the Li/electrolyte interface is poorly controlled, resulting in unsatisfied cycling stability. Modification of a S cathode thus is necessary to overcome its intrinsic deficiencies. First of all, the addition of an electrical conducting agent is always required to improve the conductivity of S. Meanwhile, the size of S should be minimized to shorten the diffusion length of both Li-ions and electrons, as well as to ease the volume expansion of S. In addition, surface protection of S is recommended to inhibit the dissolution of polysulfides. Carbon materials, which have relatively high conductivity and variable structures and morphologies, are extensively used in the modification of S. Recently, graphene, as a new member of carbon materials, is beginning to play an important role in the application of energy storage due to its ultrahigh electrical conductivity, large surface area, and flexible 2-D nanostructure. These features also endow graphene with great potential in the modification of the S cathode for high-performance Li-S batteries.

4.2.2 PROGRESS OF THE APPLICATION OF GRAPHENE IN LI-S BATTERIES

The first report on graphene/S composite materials for Li-S batteries was published in 2011 by Wang and coworkers [1]. Their experiment was quite simple. Graphene nanosheets were first mixed with the element S at a weight ratio of 1/1.5. The mixture was then heated at 200°C to allow the infiltration of melted S into graphene layers. Final heating of the mixture at 300°C resulted in the coating of S on graphene sheets. EDS mapping showed homogeneous distribution of both S and C in the composite; however, S was irregularly coated on graphene sheets as revealed by SEM images. In addition, the weight percentage of S was quite low (22%) in the graphene/S composite. The initial discharge capacity of the composite based on the mass of S at a current density of 50 mA g^{-1} was measured to be 1611 mAh g^{-1}, which is ~96% of the theoretic value and was higher than pure S (1100 mAh g^{-1}), suggesting high utilization of S by modification with graphene. Nevertheless, the cycle stability of the composite was not improved much. The discharge capacity quickly faded to less than 600 mAh g^{-1} after 40 cycles, probably due to the fact that S was not firmly anchored on the graphene sheets and was directly exposed to electrolytes. Therefore, better microstructure of graphene/S composite materials by more delicate structural design is required to achieve better electrochemical performance.

Later, graphene/S composite material with a novel structure and much improved electrochemical performance was reported by Cui and Dai's research group [2]. The main idea of their research was to wrap S particles in graphene sheets in order to accommodate the volume expansion of S during charge/discharge cycles, trap soluble polysulfide intermediates, as well as to enhance the electrical conductivity of S. As illustrated in Figure 4.1, S particles were synthesized by the reaction of sodium thiosulfate with HCl in the aqueous solution in the presence of a surfactant (Triton X-100). The surfactant could control the growth S particles to limit their size in

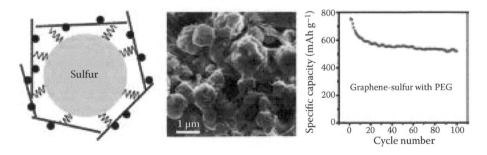

FIGURE 4.1   The structure model, SEM image, and cycle performance of graphene-wrapped sulfur cathode material. (Reprinted with permission from Ref. 2, 2644–2647. Copyright 2011 American Chemical Society.)

the submicrometer region. As-prepared S particles were then mixed with mildly oxidized graphene oxide (GO) loaded with carbon black nanoparticles to produce the final product. The infinity between polyethylene glycol moieties in Triton X-100 and GO sheets resulted in the encapsulation of S particles by GO. The addition of carbon black was aimed to further enhance the electrical conductivity of the composite electrode materials. As revealed by electron microscopy images, the S particles in the graphene/S composite were uniformly and tightly wrapped by thin GO sheets. The weight content of S was measured to be ~70%. The composite electrode material exhibited excellent electrochemical performance. It could deliver an initial discharge capacity higher than 1000 mAh g^{-1}. When cycled at a rate of 0.2 C, an initial capacity of 750 mAh g^{-1} was obtained, followed by a decrease of the capacity to ~600 mAh g^{-1} after 10 cycles. Within the subsequent 90 cycles, the capacity only decreased by 13%. Even better cycle stability was achieved at a higher rate of 0.5 C. There was only 9% capacity loss from the 10th to 100th cycles. For comparison, the performance of surfactant-coated S particles without graphene coating was also measured. Its capacity dramatically decreased from ~700 mAh g^{-1} to ~330 mAh g^{-1} within 20 cycles. In another control experiment, S particles mixed with graphene sheet but without surfactant coating also showed poor cycle stability. These results clearly indicated that both graphene and PEG-containing surfactant played important roles in improving the cycle life of S. The surfactant could serve as a buffer to accommodate the volume change and stress of S during cycling. Meanwhile, PEG chains were also able to capture polysulfides. The graphene coating could further immobilize S, and the carbon-black-coated graphene sheets afforded electrical conductivity to S particles. Consequently, the reported S/graphene composite material exhibited excellent electrochemical performance.

A few months later, Ji et al. proposed another high-performance S/graphene composite material with different microstructure [3]. In their work, formation of strong chemical bonds between S and GO was realized, which helped to firmly immobilize S on graphene sheets, giving rise to outstanding cycle stability. To obtain the S/GO composite material, sodium sulfide (Na_2S_x), which was used as the source of S, was added to the aqueous suspension of GO in the presence of the surfactant cetyltrimethylammonium bromide. The blended solution was sonicated and then titrated into formic acid (HCOOH) solution. The reaction between S_x^{2-} and H^+ resulted in the formation of S on graphene sheets. The as-prepared composites were then heat-treated at 155°C in Ar. It was observed that before heat treatment, nano-sized S particles with irregular morphologies were randomly dispersed on graphene sheets and the S content in the sample was measured to be 87 wt%. After heat treatment, a thin layer of S with a thickness of tens of nanometers was homogeneously dispersed on the surface of graphene, and the weight content of S decreased to 66 wt%. It was speculated that during the heating process, bulk S melted and diffused over the entire surface of graphene due to the strong adsorption effects derived from the large surface area of GO and its abundant functional groups. Partial S particles that were not directly attached to GO were removed by evaporation of S, leading to the reduced S content in the final product. Meanwhile, partial functional groups on GO sheets were also removed during this low-temperature heating process, giving rise to dramatic increase of the electrical conductivity of GO by ~2000 times. The remaining functional groups had strong absorbing abilities to anchor S and immobilize polysulfides, which was favorable for the enhancement of the stability of the S cathode. The authors also carried out *ab initio* calculations to clarify the role of these functional groups. As shown in Figure 4.2, both epoxy and hydroxyl groups could enhance the binding of S to the C-C bonds because of the induced ripples by these functional groups. Soft x-ray absorption spectroscopy (XAS) was also used to investigate the chemical bonding in the sample. The comparison of the C K-edge XAS spectra between GO and S/GO composite revealed that the sharpness of both π^* state and excitonic state increased in the S/GO sample compared with the GO sample, which suggested that the ordering of sp^2 hybridized carbon was better formatted after S incorporation. Meanwhile, the signal possibly from the C-O bond was weakened when S was loaded, also suggesting the strong interaction between S and oxygenous groups in GO. In addition, a new signal corresponding to the C-S σ^* excitation was observed in the S/GO nanocomposite, indicating the formation of a strong covalent bond between S and graphene. Eletrochemical measurement of the S/GO nanocomposite showed that at a rate of 0.02 C, it could deliver a high initial discharge capacity of 1320 mAh g^{-1} (based on the mass of S), ~79% of the theoretical value of S and the corresponding Coulombic efficiency in the first charge/discharge cycle was 96.4%. A reversible capacity of 1247 mAh g^{-1} (corresponding to a retention rate of 94.5%) could still be retained in the second cycle. Afterward, the cell was cycled at 0.1 C. The reversible capacity remained at ~1000 mAh g^{-1} in the first cycle and decreased to ~950 mAh g^{-1} in the second cycle. In the following 50 cycles, no capacity decay occurred, suggesting outstanding cycle stability of the S/GO composite. The good reversibility and stability of the nanocomposite could first be ascribed to the strong interaction between S and GO. The functional groups on GO could form strong chemical bonds with S, which effectively immobilized S and prevented the dissolution of polysulfides. Meanwhile, the intimate contact between the two components was favorable for good electron and Li-ion accessibility. In addition, flexible GO sheets with a large surface area could also accommodate the volume change of S during charge/discharge cycles.

The above pioneering works on the modification of S with graphene exhibited significant potential of the application of graphene in Li-S batteries, and S/graphene composite electrode materials with various microstructures and prepared by diverse

FIGURE 4.2 Representative pattern of GO immobilizing S. The hydroxyl groups enhance the binding of S to the C atoms due to the induced ripples by epoxy or hydroxyl group. Yellow, red, and white balls denote S, O, and H atoms, respectively, while the others are C atoms. (Reprinted with permission from Ref. 3, 18522–18525. Copyright 2011 American Chemical Society.)

methods were subsequently reported. For example, a graphene-enveloped S cathode was synthesized by a one-pot reaction [4]. The synthesis relied on *in situ* oxidation of polysulfides in the acidic solution and with the presence of GO, resulting in the formation of micrometer-sized S particles. Meanwhile, S particles were simultaneously wrapped by reduced GO sheets that formed a highly conductive network and could trap polysulfides. A high S content of 87 wt% could be obtained in the composite. An initial capacity of 705 mAh g^{-1} could be delivered and a relatively high Coulombic efficiency of 93% could be maintained after 50 cycles for the S/graphene composite. In another report, graphene-encapsulated S was prepared in a quasi-emulsion-templated method in an oil/water system [5]. In the experiment, the oil phase containing a CS_2 solution of S was added dropwise to the aqueous suspension of GO under ultrasonication. After CS_2 was completely evaporated from the mixture, hydrazine was added to reduce GO to graphene, and the final product was obtained by centrifugation. The as-prepared S/graphene composite had a sacculelike morphology. The S encapsulated by graphene sheets was composed of small particles having the sizes of 10–100 nm. The composite showed good rate capability and cycle stability. Even at a high rate of 4 C, it could still deliver a relatively high reversible capacity of 697.5 mAh g^{-1}. When the cell was cycled at 1 C 60 times, capacity retention of 85.8% could be maintained. Wang et al. proposed an interleaved nancomposite composed of S embedded in stacked graphene sheets [6]. The stacked graphene nanosheets were obtained by thermal expansion of GO at a high temperature of 1050°C followed by reduction with H_2. Because there are plenty of voids and nanocavities in stacked graphene sheets, S could be easily impregnated by simply heating the mixture of graphene and S. At a proper weight ratio between S and graphene, small S nanoparticles could be homogeneously dispersed on graphene sheets and no large bulk S particles were formed. Meanwhile, the interlayer spacing and large quantities of edge and defect sites in the graphene framework could effectively confine and S and polysulfide intermediates during electrochemical cyclings. As a result, the nanocomposite exhibited much improved rate capability and cycle stability. With similar structural design, Li et al. also used thermally expanded GO as a substrate for S loading [7]. However, in their experiment, the authors took an extra step of coating the surface of the composite particles with reduced GO sheets in order to further inhibit the dissolution of polysulfides. As a result, a relatively high capacity retention of 63% after 200 cycles at 1 C was achieved. Park and coworkers reported the synthesis of S-impregnated graphene cathode by a one-step synthesis procedure [8]. Graphene sheets were treated by hydrogen fluoride (HF) to create extra active sites of the nucleation of S particles before they were mixed with sodium thiosulfate ($Na_2S_2O_3$) solution. The addition of H_2SO_4 in the solution then initiated the nucleation and growth of S. Because of abundant active sites in graphene nanosheets, S particles with sizes of a few microns were grown in the interior spaces between randomly dispersed graphene sheets. The signal corresponding to C=S and C-S bonds observed in both Fourier transfer (FT)-IR and x-ray photoelectron spectroscopy (XPS) data indicated that S particles were chemically bonded to graphene, which could enhance the stability of S during charge/discharge cycles. The composite cathode material could deliver a high initial discharge capacity of 1237 mAh g^{-1} with an initial Coulombic efficiency of 98.4% and had a good cyclic retention of 67% after

50 cycles. The morphological characterization after charge/discharge cycles showed that the structure of the S-impregnated graphene composite was well maintained and the separator in the cell still reserved the porous morphology, which implied that S and polysulfides could be successfully trapped in the interior space between graphene sheets. A two-step hydrothermal method was employed by Wei and coworkers to prepare S/graphene nanohybrids [9]. In the first step, GO in the aqueous solution was *in situ* reduced by Na_2S to form graphene with very low S content. Afterward, a large amount of S was deposited on graphene by a hydrothermal process based on the reaction between Na_2SO_3 and Na_2S. An initial capacity of 1053 mAh g^{-1} could be delivered for the hybrid, and after 100 cycles at a current density of 1 A g^{-1} the cell could reach a reversible capacity of 662 mAh g^{-1}.

Rong and coworkers recently developed a general method of coating GO sheets on particles that was successfully applied to the preparation of S/graphene nanocomposites with core-shell structure [10]. The key to the uniform coating of GO on particles was the minimization of the surface energy of GO in the solution containing ions. It is well known that GO is negatively charged in the aqueous solution because it is rich in hydroxyl and carboxylic groups, which also helps to stabilize GO in water by electrostatic repulsion among adjacent GO sheets. However, the balance will be easily disturbed if positively charged ions are added because of the screening of electrostatic repulsion, and finally GO sheets coagulate to reduce the surface energy. If other particles coexist in the solution, GO sheets can also minimize their surface energy by coating on the particles because contact of the inner side of GO sheets with the solvent can be completely eliminated. As a result, core-shell structure can be obtained. By employing this concept, the authors succeeded in the preparation of core-shell S/GO composites. By simply mixing S and GO (1:1 weight ratio) in HCl solution, S particles with different particle sizes could be tightly coated by GO sheets, as shown in Figure 4.3. In this core shell structure nanocomposite, GO sheets could effectively limit the dissolution of polysulfides. Meanwhile, the wrinkles on the graphene shell could provide extra space for the expansion of S when lithiated, and prevented the electrode from disruption. As a result, the S/GO composite cathode material exhibited outstanding electrochemical performance. The charge transfer resistance of the core-shell particle was only one-eighth of that of bare S particles, and the voltage difference between charge and discharge plateaus of S/GO composite was also much smaller than that of bare S. The composite cathode material could deliver a stable reversible capacity of 600 mAh g^{-1} (corresponding to 1200 mAh g^{-1} based on the net weight of S) at a current density of 100 mA g^{-1}. It was surprising to find that the S/GO composite maintained a high capacity of 400 mAh g^{-1} (800 mAh g^{-1} in relation to the weight of S) after ultralong cycles of 1000 times at the rate of 1 A g^{-1} and the capacity loss per cycle was only 0.02% over 1000 cycles. The extremely excellent cycle stability of the S/GO composite could be mainly ascribed to the tight coating of GO on S particles.

The core-shell structured S/graphene composite cathode material was also prepared by a one-pot chemical-reaction deposition route as reported by Xu and coworkers [11]. S was produced by the reaction between $(NH_4)_2S_2O_3$ and HCl in the presence of GO, and GO was then chemically reduced by urea. The hydrophobic characteristic of reduced GO was believed to be the driving force to the formation of core-shell

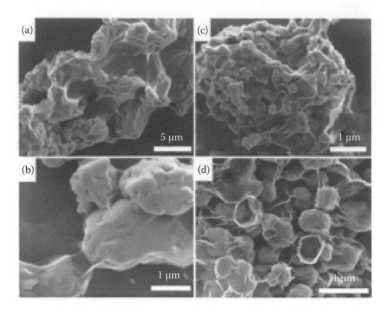

FIGURE 4.3 SEM images of graphene coated on sulfur particles with diameters between 1 and 10 µm (a, b) and sulfur particles with diameters of ~500 nm (c, d). (Reprinted with permission from Ref. 10, 473–479. Copyright 2013 American Chemical Society.)

structure in aqueous solutions. As shown in Figure 4.4, in contrast with the previous core-shell case in which GO sheets were tightly wrapped on S particles, there were relatively large void spaces between the graphene shell and S core, which was expected to be beneficial to buffer the volume expansion of S during lithiation. The graphene shell was quite robust. When the S particles inside were completely evaporated at high temperatures, the graphene shell was still quite intact with little cracks. The graphene shell also affected the growth of S particles. If no GO was added in the experiment, the size of S particles obtained could reach ~10 µm, whereas the average size of S particles encapsulated in graphene shell was only ~2 µm. The core-shell S/graphene composite showed excellent rate capability. Even at a high rate of 6 C, a relatively high reversible capacity of 480 mAh g^{-1} could still be delivered, which is about 47% of the capacity at 0.2 C. The cycle performance of S could also be significantly enhanced by the core-shell structure. A capacity retention of 86% was achieved when cycled at 0.75 C after 160 times. Even higher retention of 94.2% after 500 cycles could be obtained if the rate was increased to 3 C.

Similar core-shell S/graphene composite particles were also prepared by Zhao et al. using a similar reaction process as in Xu's paper [12]. The main difference was that the authors used Na_2S_x as the precursor to prepare S, and GO was reduced by hydrogen iodide (HI) to achieve higher electrical conductivity. The volume conductivity of the sample before HI reduction was measured to be 3 MΩ, while it was dramatically reduced to 73 Ω after reduction, indicating the strong reducibility of HI. The high electrical conductivity of the S/graphene composite cathode material also effectively lowered the internal resistance of the cell. The charge transfer resistance

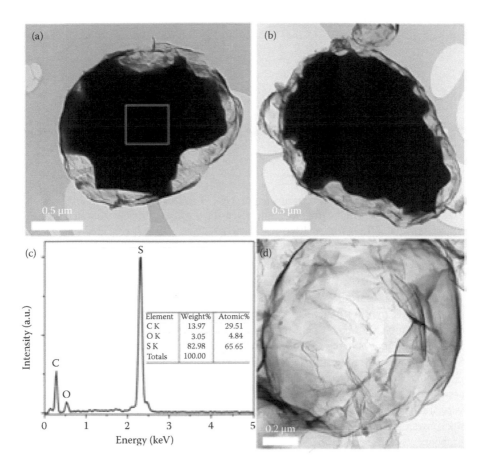

FIGURE 4.4 (a, b) TEM images of graphene-encapsulated S particles. (c) Energy-dispersive spectroscopic image of the selected region in (a). (d) TEM image of graphene encapsulated after heat-treatment at 250°C in N$_2$. (From Xu, H., Deng, Y., Shi, Z. et al., Graphene-encapsulated sulfur [GES] composites with a core–shell structure as superior cathode materials for lithium–sulfur batteries. *J. Mater. Chem. A* 1:15142–15149, 2013. Reproduced with permisson of The Royal Society of Chemistry.)

of the S/reduced GO samples was ~30% of the S/GO sample. As a result, both rate capability and cycle stability were significantly improved.

Flexibility is one of the most attractive characteristics of 2-D graphene sheets, which brings about lots of possibilities in the design and synthesis of graphene-modified S cathode with unique structures that cannot be obtained by traditional carbon materials. For instance, graphene/S hybrid with fibrous structure was reported by Zhou et al. [13]. The fibrous hybrid was prepared by hydrothermal treatment of the mixture of aqueous suspension of GO and CS$_2$ solution of S, as illustrated in Figure 4.5. A certain amount of alcohol had to be added to the mixture in order to improve the miscibility of the two solutions to form homogeneous S/graphene composite. During the hydrothermal process, GO sheets were rolled up to form fibrous

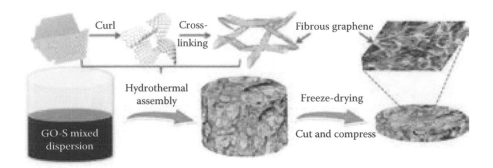

FIGURE 4.5 Illustration of the formation process of the S/graphene hybrid and schematic of fabrication of a self-supporting electrode. (Reprinted with permission from Ref. 13, 5367–5375. Copyright 2013 American Chemical Society.)

structures and simultaneously cross-linked to form interconnected porous network. The reduction of GO also occurred at the same time. Meanwhile, S nanocrystals precipitated on reduced GO from CS_2 solution. The cross-linking of graphene resulted in the formation of a bulk hybrid of a size in the centimeter scale rather than powders. Structural characterization showed that the fibrous graphene in the bulk hybrid had widths of 1–2 µm and large quantities of micrometer-scale pores were distributed in the hybrid. S nanocrystals with diameters of 5–10 nm were homogeneously and strongly adhered on graphene. It was noted that the oxygen groups on GO were critical to anchor and control the size of S. As a comparison, if intercalation-exfoliated graphene sheets with very small content of oxygen were used instead of GO, S tended to agglomerate to form micrometer-size particles. The formation of covalent bonds between S nanocrystals and graphene was observed by XPS results, which could effectively immobilize S and polysulfides during charge/discharge processes. In order to measure the electrochemical performance of the S/graphene composite cathode material, the bulk hybrid was cut into thin slices, compressed, and shaped into circular pellets, and then directly used as the electrode without additional binders and conductive additives. At a current density of 0.3 A g^{-1}, the cell could deliver an initial capacity of 731 mAh g^{-1} based on the total mass of S and graphene, corresponding to a capacity of 1160 mAh g^{-1} for S. When the current density was increased to 1.5 and 4.5 A g^{-1}, relatively high reversible capacities of 422 and 246 mAh g^{-1} (based on the total mass of S and graphene) could be obtained, respectively. After the cell was cycled at 0.75 mA g^{-1}, a capacity as high as 541 mAh g^{-1} was maintained. The excellent electrochemical performance of the hybrid could be ascribed to its unique structures. The highly conductive and interwoven fibrous graphene and ultrasmall S nanoparticles significantly enhanced the charge and ion transfer in the electrode, while the strong interaction between graphene and S through oxygen-containing groups effectively reduced the dissolution of polysulfides.

In another report, S immobilized in mesoporous graphene paper as a flexible electrode was proposed [14]. The mesoporous graphene paper was prepared using a hard template method. By filtrating the aqueous dispersion containing GO sheets, prepre-pared silica nanoparticles as the hard template, and carbon nanofibers, a thin paper was obtained. After calcination in Ar, the paper was washed with NaOH to remove

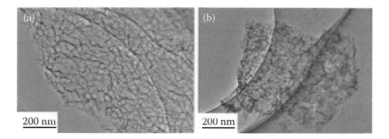

FIGURE 4.6 TEM images of mesoporous graphene paper (a) and S-loaded mesoporous graphene paper (b). (From Huang, X., Sun, B., Li, K. et al., Mesoporous graphene paper immobilized sulfur as a flexible electrode for lithium–sulfur batteries. *J. Mater. Chem. A* 1:13484–13489, 2013. Reproduced by permission of The Royal Society of Chemistry.)

silica nanoparticles and the mesoporous graphene paper with high flexibility was then obtained. TEM characterization revealed that the paper had a uniform pore diameter of ~50 nm (Figure 4.6). Carbon fibers could also be observed in the sample, which could bridge the porous graphene sheets into an intact and free-standing paper. The porous paper had a relatively high specific surface area of 524.8 m^2 g^{-1} and pore volume of 1.25 cm^3 g^{-1}. The porous graphene paper was then applied as the substrate for the loading of S by a vapor treatment approach. S vapor formed at an elevated temperature infiltrated into the porous paper and then deposited as solid S in the mesopores when cooled down. After S deposition, both the surface area and pore volume were dramatically reduced. Both XRD and selected area electron diffraction results suggest that S in the graphene paper had an amorphous nature. The S/graphene composite paper could be used directly as the cathode in the Li-S cell. Galvanostatic charge/discharge tests showed that the paper electrode could deliver an initial discharge capacity of 1393 mAh g^{-1} and charge capacity of 1288 mAh g^{-1} corresponding to a relatively high initial Coulombic efficiency of 92.5%, suggesting that the mesoporous graphene paper was effective in confining S and polysulfides inside. When the cell was further cycled at 0.1 C, a reversible capacity of 689 mAh g^{-1} was maintained after 50 cycles.

Besides the templating method, another effective way of generating highly porous graphene matrices for S loading is chemical activation. Though graphene has a high theoretical specific surface area of 2630 m^2 g^{-1}, the values of the practical samples are always far smaller, which is mainly due to the inevitable restacking of graphene sheets. In order to generate higher surface area and pore volume, chemical activation that has been applied in the production of activated carbon for a long time period was adopted to the modification of graphene. The activated graphene can have an ultrahigh surface area up to 3100 m^2 g^{-1} and abundant nanopores, which is expected to be a suitable substrate for the loading of S. Based on this idea, Ding and coworkers synthesized the S/activated graphene nanocomposite as a high-performance cathode material for Li-S batteries [15]. In the experiment, hydrothermally reduced GO was infiltrated in potassium hydroxide (KOH) solution. The mixture was then dried and annealed at 800°C. KOH as the activation reagent could have chemical reactions with graphene at high temperatures to generate nanopores, which was regarded as the activation process. The as-prepared activated graphene had a specific surface area of 2313 m^2 g^{-1} and pore volume of 1.8 g cm^{-3}, which is dramatically larger than the value of the reduced graphene without activation (only 106 m^2 g^{-1}). The nanopores in the activated graphene had a mean pore diameter of ~4 nm. TEM characterization also indicated that graphene had been successfully etched by KOH, as numerous nano-sized holes could be clearly observed on the graphene sheets, giving rise to a 3-D porous network. S was then impregnated into the nanopores by simply heating the mixture of activated graphene and S. Elemental mapping revealed homogeneous distribution of S in active graphene. No large S particles or aggregates were observed. XRD results suggested the amorphous nature of S in the nanopores. After S loading, the nanocomposite still had a similar electrical conductivity compared with activated graphene, suggesting that highly dispersed S particles did not block the conductivity. The coin cell employing S/activated graphene as the cathode could deliver a high initial capacity of 1379 mAh g^{-1} at 0.2 C, and after 60 cycles, a relatively high capacity of 1007 mAh g^{-1} could still be retained. When the rate was increased to 1 C, the initial capacity was still as high as 927 mAh g^{-1} and the capacity remained at 685 mAh g^{-1} after 100 cycles, corresponding to the capacity retention of 74%. For comparison, the authors also measured the electrochemical performance of the S/reduced GO sample with no chemical activation treatment. At 0.2 C, S/reduced GO could only deliver an initial capacity of 1093 mAh g^{-1}, ~80% of the one of S/activated graphene. After 60 cycles, a large drop in the capacity of 435 mAh g^{-1} was observed for the S/reduced GO cathode, corresponding to a capacity retention of ~40%, much lower than the 73% retention for the S/activated graphene cathode. The comparison proved that the nanostructure of the graphene substrate could significantly affect the electrochemical performance of S. The advantages of activated graphene over conventional 2-D graphene sheets in improving the electrochemical performance, especially the cycle stability of S, could be mainly ascribed to the following points. First, S was embedded in the nanopores of activated graphene, which gave rise to intimate contact between S and graphene and resulted in rapid electron transfer. Meanwhile, S was homogenously dispersed as tiny nanoparticles in the activated graphene substrate because of its large surface area and nano-sized pores, which could effectively raise the electrochemical activity of S compared with large S particles. Second, the small nanopores that acted as microreactors to entrap S and polysulfides during cycling could remarkably reduce the mass loss of active material, giving rise to improved cycle stability.

Whatever the structure of the graphene/S composite is, the interaction between graphene and S is always important in affecting the electrochemical performance of the composite cathode material. To better understand the electrochemical property of the graphene/S nanocomposite, Zhang and coworkers carried out a systematic investigation on its chemical bonding and electronic structure by using a series of characterization techniques, including XPS, near-edge x-ray absorption fine structure (NEXAFS) and x-ray emission spectroscopy (XES) [16]. The nanocomposite was prepared by chemical deposition of nano-sized S on GO sheets and subsequent low-temperature heat treatment to partial reduction of GO. High-resolution XPS measurements revealed the existence of C-S and O-S species in the composite, suggesting the formation of covalent bonds between S and mildly reduced GO. The authors also compared the C K-edge and O K-edge NEXAFS spectra between mildly reduced GO/S composite and pure GO. Some additional peaks were found in the composite material, which could be assigned to C-S and O-S bonds. The O K-edge XES spectra indicated that the density of the occupied Q 2p states in the composite cathode material is different from that of GO, implying the strong interaction between O atoms of mildly reduced GO sheets and S species incorporated in the composite material. The above characterization results demonstrated the relatively strong chemical interaction between S and mildly reduced GO, which could effectively help to immobilize S on the graphene substrate.

Apart from tuning the microstructure, addition of the third component in the S/graphene composite electrode materials to achieve better electrochemical performance has been broadly investigated in recent years. Chen et al. proposed 3-D hierarchical sandwich-type architecture composed of S, graphene, and MWCNTs for high-performance electrode materials in Li-S batteries, as shown in Figure 4.7 [17]. In the first step, S was absorbed on the surface of CNTs by a thermal infusion method to form core-shell MWCNT@S structure. Then the MWCNT@S composite was dispersed in the GO solution and underwent an ultrasonication treatment to be impregnated into the interlayer galleries of GO. After reduction of GO, the graphene-MWCNT@S nanocomposite electrode material was finally obtained.

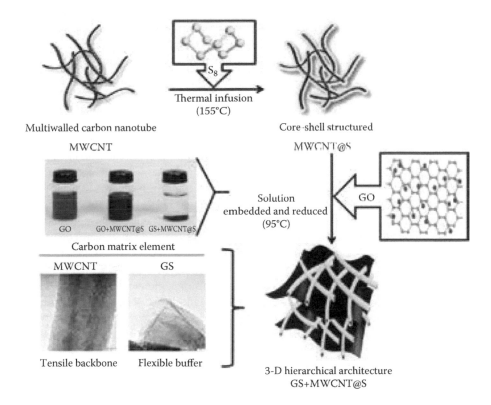

FIGURE 4.7   Schematic illustration of the fabrication and architecture of 3-D hierarchically structured graphene-MWCNT@S composite. (Reprinted with permission from Ref. 17, 4642–4649. Copyright 2013 American Chemical Society.)

Electron microscopy analysis revealed that the surface of MWCNTs became rough after incorporation of S, and a thin layer of S wrapping the outer surface of MWCNTs could be observed. The core-shell MWCNT@S composite nanofibers were interwoven and sandwiched by thin graphene sheets to form a hierarchical structure and MWCNT@S were strongly attached to graphene even after ultrasonication. The thermogravimetric analysis (TGA) results confirmed that there was a high loading of S of ~70 wt% in the composite and the nitrogen sorption analysis showed that the composite had a BET surface area of 59.6 m^2 g^{-1} and a narrow pore size distribution centered at 4.8 nm. This unique hierarchical architecture was believed to be beneficial for improving the charge/discharge performance of S mainly because of the following features. First, mechanically tensile MWCNTs backbone could prevent the restacking of graphene and enhance the structural stability of S. Second, the flexible graphene sheets could act as an elastic buffer to endure the volume change of S during charge/discharge cycles. Third, the hybrid carbon matrix composed of graphene and MWCNTs could act as a half-closed electrochemical chamber to facilitate rapid electron transfer in three dimensions and trap intermediate polysulfides from dissolution into the electrolyte. Fourth, the nanopores in the composite could facilitate electrolyte diffusion and transportation of Li-ions. The electrochemical measurement of the cell containing graphene-MWCNT@S nanocomposite confirmed its outstanding performance. The composite cathode could deliver an initial capacity of 1396 mAh g^{-1}, corresponding to 83% usage of S. The initial Coulombic efficiency of the nanocomposite was as high as 95%. When cycled at 0.2 C 100 times, it still maintained a capacity of 844 mAh g^{-1} with capacity retention of 60.5%. As a comparison, either MWCNT@S composite with no graphene or S/graphene composite with no CNTs exhibited much worse cycle stability (capacity retention less than 50%) and lower Coulombic efficiency (<80%). The authors also monitored the electrochemical impedance response at a different depth of discharge (DOD) during the first cycle. The analysis of the Nyquist plots indicated that the solution resistance did not significantly change during the discharge process, implying that the hierarchical structure was effective in preventing the dissolution of polysulfides. The charge transfer resistance of the graphene-MWCNT@S composite cathode reached the maximum at the DOD of 28% and then decreased as the discharging proceeded, which suggested that the hierarchical structure was beneficial for maintaining good electrical conduction for charge transfer. The morphology of the graphene-MWCNT@S composite could remain after 100 cycles, whereas the surface of either MWCNT@S or S/graphene composite cathode material contained large amounts of agglomerated Li$_2$S/Li$_2$S$_2$, suggesting that the porous hybrid matrix of graphene and MWCNTs could provide sufficient space to accommodate Li$_2$S/Li$_2$S$_2$.

Lu and coworkers reported sulfur-coated carbon nanofibers coaxially wrapped by graphene as high-performance electrode materials (Figure 4.8) [18], which was synthesized by a two-step approach. Carbon nanofibers (CNFs) were functionalized with hydroxyl and carboxyl groups by refluxing in HNO$_3$ before coating of S in order to generate strong interaction between sulfur and CNFs. The reaction between Na$_2$SO$_3$ and HCl then formed thin S coating layers on the surface of CNFs. Afterward, the S-CNF composite was dispersed in the aqueous solution of graphene and the pH of the solution was preadjusted to 10. Both graphene and S-CNF were negatively

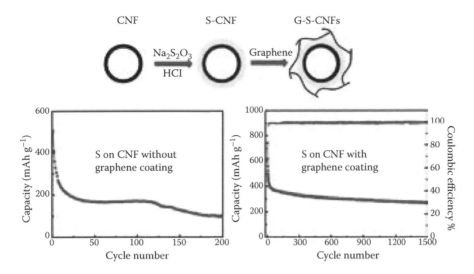

FIGURE 4.8 Fabrication process and the electrochemical performance of S-coated carbon nanofibers coaxially wrapped by graphene. (Reprinted with permission from Ref. 18, 2485–2489. Copyright 2013 American Chemical Society.)

charged at this pH value. The wrapping of graphene was then realized by gradually decreasing the pH of the solution until 2 using HCl. During this process, negatively charged S slowly turned to be positively charged, while graphene kept negatively charged. The electrostatic interaction then induced the wrapping of graphene sheets on the surface of individually S coated CNFs. TEM characterization showed that the CNFs had a mean diameter of ~90 nm, and the mean thickness of S coating layers was ~23 nm. The sulfur layer was uniform and no isolated S particles were observed. Each S-CNFs composite fiber was coaxially and intimately wrapped by graphene sheets. The graphene-S-CNFs nanocomposite could be vacuum-filtrated to form a freestanding film, which could be used directly as the cathode in the coin cell by cutting into a suitable size without any binders and conductive additives. The cell had an initial discharge capacity of 1047 mAh g^{-1} based on the weight of S at 0.1 C. The capacity decayed to 694 mAh g^{-1} after 10 cycles and was stabilized at this value in the subsequent cycles. The cell was also tested for 1500 cycles at a higher rate of 1 C. Although the capacity faded relatively fast in the initial 50 cycles, the cell was extremely stable from the 50th to the 1500th cycles with a decay rate of 0.019% per cycle. The superior cycle stability of this composite cathode material was mainly ascribed to the synergetic effect from graphene and CNFs, which was effective in immobilization of polysulfides and provided good mechanical support to accommodate the volume expansion of S. In comparison, the S/CNFs composite without graphene wrapping exhibited much worse cycle stability. Its capacity quickly decreased from ~500 to ~100 mAh g^{-1} within only 200 cycles.

Besides 1-D carbon materials such as CNTs and CNFs reported in the above researches, mesoporous carbon with large surface area and high pore volume was also used along with graphene in the modification of S cathodes. Mesoporous carbon can be obtained by using ordered mesoporous silica as the hard template. As an inverse replica of the mesoporous silica, mesoporous carbon has an ordered pore structure, uniform pore diameters of several nanometers, and ultrahigh surface area, which is suitable to be a substrate for S loading. The mesopores can effectively constraint the growth of S and provide abundant pathways for Li-ion diffusion, whereas the carbon framework can serve as a continuous conductive network for fast electron migration. However, the open end of the mesopores on the outer surface is directly exposed to the electrolyte, which cannot avoid the dissolution of intermediate polysulfides during cycling. Scientists have proposed a resolution to prevent this by coating of S-loaded mesoporous carbon particle with graphene sheets on the surface. The coating could be realized by electrostatic interactions, as reported by Zhao et al. [19]. The mesoporous carbon/S composite was first functionalized by tris(hydroxymethyl)aminomethane to be positively charged. It was then mixed with negatively charged GO sheets in the aqueous solution, giving rise to simultaneous coating of GO on the surface of the mesoporous carbon/S composite particles. After reduction of GO, the graphene/mesoporous carbon/S composite cathode material was finally produced. Because of the electrostatic interaction, each mesoporous carbon/S composite particle was intimately wrapped by thin graphene sheets (Figure 4.9). As a result, the cycle stability of this ternary composite was much improved compared with mesoporous/S composite without graphene coating. Although both cathode materials have similar initial discharge capacity of over 1200 mAh g^{-1} at 0.1 C, a relatively high reversible capacity of ~650 mAh g^{-1} was retained after 100 cycles for the graphene-coated samples, while the corresponding value for the noncoated samples was only ~400 mAh g^{-1}. Over the entire 100 cycles, the Coulombic efficiency of the cell using graphene/mesoporous carbon/S composite cathode material was over 97%, whereas the one for the mesoporous carbon/S composite was only around 90%. The comparison of the electrochemical performance clearly indicated the graphene sheets could effectively minimize the dissolution of polysulfides, giving rise to better cycle stability. Similar graphene-coated mesoporous carbon/S composite cathode was also prepared by Zhou et al. [20] and the authors also observed improved cycle stability by graphene coating.

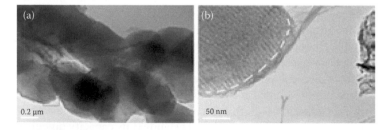

FIGURE 4.9 TEM images of graphene wrapped mesoporous C/S composite at low (a) and high (b) magnifications. (Reprinted from *Electrochim. Acta*, 113, Zhao, X. Y., Tu, J. P., Lu, Y. et al., Graphene-coated mesoporous carbon/sulfur cathode with enhanced cycling stability, 256–262, Copyright 2013, with permission from Elsevier.)

Amorphous carbon derived from pyrolysis of polymers has also been used along with graphene to modify S cathode. Pyrolyzed polyacrylonitrile (PAN)/graphene/S composite cathode material was proposed by Yin et al. [21]. In the first step, PAN/graphene composite was synthesized by *in situ* polymerization of acrylonitrile in the aqueous suspension of GO and subsequent chemical reduction of GO. PAN had spherical morphology with diameters of ~100 nm and was uniformly attached to the graphene sheets. Then the composite was mixed by S using ball milling and the mixture was heat-treated at 300°C in Ar. During the heating process, PAN was pyrolyzed and S was gradually embedded into the matrix of pyrolyzed PAN, which still possessed spherical morphology. Charge/discharge measurement showed that at an optimal graphene content of 4 wt%, the ternary composite could reach capacity retention of ~80% after 50 cycles at 0.1 C, whereas the pyrolyzed PAN/S composite without graphene only had a capacity retention of ~60% under the same conditions. The ternary composite cathode material also exhibited better rate capability. Even its reversible capacity at a high rate of 6 C was higher than the capacity of the composite without graphene at the rate of 3 C. The improved cycle stability and rate capability was mainly due to the high electrical conductivity of graphene. Graphene sheets could serve as mini current collectors to accelerate electron migration in the cathode. In addition, GO was able to reduce the size of PAN, which could shorten the transport lengths of Li-ions. In another paper published by the same research group, the authors proposed another route to the synthesis of pyrolyzed PAN/graphene/S ternary composite [22]. In this work, S was first embedded in pyrolyzed PAN nanoparticles, which were mixed with GO nanosheets in an aqueous solution. Finally, nano-sized S was deposited onto GO and pyrolyzed PAN through a solution-based chemical reaction. The as-prepared ternary composite had a dual-mode structure. Part of S was embedded in pyrolyzed PAN, and the others were deposited on graphene. Compared with single sulfur-containing mode, this dual-mode composite cathode material possessed higher S content than bare pyrolyzed PAN/S composite with no graphene, while its utilization of S and cycle stability was better than those of graphene/S composite with pyrolyzed PAN.

Similar to mesoporous carbon, metal organic framework (MOF) also has high surface area and abundant nanopores, which is suitable to be used as a host material for S cathode. However, its electrical conductivity is quite poor. Graphene thus is quite useful to improve the conductivity of MOFs. Based on this idea, Bao et al. synthesized MOF/graphene/S composite as a cathode material for Li-S batteries [23]. The MOFs they used had open channels with diameters from 1.4 to 3.0 nm and was first dispersed in an aqueous suspension of GO. After the addition of hydrazine hydrate as the reducing agent and being heated at 95°C, precipitation of MOFs/graphene composite was formed. The composite was then dispersed in the CS_2 solution of S. The final ternary composite was obtained after evaporation of CS_2. S was homogeneously distributed in the porous framework of MOFs, and the MOF crystals with a size of several micrometers were wrapped by graphene sheets. At a rate of 0.2 C, the ternary composite cathode material delivered an initial capacity of 980 mAh g^{-1} and the reversible capacity of 650 mAh g^{-1} could be retained after 50 charge/discharge cycles with a retention rate of 67%. Though the MOFs/S binary composite without graphene modification had a higher initial capacity of 1232 mAh g^{-1}, it quickly dropped to 504 mAh g^{-1} after 50 cycles, corresponding to a low-capacity retention of 41%. Consequently, graphene played an important role in improving the electrochemical performance of S loaded in MOF via its high electrical conductivity.

Polymer is another kind of functional component used to improve the electro-chemical performance of graphene/S composite cathode materials. Polymers usually have soft nature and relatively strong interaction with chemically modified graphene or S, which is able to further immobilize S and accommodate the volume expansion of S during cycling. For example, Zhou and coworkers used amylopectin, a natural polymer to modify GO/S composite [24]. S particles dispersed on GO sheets were mixed with the aqueous solution of amylopectin containing carbon black. The fol-lowing addition of ethanol in the solution then resulted in the precipitation of amylo-pectin/graphene/S/carbon black composite (Figure 4.10). The branched amylopectin molecules wrapped the GO/S composite, which closed the open channel among GO sheets and minimized the diffusion of polysulfides. Compared with GO/S, the amylopectin-wrapped sample exhibited much better cycle stability. At 0.8 C, an ini-tial capacity of ~900 mAh g^{-1} could be delivered for the polymer-coated sample, and after 100 cycles, a reversible capacity of ~600 mAh g^{-1} could still be retained. In contrast, the capacity of GO/S decayed quickly from ~800 mAh g^{-1} to less than 300 mAh g^{-1} within only 80 cycles at the same rate.

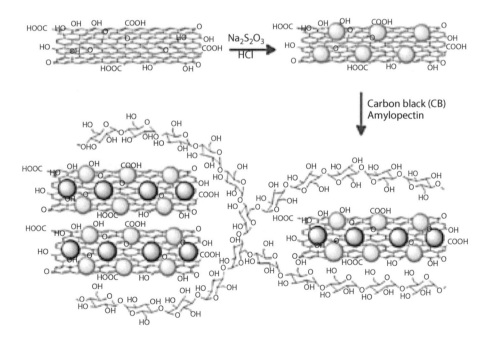

FIGURE 4.10 Schematic illustration of the synthesis route for a graphene-S-amylopectin composite, with light gray balls representing S and dark gray balls representing carbon black. (Reprinted with permission from Ref. 24, 8801–8808. Copyright 2013 American Chemical Society.)

Nafion was used in another case to coat graphene/S composite cathode material [25]. The capacity retention of the Nafion-coated sample after 50 cycles at 0.1 C increased by ~27% compared with the noncoated one. At a higher rate of 1 C, a better capacity retention of 84.3% could be achieved after 100 cycles for the Nafion/graphene/S composite cathode material. The improved cycle stability could also be ascribed to the protective effect of Nafion polymer that physically prevented the dissolution of polysulfides in the electrolyte. In addition, Nafion, which contains a large quantity of negatively charged sulfonate groups further suppressed the diffusion of negatively charged polysulfide anions through the polymer film because of static-electric repulsion. Other polymers, such as polypyrrole [26,27] and polyacrylonitrile [28] have also been successfully used to form ternary composite composed of polymer, graphene, and S, and effective prohibition of polysulfides dissolution was realized.

Though graphene is mostly used in the modification of S cathode for better electrochemical performance, exploration of other functions of graphene in Li-S batteries has also been attempted by some research groups. Wang and coworkers reported the application of graphene film as a shuttle-inhibitor interlayer [29]. In contrast to other research in which graphene was incorporated with S to form a composite cathode, in this paper, a freestanding graphene film that was sandwiched between the S cathode and the separator was used to inhibit the diffusion of dissolved polysulfides to the anode. The graphene film could be easily prepared by vacuum filtration of GO suspension and subsequent heat reduction. However, graphene sheets were tightly stacked in the film, which made it difficult for the entering and diffusion of electrolyte. Thus, the authors added some carbon black nanoparticles in the graphene film to enlarge the gaps between graphene sheets and generate more channels for the permeation of electrolytes and polysulfides. With a weight ratio of graphene/carbon black = 2/1, the cell using this interlayer exhibited excellent cycle stability. It delivered an initial capacity of 1260 mAh g^{-1} and maintained at 894 mAh g^{-1} after 100 cycles, corresponding to capacity retention of ~70%. Elemental mapping showed a homogeneous distribution of S in the film after cycles, suggesting the capture of polysulfides of the film during cycling. The FTIR spectra revealed that the C-O and C=O bonds disappeared while the S-S and CO_3^{2-} bonds emerged after cycling, which the author believed was due to the oxidation of carboxyl and carbonyl groups into carbonate groups with the assistance of polysulfides. The above results demonstrated that the graphene film was effective to capture and immobilize polysulfides by its oxygen functional groups, which was could be used as a shuttle-inhibitor to improve the cycle stability of Li-S batteries.

In another report, graphene was applied as a coating layer on polymer substrate to form metal-free current collectors for Li-S batteries [30]. Graphene was dispersed in DMF and then cast on polyethylene terephthalate (PET). Compared with metal current collectors, graphene-coated PET film has a much lower density of 1.37 m^2 g^{-1}, which is beneficial to raise the energy density of the battery, and it can also avoid the corrosion of conventional metal current collectors during long-term charge/discharge cycles because of the electrochemical stability of graphene. The prototype Li-S battery using graphene-coated PET as the current collector had an energy density of

452 Wh kg^{-1} excluding the package weight. The capacity retention of 96.8% after 30 cycles could be achieved for the battery.

4.3 APPLICATION OF GRAPHENE IN LI-AIR BATTERIES

4.3.1 INTRODUCTION

Li-air battery is another type of high-energy density energy storage system. Li metal is employed as the anode in the Li-air battery. On discharge, Li is oxidized to release Li$^+$ into the electrolyte, and the reverse process occurs on charge. The cathode of the Li-air battery usually adopts porous and conductive materials, and air, or more specifically, oxygen, is employed as the "fuel" to generate electricity on the porous materials. During the discharge process, oxygen enters the porous cathode, dissolves in the electrolyte, and then is catalytically reduced on the surface of the porous materials. This catalytic reduction reaction is strongly dependent on the properties of the electrolyte. When a suitable nonaqueous electrolyte is employed, O$_2$ is reduced to form O$_2^{2-}$, which subsequently combines with Li$^+$ in the electrolyte from the Li anode to form Li$_2$O$_2$ as the discharge product. Li$_2$O$_2$ then decomposes on charge. If aqueous electrolyte is used, OH$^-$ is formed instead of O$_2^{2-}$, which results in the formation of LiOH as the discharge product. LiOH is then oxidized on charge. Because of the significant difference of the electrochemical reaction and discharge products between nonaqueous and aqueous electrolytes, Li-air batteries are usually divided into the categories of nonaqueous Li-air batteries and aqueous Li-air batteries. As oxygen is inexhaustible in the air, the capacity of a Li-air cell is theoretically only dependent on the mass of the Li used in the anode. Therefore, Li-batteries have ultrahigh energy density up to more than 10,000 Wh kg^{-1}. The practical performance, including capacity and the reversibility of a Li-air cell, is dominantly affected by the catalytic activity of oxygen reduction and regeneration in the cathode no matter what kind of electrolyte is used. Therefore, design and preparation of a high-performance cathode has become the major task in the research of Li-air batteries. Porous carbon materials have been widely used as the catalyst of catalyst supporter in the cathode. As a new type of carbon material, graphene has shown great advantages over traditional carbon materials in improving the catalytic activity in the cathode as reported in some of the recent publications, which will briefly be discussed in Section 4.3.2.

4.3.2 PROGRESS OF THE APPLICATION OF GRAPHENE IN LI-AIR BATTERIES

4.3.2.1 Nonaqueous Li-Air Batteries

The oxygen reduction reaction (ORR) taking place in the cathode of the nonaqueous Li-air batteries is significantly affected by the composite and structure of the air electrode. To improve the kinetics and enhance the energy density of the Li-air battery, the porous structure of the cathode should be optimized to realize rapid oxygen diffusion, accelerate the catalytic reaction, and prevent excess growth of discharge products (mainly Li$_2$O$_2$). Carbon materials such as commercial carbon black are commonly used as the cathode materials of nonaqueous Li-air batteries. The emergence of graphene provides another option in the design and synthesis of high-performance carbon cathodes because of the unique and excellent physico-chemical properties of graphene.

The first publication on a graphene-based air electrode for nonaqueous Li-air batteries was reported by Li and coworkers in 2011 [31]. The cell using the graphene electrode could deliver a capacity of 8705.9 mAh g^{-1}, which was 4–8 times higher than the cells using commercial carbon black as the air electrode and was the highest capacity ever reported for any carbon-based electrode. It could be mainly ascribed to the different pore structure of the graphene electrode. Comparing with carbon black, graphene had a wide pore size distribution and much higher porosity in the meso-pore range (2–50 nm). These mesopores were believed to play important roles in the diffusion of O_2 and electrolyte. It was also observed that more discharge products were deposited on the edges of graphene sheets than on the basal planes, as shown in Figure 4.11, suggesting higher catalytic activity of the ORR reaction on the edges, which was probably due to large amounts of unsaturated and functionalized carbon atoms on the edge sites. The advantage of graphene over carbon black in the application as the air electrode was also confirmed later by Sun and coworkers [32].

Shortly afterward, scientists from the Pacific Northwest National Laboratory improved the performance of a graphene-based air electrode by designing a hierarchical porous structure using functionalized graphene sheets, as shown in Figure 4.12 [33]. The functionalized graphene, which was prepared by thermal expansion and reduction of GO, was first dispersed in a microemulsion solution containing the binder materials polytetrafluoroethylene (PTFE). The mixture was then casted and dried. Graphene sheets aggregated to form a "broken egg" structure as described

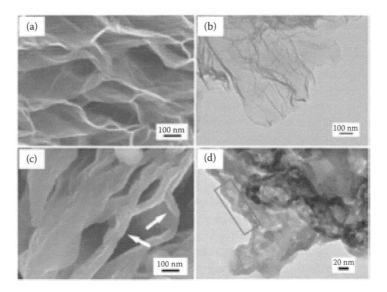

FIGURE 4.11 SEM and TEM images of GNSs electrodes before (a and b) and after (c and d) discharge. (From Li, Y., Wang, J., Li, X. et al., Superior energy capacity of graphene nanosheets for a nonaqueous lithium-oxygen battery. *Chem. Commun.* 47:9438–9440, 2011. Reproduced by permission of The Royal Society of Chemistry.)

(a) $2Li^+ + O_2 + 2e^- \longrightarrow Li_2O_2$

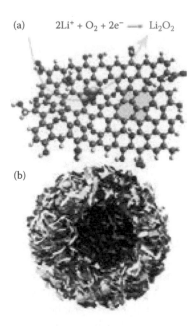

(b)

FIGURE 4.12 Schematic structure of a functionalized graphene sheet (a) with an ideal bimodal porous structure (b), which is highly desirable for Li-air battery operation. (Reprinted with permission from Ref. 33, 5071–5078. Copyright 2011 American Chemical Society.)

by the authors, where the "shell" consisted of numerous nanoscale pores consti-tuted by randomly aggregated graphene nanosheets. The broken egg structure was then loosely packed to generate large interconnected tunnels that went through the entire electrode. This unique hierarchically porous structure was mainly ascribed to the microemulsion of PTFE. During the mixing of graphene with PTFE micro-emulsion, significant foaming occurred, which resulted in the formation of large quantities of bubbles whose diameters ranged from tens of nanometers to tens of micrometers, and in many cases, one large bubble contained lots of smaller bubbles. The hydrophobic graphene sheets tended to assemble at the surfactant-rich surface of the bubbles (liquid–gas interface). After drying, porous graphene as a replica of the bubbles was formed. The hierarchical pore structure was considered as an ideal structure for the air electrode. The large tunnels among the broken egg structures could effectively help the fast diffusion of oxygen while the small pores in the shells provided triphase (solid–liquid–gas) regions for the ORR. When the Li-air cell using the porous graphene as the air electrode was tested in pure O_2 at an initial pressure of ~2 atm, the discharge capacity could reach as high as 15,000 mAh g^{-1} with a voltage plateau of ~2.7 V, corresponding to a high energy density of 39,714 Wh kg^{-1} (based on the mass of carbon), which was the highest record ever reported for the nonaque-ous Li-air batteries at that time. To evaluate the practical performance, the cell was also tested in a pouch-type configuration and operated in an ambient environment with an O_2 partial pressure of 0.21 atm and 20% relative humidity. The cell could still reach a discharge capacity of 5000 mAh g^{-1} at a current density of 0.1 mA g^{-1}.

It was also observed that Li_2O_2 (the discharge product) deposited on graphene were nanoscale particles and isolated, rather than thick layers of large particles formed in conventional carbon electrodes. It was interesting to note that the specific capacity of the cell was significantly affected by the defects in the graphene sheets. The authors measured the electrochemical performance of the air electrode composed of graphene with different C/O atomic ratios of 14 and 100, respectively, and found that the capacity of the former was approximately two times higher than the latter. The Li_2O_2 particles deposited on the former sample were also smaller and more homogeneous than on the latter one. Theoretical calculation was then applied to investigate the deposition mechanism of Li_2O_2 on the graphene surface, and the effects of the defect sites and functional groups were taken into great account. It was concluded that the interaction between Li_2O_2 monomers with defects and functional groups was much stronger than the one between Li_2O_2 monomers and perfect graphene. Consequently, $(Li_2O_2)_n$ clusters preferred to be isolated on the defect sites rather than aggregated to form larger Li_2O_2 particles. The calculation results were in good accordance with the observed morphology of Li_2O_2 particles on graphene sheets with different C/O ratios. The limited size of Li_2O_2 particles formed on graphene would not block the oxygen transportation pathways, and also prevented the increase in electrode impedance, which was also a significant factor that improved the performance of the graphene-based air electrode.

Graphene-based hierarchically porous air electrode was also reported by Wang et al. using a different preparation method, and its structure was illustrated in Figure 4.13 [34]. In their experiments, a mixed sol containing GO, resorcinol, and formaldehyde was dropped into the Ni foam disks and then cured in an oven. The gel formed in the Ni foam was then dehydrated by freeze-drying and finally carbonized in N_2 atmosphere. During the synthesis, GO acted as the framework of the 3-D gel. Meanwhile, the carboxyl (COOH) functional groups on the GO sheets could provide a weak acidic environment to etch Ni to form nickel oxyhydroxide (NiOOH) in the gel, which then helped to adhere the gel to the Ni foam. NiOOH decomposed to nickel(II) oxide (NiO) during the carbonization process and was further reduced by carbon to form Ni particles. Meanwhile, the consumption of carbon when reducing NiO generated nanopores around Ni. In this carefully designed air electrode, Ni foam as the conductive skeleton was uniformly covered with carbon-coating layers. The carbon species had a sheet-like morphology because of the introduction of GO. The carbon sheets were loosely packed and aligned roughly perpendicular to the surface of Ni foam skeleton. Large tunnels existed among carbon sheets. High-magnification SEM images showed that a large amount of nanopores with a diameter of 20–100 nm could be observed on carbon sheets and Ni nanoparticles were embedded in the pores. This hierarchical pore structure was favorable for the fast diffusion and catalytic reduction of oxygen. As a result, the Ni/carbon composite air electrode could deliver a discharge capacity of 11,060 mAh g^{-1} at a current density of 280 mA g^{-1} in pure oxygen. At a higher current density of 2.8 A g^{-1}, a discharge capacity of 2020 mAh g^{-1} could still be retained, which showed excellent rate capability. The galvanostatic charge and discharge measurement showed that the cutoff voltage did not change dramatically within 10 cycles at a current density of 700 mA g^{-1} when the capacity was restricted to 2000 mAh g^{-1}. A SEM

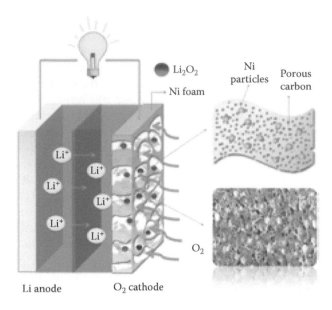

FIGURE 4.13 Structure model of a graphene-based air electrode with a hierarchically porous structure. (From Wang, Z.-L., Xu, D., Xu, J.-J. et al.: GO gel-derived, free-standing, hierarchically porous carbon for high-capacity and high-rate rechargeable Li-O$_2$ batteries. *Adv. Func. Mater.* 2012. 22. 3699–3705. Copyright Wiley-VCH Verlag GmbH & Co. KGaA, Weinheim. Reproduced with permission.)

image showed that the Li$_2$O$_2$ formed after discharge had toroidal shape with particle size of ~400 nm. After charging, the Li$_2$O$_2$ particles almost disappeared, suggesting a highly reversible process.

Besides pure graphene, graphene doped with heteroatoms were also applied in the Li-air batteries. Nitrogen-doped graphene as the air electrode was first reported by Li and coworkers [35]. The sample was prepared by heating reduced GO in the mixed atmosphere of ammonia and Ar. The atomic percentage of N in the sample was measured to be 2.8%. The quantity of defects increased after N-doping. Compared with nondoped graphene, the discharge capacity of the N-doped graphene electrode was increased by 40% and N-doped graphene also showed higher discharge plateau, indicating higher catalytic activity of ORR of the N-doped sample, which was possibly due to the higher binding energy of oxygen after N doping. Though a relatively high capacity was achieved, the discharge voltage plateau only reached 2.58 V, implying high ORR overpotential. In another report, Wu et al. proposed an alternate way to prepare N-doped graphene-rich catalyst for oxygen oxidation, and excellent electrochemical performance was achieved [36]. In contrast with a conventional N doping process as adopted in the previous paper, the authors of this paper employed the method of graphitization of heteroatom polymers under the catalysis of a Co species and using CNTs as a supporting template, as illustrated in Figure 4.14. Aniline, which contains both carbon and nitrogen atoms, were used as the precursor. It was

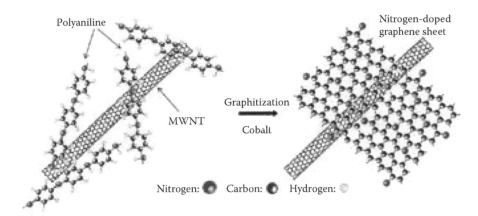

Nitrogen: Carbon: Hydrogen:

FIGURE 4.14 Scheme of the formation for nitrogen-doped graphene sheets derived from polyaniline and Co precursors using CNT as a template. (Reprinted with permission from Ref. 36, 9764–9776. Copyright 2012 American Chemical Society.)

mixed with acid-treated CNTs and Co^{2+} in an aqueous solution. After *in situ* polymerization, the mixture was dried and ball-milled, and then heat-treated to obtain the final product. Large amounts of thin graphene sheets could be observed in the sample, which wrapped around CNTs. Both CNTs and Co played significant roles in the formation of this unique electrode material. If CNTs were replaced by carbon black, agglomerated bubble-like graphene structures were obtained instead of stretched and large-size graphene sheets in the existence of CNTs, and the specific surface area of the former samples was less than half of the latter one. In the absence of Co, only amorphous carbon species were obtained, indicating that Co was required to form highly graphitized carbon nanostructures. As pyrrolic N was usually the dominant species when using ammonia as the doping agent, a high level of quaternary and pyridinic N species were observed in this Co catalyzed sample. The Co-CNT-graphene composite electrode had a relatively high average discharge voltage plateau of 2.85 V, indicating greatly reduced ORR overpotential. It could also deliver an initial discharge capacity of 3700 mAh g^{-1} at a current density of 50 mA g^{-1}. When the current density was raised to 800 mA g^{-1} the capacity of 1200 mAh g^{-1} could still be maintained. The cell also exhibited excellent cycle stability. Within the initial 20 cycles, at a current density of 400 mA g^{-1}, no significant capacity loss was observed. When further cycling to 30 and 50 cycles, capacity loss of 8.4% and 20.4% resulted, respectively. The degradation of the electrode was probably due to the formation of insoluble solid particles (such as Li_2O) during cycling that may block the catalytic sites and O_2 transporting channels. The enormously improved electrochemical performance of this novel cathode compared with traditional carbon black and Pt/C catalyst was mainly ascribed to its unique structure and the N-doping functionalities.

N doping was also combined with chemical activation to produce N-doped graphene with a large surface area and high porosity as a cathode catalyst for Li-air batteries [37]. The sample was prepared by chemical activation of thermally reduced

and exfoliated graphene with KOH and subsequent N doping by heat treatment of
the activated graphene with ammonia. It had an ultrahigh specific surface area of
2980 m^2 g^{-1}, which was even higher than the theoretical value of graphene. The
atomic content of N was measured to be 1.57%. When tested in a nonaqueous Li-air
cell, the N-doped and activated graphene electrode could deliver an initial discharge
capacity of 11,746 mAh g^{-1}, which was 42% higher than that of carbon black. Its
discharge voltage plateau was also 0.16 V higher than that of carbon black.

The effect of N doping in increasing the catalytic activity of ORR was also investigated by theoretical calculation [38]. First, principle calculation was carried out to
study the oxygen adsorption and dissociation of graphene and N-doped graphene.
It was found that the adsorption O$_2$ molecules were energetically favorable on graphene sheets as charge transfer occurred between two components. N doping could
further enhance the adsorption of O$_2$ because a N atom has one more electron than a
C atom. The energy barrier to the dissociation O$_2$ on graphene was calculated to be
2.39 eV, smaller than the binding energy of O$_2$, suggesting that O$_2$ dissociation could
be catalyzed directly by graphene. After N doping, the energy barrier could be further decreased to 1.20 eV, indicating higher catalytic activity of N-doped graphene
compared with a pristine one.

S-doped graphene was also reported as the cathode for Li-air batteries [39]. It was
prepared by simply mixing graphene with p-toluenesulfonic acid in acetone and heat
treatment of the dried mixture at 900°C. S was homogenously distributed on graphene sheets after doping as revealed by energy dispersive x-ray (EDX) mapping. The
formation of C-S-C and C-SO$_x$ covalent bonds was confirmed by XPS and XANES
spectra, indicating that S was successfully doped into the carbon framework of graphene. The atomic percentage of S was measured to be 1.9%. Comparing the initial
discharge capacity of 8700 mAh g^{-1} of pristine graphene-based electrode, the S-doped
graphene cathode only had an initial discharge capacity of 4300 mAh g^{-1}. However,
its subsequent charge capacity reached 4100 mAh g^{-1}, which was significantly greater
than 170 mAh g^{-1} of the pristine graphene. Moreover, the discharge capacity in the
second cycle of the S-doped cathode could still reach 3500 mAh g^{-1}, while the corresponding value for the pristine graphene was only 220 mAh g^{-1}. The extremely
enhanced cycle stability could be mainly ascribed to the influence S doping in the
microstructure of the discharge product. Irregular Li$_2$O$_2$ particles were formed on
pristine graphene, whereas Li$_2$O$_2$ nanorods with a mean diameter of ~100 nm grown
on S-doped graphene were observed after discharge. The authors also monitored the
growth of such nanorods by controlling the discharge depth of the cell, as shown in
Figure 4.15. At the discharge voltage of 2.6 V, nanorods had already grown on most
of the surface of S-doped graphene and had diameters of ~35 nm. As the discharge
proceeded, the nanorods gradually covered the entrie surface of the electrode and
their diameters continuously increased until 100 nm at a voltage of 2 V. The authors
also observed that the morphology of the discharge products varied when the current
density was changed. The formation of nanorods could provide sufficient channels
for the steady electrochemical reaction in the air electrode, thus giving rise to much
improved cycle stability compared with large particles that easily blocked the reaction channels in the case of pristine graphene. However, the role that S played in the
unique morphology of the discharge product is still not fully understood.

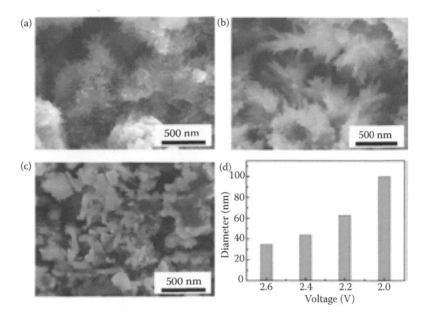

FIGURE 4.15 SEM images of discharged S/graphene electrodes at (a) 2.6 V, (b) 2.4 V, and (c) 2.2 V at a current density of 75 mA g^{-1}. (d) Mean diameters of the discharge products. (From Li, Y., Wang, J., Li, X. et al., Discharge product morphology and increased charge performance of lithium–oxygen batteries with graphene nanosheet electrodes: The effect of sulphur doping. *J. Mater. Chem.* 22:20170–20174, 2012. Reproduced by permission of The Royal Society of Chemistry.)

In order to achieve better catalytic activity in the air electrode of Li-air batteries, graphene-based hybrid catalysts have been proposed in recent years. By incorporation of graphene with traditional catalysts for ORR, such as transition metal oxides and noble metals, a series of hybrid catalysts with excellent electrochemical performance were successfully prepared. MnO_2 is a good candidate for the catalytic reaction of oxygen in the cathode of Li-air batteries due to its low cost, environmental benignity, and high catalytic activity. Consequently, several research groups have prepared MnO_2/graphene composite catalyst for nonaqueous Li-air batteries. For example, Cao and coworkers presented the α-MnO_2 nanorods/graphene hybrid as a high-performance catalyst [40]. The hybrid was prepared by the redox reaction between reduced GO and $KMnO_4$ in an acidic solution. Reduced GO sheets were first treated by moderate acid to generate oxygen-containing groups (e.g., carboxyl, carbonyl, and hydroxyl groups) on the surface, which was then mixed with $KMnO_4$. The α-MnO_2 nanorods were grown *in situ* on graphene sheets by the reaction between MnO_4^- and Mn^{2+} resulting from the reaction of MnO_4^- with carbon. The α-MnO_2 nanorods with a mean diameter of 10 nm and lengths of 50–80 nm were uniformly distributed on graphene sheets. The α-MnO_2-loaded graphene sheets were interconnected to form a porous structure. The hybrid cathode delivered a reversible capacity of 11,520 mAh g^{-1} (based on the mass of carbon) at a current density of 200 mA g^{-1}, and it had a discharge voltage plateau of ~2.86 V, suggesting excellent electrocatalytic activity of the ORR process. For comparison, a simple mixture of α-MnO_2 nanorods and graphene was prepared and its electrochemical performance was also measured. It could only deliver a reversible capacity of 7200 mAh g^{-1}, which was ~62.5% of the value of the hybrid. Its discharge plateau was also 0.15 V lower than that of the hybrid. The advantage of the hybrid electrode over the mixture could be ascribed to the following factors. First, in the hybrid, α-MnO_2 nanorods were firmly attached to graphene sheets, resulting in better conductance than the mixture. It was measured that the electrical conductivity of the hybrid was about twice as much as that of the mixture. Second, the hybrid possessed larger pore volume and average pore size than the mixture, which could provide more space to accommodate discharge products. The α-MnO_2/graphene hybrid also showed good cycle stability. At a current density of 300 mA g^{-1}, the cell had steady charge and discharge capacities over 25 cycles, and its overpotential did not increase during cycling. MnO_2/graphene hybrid air electrode was also reported by other research groups, such as a similar α-MnO_2 nanorod/graphene composite cathode [41], MnO_2 nanoparticle/graphene hybrid electrode [42], and MnO_2 nanotube/N-doped graphene composite cathode [43]. All these studies confirmed that the synergetic effect between graphene and MnO_2 was effective in improving the catalytic activity of the ORR process.

By substituting the Mn^{3+} in the spinel MnO_2 with Co^{3+}, $MnCo_2O_4$/graphene was synthesized and used as a cathode catalyst for Li-air batteries [44]. The hybrid was prepared by a solvothermal method, and covalent bonds were formed between $MnCo_2O_4$ and graphene, which induced effective electrochemical coupling between two components and thus excellent electrochemical performance. The hybrid electrode exhibited an ultrahigh discharge voltage of 2.95 V, which was among the lowest overpotentials ever reported. It also showed good cycle stability. A discharge capacity of 1000 mAh g^{-1} could be delivered for 40 cycles without significant increase in the overpotential.

Graphene hybrids with other transition metal oxides as the cathode have also been reported. For example, Fe_2O_3 nanoclusters/graphene composite was prepared and exhibited increased discharge capacity and better round-trip efficiency than pure graphene electrode [45]. Its excellent electrochemical performance was ascribed to the combination of fast electron transport induced by highly conductive graphene sheets and high electrocatalytic activity of O reduction provided by Fe_2O_3 nanoclusters. In another report, Co_3O_4/graphene hybrid was proposed as a high-performance air electrode [46]. Graphene-supported Co_3O_4 nanoparticles were well dispersed with no severe agglomeration that occurred in the simple mixture of Co_3O_4 and carbon black. As a result, it exhibited much improved cycle stability. Zirconium (Zr)-doped CeO/graphene hybrid was also reported as the cathode for Li-air batteries [47]. With a 10% weight loading of doped CeO on graphene, the discharge capacity of the cell could be increased by three times.

Noble metal is another important type of catalyst used in the electrochemical catalysis of oxygen reduction. In some recent publications, noble metal/graphene hybrid catalyst was prepared and used as a high-performance air electrode in the Li-air batteries. Jung and coworkers reported graphene-supported Ru-based electrocatalysts for Li-air batteries (Figure 4.16) [48]. Two types of Ru-based nanomaterials loaded on graphene were presented in the paper. One was metallic Ru synthesized

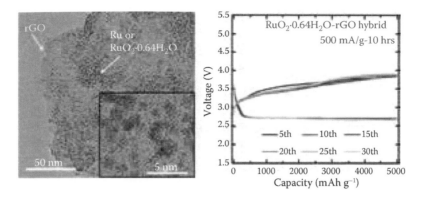

FIGURE 4.16   TEM characterization and charge/discharge performance of Ru-based electrocatalyst supported on graphene. (Reprinted with permission from Ref. 48, 3532–3539. Copyright 2013 American Chemical Society.)

by reduction of $RuCl_3$ in ethylene glycol in the presence of GO and the other was hydrated RuO_2 produced by the hydrothermal method. Both Ru and hydrated RuO_2 had an ultrasmall particle size of ~2 nm and were homogeneously distributed on graphene. When using $LiCF_3SO_3$-tetraethylene glycol dimethyl ether as the electrolyte, the electrocatalytic activity of pure graphene was comparable to commercial carbon black, which has a charge potential of 4.3 V. However, the addition of Ru-based nanoparticles could significantly improve the electrochemical performance. The charge potential could be reduced to 3.9 V for the Ru/graphene hybrid and further reduced to 3.7 V for the hydrated RuO_2/graphene composite. The cell employing hydrated RuO_2/graphene cathode also maintained stable cycling performance for over 30 cycles.

Pt/graphene hybrid as the catalyst in the cathode has also been demonstrated. Yang et al. synthesized Pt/graphene composite by liquid phase pulsed laser ablation [49]. The obtained Pt nanoparticles anchored on graphene sheets had a mean diameter of 3.8 nm and the weight ratio of Pt in the hybrid was about 10%. At a current density of 70 mA g^{-1}, the hybrid electrode could deliver a discharge capacity of 4820 mAh g^{-1}, which was 70% higher than the value for pure graphene electrode. The voltage gap between charge and discharge curves for the Pt/graphene hybrid was also smaller than that for pure graphene, suggesting better electrocatalytic activity after Pt loading.

4.3.2.2 Aqueous Li-Air Batteries

As its name implies, aqueous Li-air batteries use aqueous electrolytes instead of organic solvents applied in nonaqueous Li-air batteries. Because of this difference, the mechanism of the oxygen reduction reaction is different between aqueous and nonaqueous Li-air batteries. The discharge product in an aqueous Li-air battery is normally LiOH, rather than Li_2O_2 in a nonaqueous one. As lithium is sensitive to water, a separator that can obstruct diffusion of water while maintaining high Li-ion conductivity is required in aqueous Li-air batteries. Lithium superion-conductive

glass film (LISICON) as solid-state a Li-ion conductor is usually applied as the separator. Haoshen Zhou from the National Institute of Advanced Industrial Science and
Technology in Japan has made some important contributions in aqueous Li-air batteries. Recently, Wang and Zhou [50] proposed a hybrid system combining nonaqueous electrolyte in the Li anode and aqueous electrolyte in the air electrode. They
further applied graphene in the air electrode of this hybrid Li-air battery to gain
better performance. Due to the similarity in the air electrode between this hybrid
system with traditional aqueous Li-air batteries, we ascribe this hybrid Li-air battery
to a type of nonaqueous Li-air batteries here.

In 2011, Zhou and his research group proposed graphene as a pure carbon electrocatalyst for the first time in nonaqueous Li-air batteries. The structure of the battery is shown in Figure 4.17 [51]. Graphene was prepared by chemical reduction of
GO by hydrazine hydrate, and it had a specific surface area of 342.6 m^2 g^{-1}. The
graphene-based catalyst exhibited an initial discharge voltage of 3.0 V at a current
density of 0.5 mA cm^{-2}, which was only slightly lower than the value of conventional Pt-based electrocatalyst (3.05 V), suggesting excellent electrocatalytic activity
of this pure carbon catalyst. In comparison, the carbon-black-based catalyst only
had a low discharge capacity of 2.78 V. The significant difference in the catalytic
activities between two carbon materials was mainly ascribed to the different structures of graphene and carbon black. The authors proposed the expected reaction

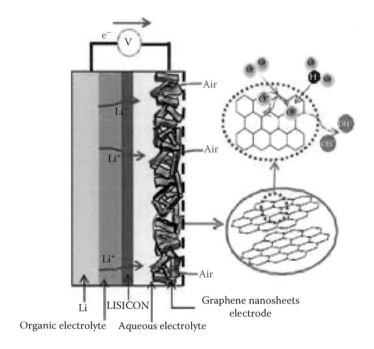

FIGURE 4.17 Structure of an aqueous Li-air battery based on graphene as an air electrode.
(Reprinted with permission from Ref. 50, 3020–3026. Copyright 2011 American Chemical
Society.)

pathway for the graphene-based air electrode, which is shown in Figure 4.17. The edge or defect sites in graphene sheets dissociated O_2 into atomic oxygen which then migrated to the surface of graphene and formed hydroxide ions by reaction with H_2O in the aqueous electrolyte. The authors also measured the cycling performance of the cell. The voltage gap between charge and discharge curves in the first cycle of the graphene-based air electrode was 0.56 V, which was relatively small for Li-air batteries. However, the gap gradually increased to ~1.2 V after 50 cycles. The rising voltage gap was probably due to the corrosion of graphene promoted by vacancies and defects in chemically reduced graphene sheets. It was also noted that the difference of the charge voltage between the first and 50th cycles was two times higher than the corresponding value of the discharge voltage, indicating that the corrosion was more serious in the charge process, probably as a result of oxidation of graphene by oxygen atoms released in the charge process. The cycling stability of the cell could be effectively enhanced by heat treatment of the graphene sheets. By calcining the chemically reduced GO sheets in flow H_2/Ar at 950°C for 30 min, the voltage gap after 50 cycles was reduced to ~0.8 V, suggesting good durability of heat-treated graphene. It could be mainly ascribed to the partial removal of defects and oxygen-functional groups and restoration of carbon framework at high temperatures, which could be confirmed by XPS spectra. Although the detailed mechanism of the ORR on graphene was not quite clear, this research showed great application potential of graphene as a high-performance metal-free catalyst in aqueous Li-air batteries.

The same research group then carried out nitrogen doping of graphene, trying to improve the catalytic activity of a graphene-based air electrode [52]. The doping process was conducted by the treatment of graphene in flowing NH_3 at high temperatures. The shift of the G band and broadening of the D band in the Raman spectra after N-doping suggested successful incorporation of N atoms in the carbon framework. XPS spectra further verified the configuration of N atoms in the N-doped sample. Three types of N atoms, including pyridine, pyrrole, and graphitic N atoms were observed. Though their content varied along with the N-doping temperature, pyridine-type N always dominated. In the acidic electrolyte, the discharge voltage for the N-doped graphene was 3.6 V, which was 0.15 V lower than Pt catalyst but 0.30 V higher than pristine graphene, suggesting improved catalytic activity by N doping. It was also observed that the discharge voltage plateau decreased when the doping temperature decreased. The voltage of the sample annealed at 850°C was 0.3 V higher than that of the sample annealed at 600°C. The enhanced activity along with the increase of the annealing temperature was due to the higher N content, especially the pyridine-type N content at higher temperatures. Besides heteroatom doping, graphene-supported nonnoble metal catalyst was also reported by this research group [53]. They prepared Fe phthalocynine/graphene composite as the air electrode for Li-air batteries. Compared with Fe phthalocynine combined with carbon nanotube and carbon black, graphene-supported catalyst exhibited higher discharge voltage and more stable cycling performance.

Zhou and coworkers proposed a novel air electrode by simple pencil-drawing on the LISICON electrolyte [54]. In contrast to traditional aqueous Li-air batteries, the air electrode reported in this paper was a complete solid with no liquid electrolyte. Because of the graphite-rich ingredients in the pencil lead, the trace on the LISICON film was composed of mechanically exfoliated thin graphite nanosheets. Some multilayer graphene sheets were detected by electron microscopy. The intensity ratio of the D band to the G band in the Raman spectrum of the pencil-trace was significantly higher than that of graphite, also indicating the formation of multilayered graphene sheets during the drawing process. At a current density of 0.1 A g^{-1}, the cell had a discharge plateau in the voltage range between 2.8 and 2.5 V and a discharge capacity of 950 mAh g^{-1} to an end voltage of 2.0 V. The cell also exhibited relatively steady capacity within 15 cycles. Although the electrochemical performance of the cell was still far below the demands for practical applications, this report provided a novel and simple preparation method of the air electrode, which might also bring about the design and fabrication of all solid state Li-air batteries.

REFERENCES

1. Wang, J.-Z., Lu, L., Choucair, M. et al. 2011. Sulfur-graphene composite for rechargeable lithium batteries. *J. Power Sources* 196:7030–7034.
2. Wang, H., Yang, Y., Liang, Y. et al. 2011. Graphene-wrapped sulfur particles as a rechargeable lithium–sulfur battery cathode material with high capacity and cycling stability. *Nano Lett.* 11:2644–2647.
3. Ji, L., Rao, M., Zheng, H. et al. 2011. Graphene oxide as a sulfur immobilizer in high performance lithium/sulfur cells. *J. Am. Chem. Soc.* 133:18522–18525.
4. Evers, S. and Nazar, L. F. 2012. Graphene-enveloped sulfur in a one pot reaction: A cathode with good Coulombic efficiency and high practical sulfur content. *Chem. Commun.* 48:1233–1235.
5. Zhang, F.-F., Zhang, X.-B., Dong, Y.-H. et al. 2012. Facile and effective synthesis of reduced graphene oxide encapsulated sulfur via oil/water system for high performance lithium sulfur cells. *J. Mater. Chem.* 22:11452–11454.
6. Wang, Y.-X., Huang, L., Sun, L.-C. et al. 2012. Facile synthesis of a interleaved expanded graphite-embedded sulphur nanocomposite as cathode of Li–S batteries with excellent lithium storage performance. *J. Mater. Chem.* 22:4744–4750.
7. Li, N., Zheng, M., Lu, H. et al. 2012. High-rate lithium–sulfur batteries promoted by reduced graphene oxide coating. *Chem. Commun.* 48:4106–4108.
8. Park, M.-S., Yu, J.-S., Kim, K. J. et al. 2012. One-step synthesis of a sulfur-impregnated graphene cathode for lithium–sulfur batteries. *Phys. Chem. Chem. Phys.* 14:6796–6804.
9. Wei, Z. K., Chen, J. J., Qin, L. L. et al. 2012. Two-step hydrothermal method for synthesis of sulfur-graphene hybrid and its application in lithium sulfur batteries. *J. Electrochem. Soc.* 159:A1236–A1239.
10. Rong, J., Ge, M., Fang, X. et al. 2014. Solution ionic strength engineering as a generic strategy to coat graphene oxide (GO) on various functional particles and its application in high-performance lithium–sulfur (Li–S) batteries. *Nano Lett.* 14:473–479.
11. Xu, H., Deng, Y., Shi, Z. et al. 2013. Graphene-encapsulated sulfur (GES) composites with a core–shell structure as superior cathode materials for lithium–sulfur batteries. *J. Mater. Chem. A* 1:15142–15149.
12. Zhao, H., Peng, Z., Wang, W. et al. 2014. Reduced graphene oxide with ultrahigh conductivity as carbon coating layer for high performance sulfur@reduced graphene oxide cathode. *J. Power Sources* 245:529–536.
13. Zhou, G. M., Yin, L. C., Wang, D. W. et al. 2013. Fibrous hybrid of graphene and sulfur nanocrystals for high-performance lithium-sulfur batteries. *ACS Nano* 7:5367–5375.
14. Huang, X., Sun, B., Li, K. et al. 2013. Mesoporous graphene paper immobilised sulfur as a flexible electrode for lithium–sulfur batteries. *J. Mater. Chem. A* 1:13484–13489.

15. Ding, B., Yuan, C., Shen, L. et al. 2013. Chemically tailoring the nanostructure of graphene nanosheets to confine sulfur for high-performance lithium-sulfur batteries. *J. Mater. Chem. A* 1:1096–1101.
16. Zhang, L., Ji, L., Glans, P.-A. et al. 2012. Electronic structure and chemical bonding of a graphene oxide–sulfur nanocomposite for use in superior performance lithium–sulfur cells. *Phys. Chem. Chem. Phys.* 14:13670–13675.
17. Chen, R., Zhao, T., Lu, J. et al. 2013. Graphene-based three-dimensional hierarchical sandwich-type architecture for high-performance Li/S batteries. *Nano Lett.* 13:4642–4649.
18. Lu, S., Cheng, Y., Wu, X. et al. 2013. Significantly improved long-cycle stability in high-rate Li–S batteries enabled by coaxial graphene wrapping over sulfur-coated carbon nanofibers. *Nano Lett.* 13:2485–2489.
19. Zhao, X. Y., Tu, J. P., Lu, Y. et al. 2013. Graphene-coated mesoporous carbon/sulfur cathode with enhanced cycling stability. *Electrochim. Acta* 113:256–262.
20. Zhou, X., Xie, J., Yang, J. et al. 2013. Improving the performance of lithium–sulfur batteries by graphene coating. *J. Power Sources* 243:993–1000.
21. Yin, L., Wang, J., Lin, F. et al. 2012. Polyacrylonitrile/graphene composite as a precursor to a sulfur-based cathode material for high-rate rechargeable Li–S batteries. *Energy Environ. Sci.* 5:6966–6972.
22. Yin, L., Wang, J., Yu, X. et al. 2012. Dual-mode sulfur-based cathode materials for rechargeable Li–S batteries. *Chem. Commun.* 48:7868–7870.
23. Bao, W., Zhang, Z., Qu, Y. et al. 2014. Confine sulfur in mesoporous metal–organic framework@reduced graphene oxide for lithium sulfur battery. *J. Alloys Comp.* 582:334–340.
24. Zhou, W. D., Chen, H., Yu, Y. C. et al. 2013. Amylopectin wrapped graphene oxide/sulfur for improved cyclability of lithium-sulfur battery. *ACS Nano* 7:8801–8808.
25. Cao, Y., Li, X., Aksay, I. A. et al. 2011. Sandwich-type functionalized graphene sheet-sulfur nanocomposite for rechargeable lithium batteries. *Phys. Chem. Chem. Phys.* 13:7660–7665.
26. Zhang, Y., Zhao, Y., Konarov, A. et al. 2013. A novel nano-sulfur/polypyrrole/graphene nanocomposite cathode with a dual-layered structure for lithium rechargeable batteries. *J. Power Sources* 241:517–521.
27. Wang, W., Li, G. C., Wang, Q. et al. 2013. Sulfur-polypyrrole/graphene multi-composites as cathode for lithium-sulfur battery. *J. Electrochem. Soc.* 160:A805–A810.
28. Li, J., Li, K., Li, M. et al. 2014. A sulfur–polyacrylonitrile/graphene composite cathode for lithium batteries with excellent cyclability. *J. Power Sources* 252:107–112.
29. Wang, X., Wang, Z. and Chen, L. 2013. Reduced graphene oxide film as a shuttle-inhibiting interlayer in a lithium–sulfur battery. *J. Power Sources* 242:65–69.
30. Wang, L., He, X., Li, J. et al. 2013. Graphene-coated plastic film as current collector for lithium/sulfur batteries. *J. Power Sources* 239:623–627.
31. Li, Y., Wang, J., Li, X. et al. 2011. Superior energy capacity of graphene nanosheets for a nonaqueous lithium-oxygen battery. *Chem. Commun.* 47:9438–9440.
32. Sun, B., Wang, B., Su, D. et al. 2012. Graphene nanosheets as cathode catalysts for lithium-air batteries with an enhanced electrochemical performance. *Carbon* 50:727–733.
33. Xiao, J., Mei, D., Li, X. et al. 2011. Hierarchically porous graphene as a lithium–air battery electrode. *Nano Lett.* 11:5071–5078.
34. Wang, Z.-L., Xu, D., Xu, J.-J. et al. 2012. Graphene oxide gel-derived, free-standing, hierarchically porous carbon for high-capacity and high-rate rechargeable Li-O$_2$ batteries. *Adv. Func. Mater.* 22:3699–3705.
35. Li, Y., Wang, J., Li, X. et al. 2012. Nitrogen-doped graphene nanosheets as cathode materials with excellent electrocatalytic activity for high capacity lithium-oxygen batteries. *Electrochem. Commun.* 18:12–15.

36. Wu, G., Mack, N. H., Gao, W. et al. 2012. Nitrogen doped graphene-rich catalysts derived from heteroatom polymers for oxygen reduction in nonaqueous lithium-O_2 battery cathodes. *ACS Nano* 6:9764–9776.
37. Higgins, D., Chen, Z., Lee, D. U. et al. 2013. Activated and nitrogen-doped exfoliated graphene as air electrodes for metal–air battery applications. *J. Mater. Chem. A* 1:2639–2645.
38. Yan, H. J., Xu, B., Shi, S. Q. et al. 2012. First-principles study of the oxygen adsorption and dissociation on graphene and nitrogen doped graphene for Li-air batteries. *J. Appl. Phys.* 112:104316.
39. Li, Y., Wang, J., Li, X. et al. 2012. Discharge product morphology and increased charge performance of lithium–oxygen batteries with graphene nanosheet electrodes: The effect of sulphur doping. *J. Mater. Chem.* 22:20170–20174.
40. Cao, Y., Wei, Z., He, J. et al. 2012. α-MnO_2 nanorods grown in situ on graphene as catalysts for Li–O_2 batteries with excellent electrochemical performance. *Energy Environ. Sci.* 5:9765–9768.
41. Yu, Y., Zhang, B., He, Y.-B. et al. 2013. Mechanisms of capacity degradation in reduced graphene oxide/α-MnO_2 nanorod composite cathodes of Li–air batteries. *J. Mater. Chem. A* 1:1163.
42. Yang, Y., Shi, M., Li, Y. S. et al. 2012. MnO_2-graphene composite air electrode for rechargeable Li-air batteries. *J. Electrochem. Soc.* 159:A1917–A1921.
43. Park, H. W., Lee, D. U., Nazar, L. F. et al. 2012. Oxygen reduction reaction using MnO_2 nanotubes/nitrogen-doped exfoliated graphene hybrid catalyst for Li-O_2 battery applications. *J. Electrochem. Soc.* 160:A344–A350.
44. Wang, H., Yang, Y., Liang, Y. et al. 2012. Rechargeable Li–O_2 batteries with a covalently coupled $MnCo_2O_4$–graphene hybrid as an oxygen cathode catalyst. *Energy Environ. Sci.* 5:7931–7935.
45. Zhang, W., Zeng, Y., Xu, C. et al. 2012. Fe_2O_3 nanocluster-decorated graphene as O_2 electrode for high energy Li–O_2 batteries. *RSC Adv.* 2:8508–8514.
46. Lim, H.-D., Gwon, H., Kim, H. et al. 2013. Mechanism of Co_3O_4/graphene catalytic activity in Li–O_2 batteries using carbonate based electrolytes. *Electrochim. Acta* 90:63–70.
47. Ahn, C.-H., Kalubarme, R. S., Kim, Y.-H. et al. 2014. Graphene/doped ceria nano-blend for catalytic oxygen reduction in non-aqueous lithium-oxygen batteries. *Electrochim. Acta* 117:18–25.
48. Jung, H. G., Jeong, Y. S., Park, J. B. et al. 2013. Ruthenium-based electrocatalysts supported on reduced graphene oxide for lithium-air batteries. *ACS Nano* 7:3532–3539.
49. Yang, Y., Shi, M., Zhou, Q.-F. et al. 2012. Platinum nanoparticle–graphene hybrids synthesized by liquid phase pulsed laser ablation as cathode catalysts for Li-air batteries. *Electrochem. Commun.* 20:11–14.
50. Wang, Y. and Zhou, H. 2010. A lithium-air battery with a potential to continuously reduce O_2 from air for delivering energy. *J. Power Sources* 195:358–361.
51. Yoo, E. and Zhou, H. S. 2011. Li-air rechargeable battery based on metal-free graphene nanosheet catalysts. *ACS Nano* 5:3020 3026.
52. Yoo, E., Nakamura, J. and Zhou, H. 2012. N-doped graphene nanosheets for Li–air fuel cells under acidic conditions. *Energy Environ. Sci.* 5:6928–6932.
53. Yoo, E. and Zhou, H. 2013. Fe phthalocyanine supported by graphene nanosheet as catalyst in Li–air battery with the hybrid electrolyte. *J. Power Sources* 244:429–434.
54. Wang, Y. and Zhou, H. 2011. To draw an air electrode of a Li–air battery by pencil. *Energy Environ. Sci.* 4:1704–1707.

5 Applications of Graphene in Supercapacitors

Chao Zheng, Xufeng Zhou, Hailiang Cao, and Zhaoping Liu

CONTENTS

5.1   Introduction 171
5.2   Graphene-Based EDLCs.. 173
    5.2.1   Electrode Materials Based on Graphene 174
    5.2.2   Electrode Materials Based on Graphene Hybrid Composites 177
    5.2.3   Electrode Materials Based on Porous Graphene 180
    5.2.4   Electrode Materials Based on GNRs... 183
    5.2.5   Effect of the Electrolyte.. 186
5.3   Graphene-Based Electrode Materials with Pseudocapacitive Properties..... 190
    5.3.1   Electrode Materials Based on Graphene-Metal-Oxide Composites... 190
    5.3.2   Electrode Materials Based on Graphene-Conductive Polymers 193
    5.3.3   Electrode Materials Based on Graphene with Heteroatoms in
          Carbon Network.. 196
5.4   Graphene-Based Hybrid Supercapacitor 199
    5.4.1   Hybrid Supercapacitor in Aqueous Electrolyte 200
    5.4.2   Asymmetric Supercapacitor in Organic Electrolyte.......................203
5.5   Summary and Perspectives..206
Acknowledgments..207
References...208

5.1   INTRODUCTION

Supercapacitors—also called electrochemical capacitors—are a new type of electrochemical energy storage system applied for harvesting energy and delivering high power in a short time. Their main energy storage mechanism is based on charging an electrical double-layer (EDL) at the electrode-electrolyte interface of high surface area electrode materials. They have attracted attention for a variety of applications, especially in hybrid systems combining with batteries and fuel cells, because of their high power density, excellent cyclic stability, and rapid response to external loading on a powertrain [1–4]. However, their main disadvantage is the relatively low energy density (5–6 Wh/kg based on activated carbon [AC]), which is significantly lower than that of a lithium ion rechargeable battery (~150 Wh/kg). Numerous efforts have been made to resolve this problem. In these works, various types and forms of porous carbon materials such as AC [5–9], aerogels [10–14], CNTs [15–22], mesoporous carbon [23–28], and carbide-derived carbons [29–33] have been used as electrodes in EDLCs. However, the reported results are still insufficient for practical applications. As we know, several factors of carbon materials such as specific surface area (SSA), pore size distribution, electrical conductivity, and surface wettability are crucial for electrode performance. Generally, although porous carbon materials have high SSA, the low conductivity restricts its application in high-power-density supercapacitors. On the other hand, although CNT materials possess excellent conductivity, their relatively low SSA affords them a low specific capacitance [16,34]. Thus, great effort is being put into developing novel carbon-based supercapacitor electrode materials with overall high performance.

Graphene [35], a two-dimensional carbon material consisting of a single-layer of sp^2 hybridized carbon atoms, has been considered as an outstanding candidate electrode material for supercapacitors due to its unique properties, such as exceptionally high specific surface area (2630 m^2/g, higher than that of CNTs and commercial AC, and major surface of graphene is exterior surface readily accessible by electrolyte), excellent electrical conductivity, and stable chemical properties [36–38]. Many works have been reported based on graphene and modified graphene for supercapacitors. Unfortunately, the EDL capacitance value measured is far lower than the theoretical one (550 F/g) provided the entire surface is fully utilized [39], mainly because graphene sheets have the inevitable tendency to restack themselves during all procedures of graphene preparation and subsequent electrode production. Moreover, the low packing density, with a value as low as ~0.005 g/mL, is another drawback of graphene. In the past, many works were interested in materials with high specific surface area, and the specific capacitance per unit weight is mostly adopted to judge the performance of the electrode materials. However, the specific capacitance per unit volume (Wh/L) is of prime importance for practical applications as the space for the power unit is always limited [2]. The volumetric capacitance is mainly affected by the packing density of electrode materials. It is worth noting that many works have focused on designing graphene-based EDLC electrode materials with high SSA, excellent electrical conductivity, and high packing density, and their results indicate that graphene-based materials are hugely favorable for their application to EDLCs.

Compared with EDLCs, pseudocapacitors store energy through a faradic process, which involves fast and reversible redox reactions between electrolyte and electroactive materials on the electrode surface [40]. The most commonly known electroactive materials are metal oxides (e.g., RuO_2 [41–45] and MnO_2 [46–50]), conductive polymer such as polyaniline (PANI) [51–55], polypyrrole (PPY) [56–61], polythiophene (PTH) [62–64], and heteroatoms doping materials (N, B, O) [8,65–70]. Generally, pseudocapacitive materials exhibit a high pseudocapacitance up to 1000 F/g, which is far higher than that of EDL capacitance. Unfortunately, they suffer from the drawbacks of a low power density (due to poor electrical conductivity) and lack of stability during cycling. Graphene, due to the large surface area and excellent conductivity, plays an important role in graphene-pseudocapacitive material composites. In addition, due to a combination of the fast charging rate of supercapacitors and the high energy density of lithium ion batteries, novel hybrid supercapacitors have been widely developed [71–75]. In these assemblies, graphene was not only used as a positive electrode material, but also acted as an excellent conductive additive for composite electrodes.

Graphene-based materials have been considered promising electrode materials for high-performance supercapacitors. In this chapter, we will summarize the progress made so far in graphene-based electrode materials by applying EDLCs, pseudocapacitors, and hybrid supercapacitors. We will also discuss the problems that exist in each form of supercapacitor and how to solve these issues in order to achieve high-performance supercapacitors.

5.2 GRAPHENE-BASED EDLCs

EDLCs are electrochemical capacitors that store the charge electrostatically using reversible adsorption of ions of the electrolyte onto active materials that are electrochemically stable and have high accessible SSA. Charge separation occurs on polarization at the electrode–electrolyte interface, producing what Helmholtz described in 1853 as the double-layer capacitance C [76]:

$$C = \frac{\varepsilon_r \varepsilon_0 A}{d} \quad \text{or} \quad \frac{C}{A} = \frac{\varepsilon_r \varepsilon_0}{d} \tag{5.1}$$

where ε_r is the electrolyte dielectric constant, ε_0 is the dielectric constant of the vacuum, d is the effective thickness of the double layer (charge separation distance), A is the electrode surface area, and the energy stored is proportional to voltage squared according to

$$E = \frac{1}{2}CU^2 \tag{5.2}$$

According to Equations 5.1 and 5.2, there are two approaches to improve the energy density. One approach is to improve the specific capacitance, and the other approach is to broaden the potential window. Obviously, the effect of improving the potential window is much more significant. It is well known that electrode materials are the key component of supercapacitors. Electrode materials with large SSA and high stability at a high-potential window are beneficial for supercapacitors to achieve high performance. Graphene, due to its exceptionally high SSA (2630 m²/g) and stable functioning at high-potential window (>4 V), has been considered as an outstanding candidate electrode material for supercapacitors. In the past few years, research based on graphene and graphene-based materials has achieved great progress.

5.2.1 Electrode Materials Based on Graphene

Graphene discussed in this section mainly refers to RGO, which is widely prepared via suspending GO in water followed by chemically reducing it with a reducing agent such as hydrazine hydrate. Ruoff and coworkers [77] first reported graphene-based EDLCs utilizing this kind of chemically modified graphene as electrode materials. Even though individual graphene sheets agglomerate onto particles of approximately 15–25 μm in diameter during the reduction process, the as-prepared graphene particles still exhibit a relatively high SSA (705 m²/g). As a result, large specific capacitance values of 135 and 99 F/g for aqueous and organic electrolytes, respectively, were achieved by these RGO materials. In addition, a low variation of specific capacitance for increasing voltage scan rates in the range of 20 to 400 mV/s due to the high conductivity and a short and equal diffusion path length of the ions in the electrolyte. It is worth noting that there is still much room for the improvement of electrical conductivity and SSA of RGO materials. Therefore, these preliminary results illustrate the exciting potential for high-performance electrical energy storage devices based on RGO materials.

Many reducing agents including hydrazine [78–80], $NaBH_4$ [81], ethylenediamine [82], hydroquinone [83], and strong alkali [84] have been developed for effective reduction of GO in various media. However, these agents are corrosive, toxic, and even explosive, thus raising serious safety and environmental concerns. Urea has been used as an environmentally friendly reductant to prepare graphene because the decomposition of urea not only creates volatile species that mechanically expand GO, but also produces volatile reducing gases that can promote the removal of surface oxygen groups [85]. Zhao and coworkers [85] reported a simple solution-processable approach to reduce GO to stable RGO using urea as a mild reducing agent. It has been found that most of the oxygen-functional groups of pristine GO have been removed after reduction with urea for 30 h. Compared with pristine GO, the as-prepared RGO-30 has a much higher SSA (590 m²/g) (Figure 5.1) and electronic conductivity (43 S/m). As a result, the RGO-30 electrode exhibited a gravimetric capacitance of 255 F/g and a volumetric capacitance of 196 F/g, respectively. Moreover, the RGO also displayed an excellent cyclability. Apart from the chemical reduction mentioned above, a mild solvothermal method was also employed to reduce GO to form RGO supercapacitor electrodes. Ruoff and coworkers [86] found that propylene carbonate (PC) is an excellent solvent for achieving exfoliated GO dispersions, and they successfully fabricated stable suspensions of GO sheets in PC via sonication. Furthermore, they found that thermally treating the suspension at 150°C can remove a significant fraction of the oxygen-functional groups and yield an electrically conductive film composed of such RGO platelets with a conductivity value as high as 5230 S/m. Since commercial ultracapacitors commonly use tetraethylammonium tetrafluoroborate ($TEABF_4$) in PC for the electrolye, $TEABF_4$ could be simply added to this PC/RGO suspension with the resulting slurry and then used for the EDLC electrodes. A high capacitance of 112 F/g was achieved by this RGO in a PC-based electrolyte, which compares favorably with the performance of other electrode materials (80–120 F/g) using PC-based electrolytes.

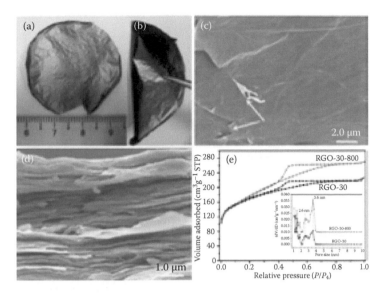

FIGURE 5.1 Photographs of RGO paper with a thickness of ~5 μm (a and b). FESEM images of the RGO papers (c and d). (e) N$_2$ adsorption isotherms and nonlocal density functional theory (NLDFT) pore size distribution (inset) of samples RGO-30 and RGO-30-800. (From Lei, Z. B., Lu, L. and Zhao, X. S., The electrocapacitive properties of graphene oxide reduced by urea. *Energy Environ. Sci.* 5:6391–6399, 2012. Reproduced by permission of The Royal Society of Chemistry.)

As mentioned in Section 5.2.1, GO can be well dispersed in aqueous solution as individual sheets; however, direct reduction of GO in solution will result in irreversibly precipitated agglomerates. It is very common that the SSA values of RGO particles are far lower than that of theoretical value (2630 m^2/g). To avoid this irreversible restacking of graphene, thermal reduction has been a popular technique to use on a large scale of RGO. Compared with chemical reduction, thermal reduction can produce few-layer graphene with less agglomeration, high SSA, and higher electrical conductivity [87–89]. Jiang and coworkers [90] prepared RGO sheets with different reduction levels by thermal reduction of GO in the temperature range of 200°C–900°C. They systematically explored the effects of interlayer spacing, oxygen content, SSA, and disorder degree on their specific capacitance. It was found that the variation of oxygen-containing groups is a main factor influencing the EDL capacitor performance. The high capacitance of 260.5 F/g at a charge/discharge current density of 0.4 A/g was obtained for the sample thermally reduced at about 200°C. In addition, microwave irradiation annealing is another convenient and rapid thermal reduction method. Ruoff and coworkers [91] prepared microwave-exfoliated and reduced GO (MEGO) by facilely and efficiently treating the GO precursor in a commercial microwave oven. Upon microwave irradiation, a large volume expansion and the as-prepared MEGO (Figure 5.2) form a wormlike morphology consisting of crumpled and curved electronically conductive graphitic sheets. Meanwhile, the MEGO powders showed a specific surface area of 463 m^2/g and a high conductivity (274 S/m). These features make it a good

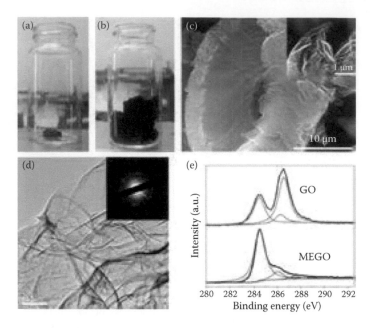

FIGURE 5.2 Optical photos of GO before (a) and after (b) treatment in a microwave oven for 1 min. (c) Typical SEM images of as-prepared MEGO by microwave irradiation with a high-magnification SEM image in the inset showing the crumpled MEGO sheets. (d) Typical TEM image of the MEGO and the corresponding electron diffraction pattern. (e) XPS Cls spectra of GO and MEGO. (Reprinted from *Carbon*, 48, Zhu, Y. W., Murali, S., Stoller, M. D. et al. Microwave assisted exfoliation and reduction of graphite oxide for ultracapacitors, 2118–2122, Copyright 2010, with permission from Elsevier.)

electrode candidate for EDLCs. As a result, EDLCs based on MEGO exhibit a specific capacitance as high as 191 F/g in KOH electrolyte. Furthermore, the simple microwave irradiation annealing process provides a promising route for the scalable and cost-effective production of graphene-based electrode materials. However, we should note that the high-temperature thermal annealing process is energy-consuming and difficult to control. Moreover, the yield is also low.

Apart from improving the reduction methods to achieve high dispensed RGO, using curved graphene sheets rather than flat ones is another effective solution to prevent the face-to-face aggregation of the graphene sheet. Fan and coworkers [92] reported a rapid, efficient, low cost, and scalable approach for the synthesis of highly corrugated graphene sheets (HCGS) by thermal reduction of GO at an elevated temperature followed by rapid cooling using liquid nitrogen. It was observed that the HCGS exhibits a porous, loose, and highly wrinkled morphology due to the violent shrinkage resulting from thermal stress (Figure 5.3) and possesses a high surface area (517.9 m²/g). As a result, the HCGS electrode displayed an ideal capacitive characteristic with the maximum specific capacitance of 349 F/g and excellent electrochemical stability. Therefore, it is highly desirable that the success in resolving the restacking problem of graphene through a simple strategy will provide considerable opportunities for the application of graphene in a buck form.

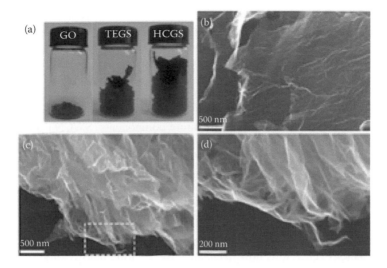

FIGURE 5.3 (a) Photographs of pristine GO and as-prepared TEGS and HCGS samples. Representative SEM images of (b) TEGS and (c and d) HCGS samples; (d) is the enlargement of the dashed frame regions in (c). (Reprinted from *Carbon*, 50, Yan, J., Liu, J. P., Fan, Z. J. et al., High-performance supercapacitor electrodes based on highly corrugated graphene sheets, 2179–2188, Copyright 2012, with permission from Elsevier.)

5.2.2 Electrode Materials Based on Graphene Hybrid Composites

Graphene has been considered as an outstanding candidate electrode material for supercapacitors due to its exceptionally high SSA (2630 m^2/g). However, the reported EDL capacitance value is far lower than the theoretical one (550 F/g) provided the entire surface is fully utilized, mainly because graphene sheets have an inevitable tendency to restack themselves during all procedures of graphene preparation and subsequent electrode production. Thus, it is a major challenge to keep graphene layers separated in order to achieve the maximum surface area for enhancing electrolyte interaction. In order to circumvent this issue, graphene-based hybrid composites have been widely reported to enhance electrochemical performance. Generally, the combination of two or more complementary materials can have a synergistic effect [93–95]. In these assemblies, the stacking of graphene sheets could be modulated by other carbon materials serving as spacers between the graphene layers, which could render the pore structure of the hybrids more accessible to the electrolyte and thus allow for higher capacitance and rate capability. Fan and coworkers [96] reported a 3-D CNT/graphene sandwich structure with CNT pillars grown in-between the graphene layers prepared by CVD. The unique structure endows the high-rate transportation of electrolyte ions and electrons throughout the electrode matrix. Nevertheless, CNT grown in sandwich structures is dominated by MWCNTs, which has a relatively low SSA. The ideal hypothesis is to prepare CNT/graphene hybrid electrode material by combination of single-walled CNT (SWCNT) and single-layer graphene. Haddon and coworkers [97] reported a hybrid structure prepared from reduced graphite oxide (rGO) and purified SWCNT that show excellent performance as supercapacitor electrode materials. The hybrid material was prepared by a simple mechanic mixture, and its morphologies (Figure 5.4) clearly show that the graphitic carbon platelets of rGO are embedded in the nanotube network, and the SWCNT bundles are coated with rGO platelets, which serve to bridge the platelets. It was found that the combination of SWCNT and rGO at a 1:1 weight ratio affords a synergistic enhancement of the specific capacitance (to 222 F/g) and energy density (to 94 Wh/kg) for supercapacitors operating with the ionic liquid, 1-butyl-3methylimidazoliu tetrafluoroborate (BMIMBF$_4$). As separate components, SWCNT and rGO have specific capacitances of 66 and 6 F/g, respectively.

Although the combination of SWCNT and graphene has achieved excellent performance, the low yields and high cost of SWCNT prevent it from being widely applied. In addition, both SWCNT and graphene have a relatively low packing density. As mentioned above, the volumetric capacitance is mainly affected by the packing density of electrode materials. Usually, it is difficult to achieve high volumetric density and high surface area simultaneously for graphene due to its unique sheet-like morphology. Therefore, how to enhance the packing density of graphene while still maintaining a high surface area is highly desirable and imperative.

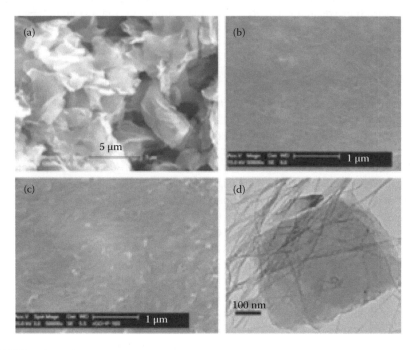

FIGURE 5.4 Microscopic characterization of the materials. SEM images of (a) rGO, (b) purified SWCNTs, and (c) hybrid material comprised of rGO and SWCNTs in a 1:1 weight ratio. (d) TEM image of the hybrid material showing few-layer graphene sheets covering a network of SWCNTs. (Jha, N., Ramesh, P., Bekyarova, E. et al.: High energy density supercapacitor based on a hybrid carbon nanotube-reduced graphite oxide architecture. *Adv. Energy Mater.* 2012. 2. 438–444. Copyright Wiley-VCH Verlag GmbH & Co. KGaA, Weinheim. Reproduced with permission.)

AC has been commercially used as supercapacitor electrode materials due to its well-developed microstructure, high SSA, relatively high packing density, and low cost. However, AC particles usually have sizes up to several tens of micrometers, which results in a long diffusion pathway for ions as well as relatively low conductivity. These disadvantages make AC unfavorable in the process of rapid charge/discharge and for the requirement of excellent cyclability. To solve this problem, our group synthesized porous graphene/AC nanosheet composite via hydrothermal carbonization and subsequent two-step activation with KOH. In the composite, a layer of porous AC is coated on graphene sheets, which not only inhibits agglomeration and increases surface area, but also enhances packing density. On the other hand, integrating graphene into AC matrix will notably increase conductivity of AC. In addition, the nanosheet composite will reduce the diffusion pathway significantly. As a result, the nanosheet composite with relatively high packing density exhibits high capacitance and excellent cycling stability. Recently, Chen and coworkers [98] reported a highly porous 3-D graphene-based bulk material with exceptional high surface area and excellent conductivity for supercapacitors. They introduced a simple and industrially scalable approach (Figure 5.5) using two standard industry steps: (1) *in situ* hydrothermal polymerization/carbonization of the mixture of cheap biomass or industry carbon sources with GO to first get the 3-D hybrid precursor materials; and (2) a chemical activation step to achieve the exceptional high surface area (up to 3523 m²/g) and excellent conductivity (up to 303 S/m). The as-prepared materials exhibited a specific capacitance of 202 F/g in 1 M tetraethylammonium

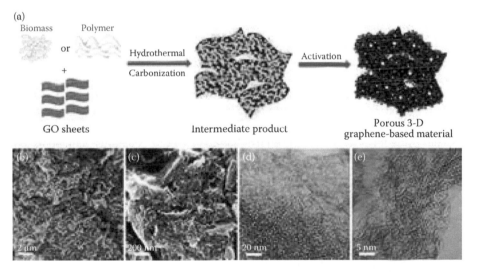

FIGURE 5.5 (a) Schematic of the simple and green process of synthesizing porous 3-D graphene-based materials. (b) Low-magnification and (c) high-resolution SEM images of products from the mixtures of PF and GO with optimized ratios, which exhibited sponge-like morphology and porous structure. (d) Low-magnification and (e) high-resolution TEM images of products from the mixtures of PF and GO with optimized ratios, which also showed a dense 3-D pore structure with a highly curved or wrinkled surface. (Reprinted by permission from Macmillan Publishers Ltd. *Sci. Rep.*, Ref. 98, copyright 2013.)

tetrafluoroborate (TEABF$_4$)/AN and 231 F/g in neat 1-ethyl-3-methyl imidazolium tetrafluoroborate (EMIMBF$_4$) electrolyte systems, respectively, at a current density of 1 A/g. The corresponding volumetric capacitance was 80 F/cm^3 and 92 F/cm^3, based on the volume of electrode films. These results are significantly higher than those of commercial AC and are the best results so far for bulk carbon materials. Furthermore, the materials have excellent cycling stability in various electrolyte systems; for instance, after 5000 times of charging/recharging at a current density of 1 A/g, the devices still keep >99% capacitance in 1 M TEABF$_4$/AN.

In view of various factors including electrochemical performance, cost, and manufacture process, porous graphene/AC composite can be considered as one of the substitutes for commercial AC. Although similar chemical activation and supercapacitor manufacturing of commercial AC were introduced into the process of synthesizing porous graphene/AC composite, hydrothermal polymerization/carbonization still hinders large-scale applications. In the future, seeking an effective and easily scaled-up process to prepare hybrid precursor materials will be very important.

5.2.3 ELECTRODE MATERIALS BASED ON POROUS GRAPHENE

As mentioned above, the performance of graphene powder is far lower than that of the expected result, which mainly is ascribed to two factors: (1) dominant single-layer graphene is very difficult to prepare, and (2) graphene sheets have an inevitable tendency to restack themselves. As a result, the SSA of graphene powder is generally in the range of 500–700 m^2/g. The development of graphene-based hybrid materials has been demonstrated as an effective approach to circumvent this issue. Another attractive method toward the improvement of graphene SSA is preparing porous graphene. Porous graphene is a collection of graphene-related materials with nanopores in the plane that has also attracted increasing attention recently. Many methods, including electron beam irradiation [99], metal-catalyzed etching [100], template synthesis [101,102], and chemical activation [103–106] have been developed. Electron beam irradiation of suspended graphene sheets is a typical way of producing porous graphene. Using focused electron beam irradiation in a TEM apparatus, nanopores and other nanoscale patterns of arbitrary design can be introduced into graphene sheets. Metal-catalyzed etching generally uses metal particles (e.g., Ni, Cr, or Cu) as chemical scissors to etch graphene. In the etching procedure, the metal particles serve as catalysts to break C-C bonds and as solvents for etched C atoms.

Ning and coworkers [101] reported a template CVD approach to produce nanomesh graphene with a high surface area and well-controlled structure (Figure 5.6). By using porous MgO layers with polygonal shapes as a template, the nanomesh graphene could be produced to have only one or two graphene layers, the polygonal morphologies similar to the porous MgO layers, and its pore size is around 10 nm. Furthermore, the porous nanomesh graphene has a large SSA up to 1654 m^2/g and total pore volume up to 2.35 cm^3/g. As result, the specific capacitance of 255 F/g obtained at 10 mV/s in 6 M KOH aqueous electrolyte, significantly, the capacitance loss was only 21% with a scan rate variation from 10 to 500 mV/s. Furthermore, the nanomesh graphene electrode showed an excellent long cycle life with 94.1% specific

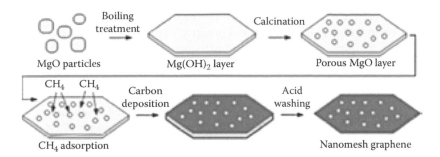

FIGURE 5.6 Illustration of the formation of polygonal nanomesh graphene. (From Ning, G. Q., Fan, Z. J., and Wang, G. et al., Gram-scale synthesis of nanomesh graphene with high surface area and its application in supercapacitor electrodes. *Chem. Commun.* 47:5976–5978, 2011. Reproduced by permission of The Royal Society of Chemistry.)

capacitance retained after 2000 cycles. This synthesis approach opens opportunities for the mass production of porous graphene at low cost and is useful for applications that need graphene as a bulk material.

Chemical activation has been widely used to obtain porous carbon-based materials for the application of supercapacitor electrodes. Chemical activation is generally made by mixing carbon precursor with chemical activating agents (KOH [107–109], $ZnCl_2$ [110,111], H_3PO_4 [112,113], etc.), followed by carbonization at 400°C–900°C. This process gives rise to porous carbons with a high SSA of over 2000 m²/g. Among the various chemical reagents, KOH is extensively used since it can result in ACs with defined micropore size distribution and very high SSA of up to 3000 m²/g. Now that chemical activation has gone far beyond the synthesizing of porous ACs, it has been successfully introduced to enhance the performance of graphene. One of the most impressive works was reported by Ruoff and coworkers [106], in which the precursor, MEGO, was chemically activated with KOH to achieve a SSA value of up to 3100 m²/g. The activation process etches the MEGO and has generated a 3-D distribution of what are referred to as mesopores. Remarkably, the activation with KOH yields a continuous 3-D network of pores of extremely small size, ranging from <1 to 10 nm. The performance of two-electrode symmetrical supercapacitor cells based on a-MEGO (SSA ~2400 m²/g) in BMIMBF₄/AN electrolyte was measured and exhibited a specific capacitance of 166 F/g at 5.7 A/g. Using a working voltage of 3.5 V, the energy density is ~70 Wh/kg for the a-MEGO in the cell. The value is four times higher than existing AC-based supercapacitors, two times higher than that reported for carbon-oxide hybrid electrochemical devices, and nearly equal to the energy density of lead acid batteries. In addition, the a-MEGO exhibited excellent cycle stability. After 10,000 constant current charge/discharge cycles at 2.5 A/g in neat BMIMBF₄ electrolyte, 97% of its capacitance was remained. Thus, by using this simple activation process already commercially demonstrated for ACs, scaled a-MEGO production for advanced energy/power electrochemical electrical energy devices might be realized in a short time.

The KOH-activated mechanism of graphene is mainly accorded to that of ACs; however, the chemical composition of graphene and AC precursor is different. In order to reveal the mechanism of KOH-activated graphene and controllably prepare the desired activated graphene, our group systematically studied the nature of graphene in the porosity development of activated graphene. It was found that the active sites (non-sp^2 hybrid carbon, including defective place and carbon attached with oxygen-functional groups) play an important role in the activation process. The precursor graphene with low crystallinity and high content of double-bond oxygen-functional group is ideal for producing highly porous activated graphene.

Although the packing density of activated graphene has improved remarkably, compared to graphene it still cannot meet the requirements for practical application. One method to overcome the low packing density of the carbon-material-based electrode is to mechanically compress porous powders for supercapacitor electrodes. It is well known that porous carbon consists of macro- and micropores, and micropores are considered to make a predominant contribution to the energy storage. Macropores are damaged but micropores are mainly remained during mechanical compression, thus, the compressed samples still remain good performance. More recently, Murali and coworkers [114] reported on the use of high pressure to compress the a-MEGO for achieving high volumetric capacitance and energy density after compression. By using 25 tons compression force, a bulk density of 0.75 g/cm^3 was obtained. The electrode exhibited a high volumetric capacitance of 110 F/cm^3 using BMIMBF$_4$/AN as the electrolyte at 3.5 V. Similarly, the volumetric energy density was increased from the uncompressed samples of 23 to 48 Wh/L for the compressed samples. Unfortunately, the rate capability was severely affected due to the ion transport channels (provided by maropores) being significantly reduced after mechanical compression. Very recently, Li and coworkers [115] reported a simple capillary compression of adaptive graphene gel films in the presence of nonvolatile liquid electrolyte to obtain porous yet densely packed graphene electrodes with high ion-accessible surface area and low ion transport resistance. The chemical-converted graphene (CCG) hydrogel films obtained by filtration of CCG dispersion were exchanged with a miscible mixture of volatile and nonvolatile liquids (sulfuric acid, EMIMBF$_4$) and were then subjected to removal of the volatile liquid by vacuum evaporation. The graphene sheets in the liquid mediated CCG (EM-CCG) films stacked in a nearly face-to-face fashion, and the packing density could be increased up to ~1.33 g/cm^3. More important, the EM-CCG film electrode still remained a highly efficient ion transport channel; due to the fluid nature of liquid electrolytes, the continuous liquid network was likely to remain within the whole film during the capillary compression process. Furthermore, as the electrolyte became integrated within the film from the start of the assembly process, there was no subsequent wettability issue for these EM-CCG films, which remained a serious problem for the dried CCG film. As a result, the supercapacitors based on EM-CCG films could obtain volumetric energy densities approaching 60 Wh/L. Additionally, the EM-CCG films exhibited excellent cycle stability; over 95% of the initial capacitance remained after a 300-h constant voltage holding at 3.5 V in a neat EMIBF$_4$ electrolyte. This means that the fabrication of CCG films and subsequent compression are essentially compatible with the traditional cost-effective paper-making process and can be readily scaled up. All these attractive features make this class of graphene materials promising for large-scale real-world applications.

5.2.4   ELECTRODE MATERIALS BASED ON GNRS

Graphene, a 2-D carbon material consisting of a single layer of sp^2-hybridized carbon atoms, has been considered as an outstanding candidate electrode material for supercapacitors due to its unique properties and high SSA. Intrinsically, the sp^2-hybridized graphene exhibits two different types of surfaces: the homogeneous and smooth hexagonal plane known as the basal plane, and the edges as opposed to the basal plane (Figure 5.7). The edge site is known to contribute a several orders higher capacitance than that of the basal plane. Theoretically, it has been reported that each carbon atom of the zigzag edge graphene has a nonbonding π-electron state that is localized in the zigzag edges and that enhances the local density of states near the Fermi energy [116], which is active to combine with other reactants. Recently, Shi and coworkers [117] quantitatively investigated the capacitance contribution of the basal plane and edge for single-layer graphene. As shown in Figure 5.8, high-quality single-layer graphene was synthesized by CVD. In order to perform the edge and the basal plane selective electrochemistry for the single-layer graphene, a nonconducting thin film was selectively coated around the graphene to fabricate the graphene edge electrode and the plane electrode, respectively. The results reveal that the graphene edge exhibits four orders of magnitude higher specific capacitance, which is much faster electron transfer rate than those of graphene basal plane. Thus, to develop edge-enriched graphene is highly desirable and imperative. GNRs (GNRs), which are dominated by edge effects, are known to behave as a quasi-1-D material with a band gap.

The properties of GNRs strongly depend on the edge types and the distance between edges (the size effect). Compare with the zigzag edge, the armchair edge is energetically more stable in chemical reactivity because of a triple covalent bond between the two open-edge carbon atoms of each edge hexagonal ring. Furthermore, the band gap of GNRs is essentially governed by the ribbon width and the edge configuration. By employing quantum approaches, Louie and coworkers [118] demonstrated that

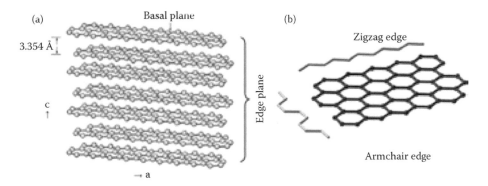

FIGURE 5.7   Schematic diagrams exhibiting (a) a 3-D graphite lattice and (b) its two-edge planes (the zigzag and armchair edges) in graphene. Edge planes are highly active due to the partial saturation of valencies. These two sites display different reactivity with reactive oxygen species, and result in different functional groups. (Reprinted from *J. Energy Chem.*, 22, Kim, Y. A., Hayashi, T., and Kim, J. H. et al., Important roles of graphene edges in carbon-based energy storage devices, 183–194, Copyright 2013, with permission from Elsevier.)

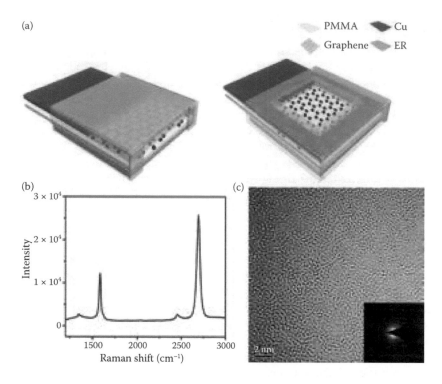

FIGURE 5.8 Configuration of plane- and edge-electrodes, and the structure of a CVD graphene sheet. (a) Schematic illustration of edge-based (left) and basal plane-based (right) electrodes. (b) Raman spectrum of a monolayer CVD graphene sheet on a SiO$_2$/Si substrate. (c) HR-TEM image of a monolayer CVD graphene sheet suspended over a microgrind; inset: a SAED pattern. (Reprinted by permission from Macmillan Publishers Ltd. *Sci. Rep.*, Ref. 117, copyright 2013.)

all GNRs with homogeneous armchair edge and zigzag edge have energy gaps that decrease nearly with increasing GNR width. Therefore, edge-enriched GNRs with small size are favorable for supercapacitors with high performance.

Cutting and unzipping CNTs into nanoribbons is an effective approach that is able to obtain large-scale production with narrow and well-defined width. Recently, numerous attempts have been dedicated to fabricate GNRs, including plasma etching [119], oxidative treatment [120], laser irradiation [121], electrochemical unzipping [122], and metal catalyzed cutting [123,124]. Wang and coworkers [125] reported oxidation with the Hummers method of CNTs and reduction with NaBH$_4$, to achieve curved graphene sheets (CGN) (Figure 5.9). It was found that the oxidation process cuts and longitudinally unzips the CNTs to form CGN, which exhibits a tubelike structure. The SSA of the obtained CGN (85 m^2/g) is much higher than that of pristine CNTs (47 m^2/g). The supercapacitor performance of the CGN was measured in 1 M KOH, 1 M H$_2$SO$_4$, and 1 M Na$_2$SO$_4$ aqueous electrolytes, respectively. A specific capacitance of as high as 256 F/g at a current of 0.3 A/g was achieved over the CGN material. Moreover, the CGN showed very impressive

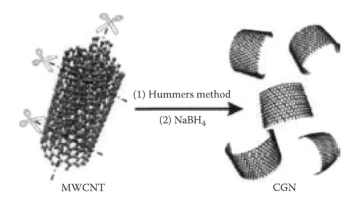

FIGURE 5.9 Preparation of CGN material by combining the Hummers method and subsequent reduction with NaBH$_4$. (Reprinted with permission from Ref. 125, 6827–6834. Copyright 2012 American Chemical Society.)

cycling stability: in 1 M KOH, the CGN electrode can keep 99% of the initial value over 5000 cycles; in 1 M H$_2$SO$_4$, the specific capacitance reaches a higher value after 5000 cycles; and in 1 M Na$_2$SO$_4$, the decay in specific capacitance after 5000 cycles is only 3%. Although the reported result of CGN cannot fulfill the demand of practical applications, the tentative research opens new approaches to synthesize edge-enriched GNRs electrode material for supercapacitors.

Up to now, large-scale one-step cutting and unzipping of CNTs has been a challenging issue, owing to the complex procedures and difficult manipulation. Furthermore, it is reasonable to expect that the GNRs derived from CNTs can exhibit much higher energy storage ability in supercapacitors. Thus, seeking an effective strategy to prepare large-scale GNRs with enriched edges and narrow size remains a major topic of interest. Recently, our group used industrially produced aligned CNTs as raw material and a mild modified Brodie method was applied to cut and unzip CNTs. The intermediate GNRs were subsequently chemically activated by using KOH to generate suitable porosity and create more edge sites (Figure 5.10). The as-prepared porous GNRs had a high SSA value of 1250 m^2/g, which is nearly 10 times as high as pristine CNTs (146 m^2/g). Using the working voltage of 3.5 V in EMIMBF$_4$ electrolyte, a specific capacitance of 130 F/g at scan rate of 1 mV/s, the energy density was calculated to be about 55.2 Wh/kg. It should be noted that although the porous GNRs do not have a relatively high SSA, they show extremely high capacitances corresponding to those of activated graphene with a SSA of 2450 m^2/g, which probably can be ascribed to the nanostructure and abundant edge sites [126].

In the next generation, broadening the potential window is one important approach to improve energy density for supercapacitors because researchers have confirmed that the energy density is dependent on the cell-voltage cubed. Nevertheless, high cell-voltage demands a stringent requirement for electrode materials. Unlike AC, which is unstable in high voltage (above 3 V), porous GNRs are considered as an outstanding candidate electrode material for supercapacitors in high cell-voltage due to their stable physicochemical properties.

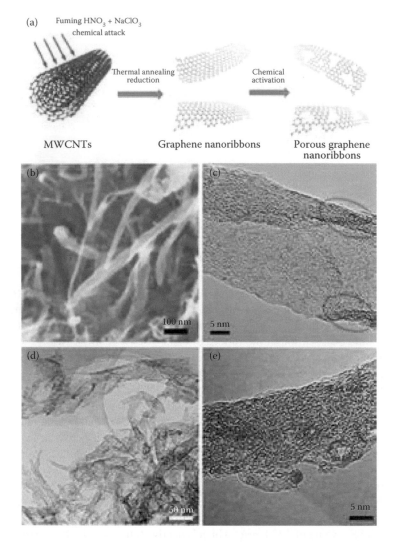

FIGURE 5.10 (a) Schematic illustration showing the process of synthesizing edge-enriched prous GNRs, (b) High-resolution SEM image of intermediate GNRs, (c) low-magnifcation, and (d) high-resolution TEM images of porous GNRs. (From Zheng, C., Zhou, X. F., Cao, H. L. et al., Edge-enriched porous graphene nanoribbons for high energy density supercapacitors. *J. Mater. Chem. A.* doi: 10.1039/C4TA00727A, 2014. Reproduced by permission of The Royal Society of Chemistry.)

5.2.5 EFFECT OF THE ELECTROLYTE

Apart from electrode materials, the electrolyte is another important component that limits the performance of supercapacitors. This has been attributed to the following facts: (1) energy density is dependent on the cell-voltage cubed, and (2) the cell voltage is limited by the electrolyte decomposition at high potentials. In order to demonstrate the dependence of energy and power on cell voltage over a wide potential range (1 to 4 V), Hata and coworkers [127] used SWCNTs as the electrode material. Unlike AC, which is unstable when the cell voltage above 3.0 V, highly graphitized SWCNT is very stable in high voltage. It was found that the energy density discharge by the SWCNT electrodes increased with the cell-voltage cubed in contrast to the conventional energy-voltage relationship ($E = 1/2 \ CV^2$) due to the linear increase of capacitance as a function of cell voltage (Figure 5.11). Both CNT and graphene have the same chemical composition; moreover, graphene has a much higher SSA than that of SWCNT. On the grounds of these results, it is believed that graphene can be considered as an outstanding candidate electrode material for supercapacitors in high cell voltage with excellent performance.

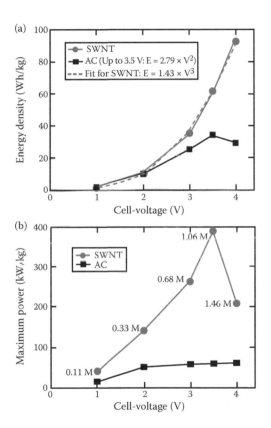

FIGURE 5.11 Energy and power performance of SWNT and AC electrodes. (a) Energy discharged by SWNT and AC electrodes. For the SWNT electrode, the energy discharged was proportional to the cell-voltage cubed. (b) Maximum power rating of SWNT and AC electrodes. For the SWNT electrode up to 3.5 V, its power performance followed the theoretical relationship; however, beyond 3.5 V, its power declined, which can be attributed to the ionic concentration of the electrolyte. (Reprinted from *Electrochem. Commun.*, 12, Izadi-Najafabadi, A., Yamada, T., Futaba, D. N. et al., Impact of cell-voltage on energy and power performance of supercapacitors with single-walled carbon nanotube electrodes, 1678–1681, Copyright 2010, with permission of Elsevier.)

Recently, numerous research efforts have been focused on the design of highly conductive, stable electrolytes with a wider voltage window. Up to now, the move from aqueous to organic electrolytes has been successful in increasing cell voltage from 1 to 2.5–2.7 V for supercapacitors. Today, the state-of-the-art is the use of organic electrolyte solutions in acetonitrile or propylene carbonate, which have a relatively wide potential window. However, organic electrolytes suffer from drawbacks involving electrolyte depletion upon charge, narrow operational temperature range, and low safety. Ionic liquids are room-temperature liquid solvent-free electrolytes, which have the tendency to replace organic electrolyte due to their wide temperature window, nonvolatility, low vapor pressure, and wide electrochemical window operating at 4–6 V. These features make it advantageous to fabricate supercapacitors with much higher energy density and to provide a much safer operation [5,128–130]. Jang and coworkers [39] reported results of a study on a mesoporous graphene structure that is ionic liquid electrolyte accessible and thus achieves an exceptionally high EDL capacitance even though ionic liquids have large molecules and high viscosity. To obtain CGNs, the GO suspension was injected into a forced conventional oven and a stream of compressed air was introduced to produce a fluidized-bed situation followed by the removal of the solvent or liquid. This CGN morphology (Figure 5.12) appears to be capable of preventing graphene sheets from closely restacking with one another when they are packed or compressed into an electrode structure, thereby maintaining a mesopore structure having a pore size in the range of 2 to 25 nm. The specific capacitances of curved graphene-based supercapacitors in an ionic liquid are typically 100–250 F/g at a high current density of 1 A/g with a discharge voltage of 4.0 V. The discharge curve is nearly a straight line, meaning a good EDL performance. Furthermore, the ionic liquid (EMIMBF$_4$) can work at a voltage up to 4.5 V, leading to an excellent energy density of 85.6 Wh/kg

(a)   (b)

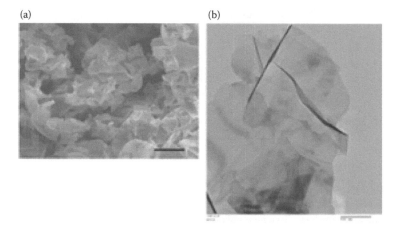

FIGURE 5.12   (a) Scanning electrode microscopy image of curved graphene sheets (scale bar 10 μm), and (b) TEM image of flat graphene sheets prepared by a conventional chemical route. This TEM image clearly shows graphene sheets overlapped together with intergraphene spacing likely <1 nm. (Reprinted with permission from Ref. 39, 4863–4868. Copyright 2010 American Chemical Society.)

at 1 A/g at room temperature (or 136 Wh/kg at 80°C). The result based on the total electrode weight of supercapacitors is comparable to that of a modern nickel metal hydride battery used in a hybrid vehicle. This breakthrough energy storage device is made possible by the high intrinsic capacitance and the exceptionally high SSA that can be readily accessed and wetted by an ionic liquid electrolyte capable of operating at a high voltage.

Another factor is the wettability. According to EDL storage mechanism, we desire the entire surface area of graphene to be electrochemically accessible by an electrolyte. Beyond the suitable pore size needed for graphene, an amphipathic graphitelike surface is just as important, which can provide high coverage that is accessible for the formation of a double layer. Usually, the reported results exhibit a lower specific capacitance than might be expected for an ideal graphene-based supercapacitor. Rao and coworkers [131] were the first to use ionic liquids (ILs) with graphene-based materials; however, the energy density is only 31.9 Wh/kg at 5 mV/s at 60°C due to the relatively low obtained specific capacitance (75 F/g). To further improve the compatibility between graphene-based materials and IL electrolyte, Suh and coworkers [132] reported a high-performance supercapacitor based on poly(ionic liquid), specifically the use of poly(1-vinyl-3-ethylimidazolium) salts bearing the bis(trifluoromethylsulfonyl)amide anion (NTf_2^- or CF_3SO_2-N-SO_2CF_3)-modified RG-O (PIL:RG-O) and an IL electrolyte (Figure 5.13). The PIL:RG-O was

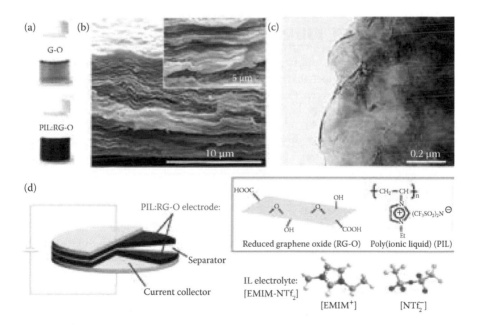

FIGURE 5.13 (a) Optical images of a suspension of a graphene oxide (G-O) in propylene carbonate (PC) and a poly(ionic liquid)-modified reduced graphene oxide (PIL:RG-O) in PC. (b) SEM and (c) TEM image of PIL:RG-O platelets. (d) Schematic diagram of the supercapacitor based on the PIL:RG-O electrodes and ionic liquid electrolyte (EMIM-NTf_2). (Reprinted with permission from Ref. 132, 436–442. Copyright 2011 American Chemical Society.)

simply produced by a reduction in propylene carbonate at an elevated temperature, which is expected to offer an advantage for supercapacitor application in that they should provide enhanced compatibility with certain IL electrolyte and improved accessibility of IL electrolyte ions into graphene electrodes. As a result, a supercapacitor assembled with such two PIL:RG-O electrodes and with EMIM-NTf$_2$ as the electrolyte exhibited a specific capacitance of 187 F/g. Nevertheless, the energy density is still low, and a maximum energy density of only 6.5 Wh/kg with a maximum power density of 2.4 kW/kg was also achieved.

It is noteworthy that although ILs seem as an attractive generation of green electrolytes, they are still very expensive. In addition, they are typically high-viscosity liquids and have low ionic conductivity at room temperature, which inevitably affect the performance of supercapacitors. Hence, further research could be carried out to design new ILs with a combination of high ionic conductivity and low molecular weight and with a wide electrochemical stability window, as well as excellent wettability for electrode materials. Currently, the reported results of supercapacitors based on IL electrolyte still fail to meet the requirements for practical applications. Its future development will be of great significance in achieving a high-performance, safe, and green energy storage system.

5.3   GRAPHENE-BASED ELECTRODE MATERIALS WITH PSEUDOCAPACITIVE PROPERTIES

In contrast to EDL capacitance, pseudocapacitance arises for thermodynamics reactions and is due to charge acceptance (Δq) and a change in potential (ΔV). The derivative $C = \mathrm{d}(\Delta q)/\mathrm{d}(\Delta V)$ corresponds to a capacitance that is referred to as the peseudocapacitance. The main difference between pseudocapacitance and EDL capacitance lies in the fact that pseudocapacitance is faradic in origin, involving fast and reversible redox reactions between the electrolyte and electroactive species on the electrode surface [133]. The most commonly known active species are RuO$_2$ [41–45], MnO$_2$ [46–50], Co$_3$O$_4$ [134–136], V$_2$O$_5$ [137–139], conductive polymers such as PANI [51–55], PPY [56–61], PTH [62–64], and heteroatoms doping materials (N, B, O) [8,65–68]. Generally, pseudocapacitive materials exhibit a high pseudocapacitance up to 1000 F/g, which is far higher than that of EDL capacitance. Unfortunately, they suffer from the drawbacks of a low power density (due to poor electrical conductivity) and lack of stability during cycling. Graphene, due to the large surface area and excellent conductivity, plays an important role in graphene-pseudocapacitive material composites. We will discuss each type of graphene-based pseudocapacitive material composites in Sections 5.3.1–5.3.3.

5.3.1   ELECTRODE MATERIALS BASED ON GRAPHENE-METAL-OXIDE COMPOSITES

Transition metal oxides are considered as promising electrode material for application in supercapacitors and have attracted much interest because their specific pseudocapacitance exceeds that of carbon materials using double-layer charge storage. The pseudocapacitive behavior of transition metal oxides is mainly ascribed to the fast and reversible electron transfer together with an electroadsorption of protons in the surface of the active materials. Among the various transition metal oxides,

MnO_2 is considered outstanding due to its environmental compatibility, low cost, and abundant availability on the earth. However, the MnO_2, the material obtained from the traditional precipitation method, has a low specific capacitance owing to its low specific surface area. Moreover, although nanoscale MnO_2 particles possess a large surface area and relatively high specific capacitance, the microstructure is easily damaged during electrochemical cycling, resulting in a relatively poor electrochemical stability. In addition, the poor electrical conductivity of MnO_2 materials also severely affect their specific capacitance. In view of these facts, graphene/MnO_2 composites were regarded as a promising strategy to improve the pseudocapacitance and cycle stability. In a hybrid system, graphene can provide a large surface area to support numerous MnO_2 nanoparticles as well as improve their conductivity; thus, the combination of graphene and MnO_2 will result in a synergistic effect.

Fan and coworkers [48] reported a rapid and facile method to prepare graphene-MnO_2 composites as electrode materials by microwave irradiation. For this composite, graphene nanosheets serve mainly as a highly conductive support that can also provide a large surface for the deposition of nanoscale MnO_2 particles (about 5–10 nm). The excellent interfacial contact and increased contact area between MnO_2 and graphene can significantly promote the electrical conductivity of the electrode due to the high electrical conductivity of graphene. Also, the easy surface accessibility of the composite by the electrolyte and the improved electrochemical utilization of MnO_2 that result from the small particle size and high surface area of the oxides could provide both the high reversible pseudocapacity and excellent capacitive retention ratio at a high charge-discharge rate. As a result, the cyclic voltammogram for graphene-MnO_2 composite electrodes used in this work retained a relatively rectangular shape in large-scale scan rates (Figure 5.14), whereas the CV curves of the carbon-MnO_2 composite usually exhibited a big distortion at a very high scan rate of 500 mV/s in previous reports. The graphene-MnO_2 composite containing 78% of MnO_2 exhibited the maximum specific capacitance of 310 F/g at 2 mV/s in 1 M Na_2SO_4 aqueous solution, and the retention ratios were still 88% and 74% at 100 and 500 mV/s, respectively.

Apart from the common composites of graphene/MnO_2 nanoparticles, nanowires [140,141], nanosheets [142,143], and flowerlike [144,145], needlelike [146], hollow sphere morphologies [47,147,148], and the composites with textile, paper, and multilayer morphologies, have all been used to enhance the capacitance of hybrid electrodes. Among them, hollow micro/nanostructure materials with a large surface area offer particular advantages to supercapacitors due to their large SSA and easy transport of species. Wu and coworkers [149] first prepared RGO-MnO_2 hollow sphere (HS) hybrid electrode materials using a solution-based ultrasonic coassembly process. GO, MnO_2 HS, and acetylene black (AB) (as an electrical conductor) were dispersed using an ultrasonic machine. They were then assembled into hybrid structures by hydrogen bonding and electrostatic interactions. After vacuum thermal treatment, GO was reduced to RGO, forming RGO-MnO_2 HS hybrid materials (Figure 5.15). Unlike other traditional techniques and active materials, this process does not suffer from the self-aggregation of RGO sheets. It facilitates excellent dispersion of active materials and electrical conductors. The as-fabricate active materials possess a porous structure for electrolyte assessment, which enhances the specific capacitance and energy density. As a result, the highest specific capacitance and energy density

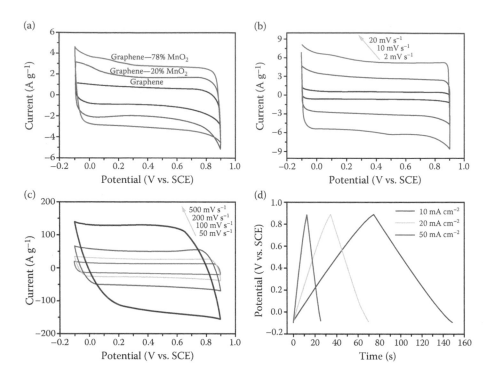

FIGURE 5.14 (a) CV curves of graphene and graphene-MnO$_2$ composites at 10 mV/s. (b and c) CV curves of graphene-78% MnO$_2$ composite at different scan rates of 2, 10, 20, 50, 100, 200, and 500 mV/s. (d) Galvanostatic charge/discharge curves of graphene-78% MnO$_2$ composite at different current densities of 10, 20, and 50 mA/cm^2 in 1 M Na$_2$SO$_4$ solution. (Reprinted from *Carbon*, 48, Yan, J., Fan, Z. J., Wei, T. et al., Fast and reversible surface redox reaction of graphene-MnO$_2$ composite as supercapacitor electrodes, 3825–3833, Copyright 2010, with permission from Elsevier.)

of as-obtained composite was found to reach 578 F/g and 69.8 Wh/kg, respectively, at a current density of 0.5 A/g. These values are far larger than those of previous graphene-MnO$_2$-based hybrid electrochemical capacitors. Moreover, the composite retains at about 83% of its original capacitance after 1000 cycles.

To date, utilization of graphene/nano-MnO$_2$ hybrid composites has successfully overcome the relatively low specific capacitance and poor cycle stability of buck MnO$_2$ electrode materials. However, it is necessary to mention that the nano-MnO$_2$ loading in graphene-based hybrid materials generally is limited, which would seriously affect the specific capacitance calculated based on the total mass of hybrid materials. This is a quite widely spread phenomenon in carbon-based pseudocapacitive composites; usually, high pseudocapacitive materials loading will result in a low utilization efficiency and poor stability for pseudocapacitive materials. In addition, it has been demonstrated that MnO$_2$ is not an optimal material for both positive and negative electrodes in symmetric capacitors. The performance of capacitors based on MnO$_2$ is limited by the two irreversible reactions Mn (IV) to Mn (II) at the negative electrode and Mn (IV) to Mn (VII) at the positive electrode. Therefore, the voltage

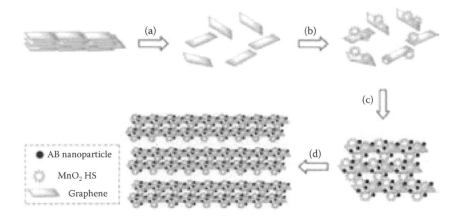

FIGURE 5.15 Schematic of the fabrication of RGO-MnO$_2$ HS hybrid ECs. (a) Exfoliation of graphite oxide into GO sheets by ultrasonic dispersion. (b) Addition of MnO$_2$ HS suspension and interaction between GO sheets and MnO$_2$ HS through hydrogen bonding. (c) Addition of AB suspension and interaction between GO-MnO$_2$ HS and AB to form GO-MnO$_2$ HS-AB hybrid via electrostatic interaction. (d) Centrifugation and vacuum drying at room temperature for 12 h, and vacuum heat treatment at 150°C for 1 h to produce RGO-MnO$_2$ HS hybrid materials. (From Chen, H., Zhou, S. X., and Chen, M. et al., Reduced graphene oxide-MnO$_2$ hollow sphere hybrid nanostructures as high-performance electrochemical capacitors. *J. Mater. Chem.* 22:25207–25216, 2012. Reproduced by permission of The Royal Society of Chemistry.)

window of MnO$_2$ symmetric capacitors cannot exceed 0.6 V due to electrode dissolution [150]. Fortunately, MnO$_2$ used as positive electrodes work up to potentials of 1.2 V versus normal hydrogen electrode (NHE) due to the high oxygen overpotential. Thus, graphene-MnO$_2$ hybrid composites were widely applied in asymmetric supercapacitors in aqueous electrolytes, and their voltage window can reach to 2.0 V [150]. Hybrid supercapacitors using graphene/MnO$_2$ composite as positive electrode in aqueous electrolytes will be discussed in more detail in Section 5.4.1.

5.3.2 Electrode Materials Based on Graphene-Conductive Polymers

Similar to transition metal oxides, conductive polymers, such as PANI, PPY, PTH, and their derivatives, have been shown to display high capacitances. Among these polymers, PANI is considered the most promising material because of its high capacitance characteristics, low cost, and ease of preparation. However, like other conductive polymers, PANI also exhibits poor stabilities during the charge/discharge process. To overcome this problem, carbon-based materials, such as AC, mesoporous carbon, and CNTs have been utilized to combine with PANI to prepare composite electrode materials. Consequently, compared with a single material, composite electrode materials exhibit high capacitances and improved stability due to the synergetic effect. Nevertheless, the obtained results are still insufficient for the desired target. Graphene has been widely applied for preparing composites with PANI to be used as supercapacitor electrodes owing to its many excellent and unique features.

GO contains a large amount of oxygen-functional groups, so it is very easy to dope into PANI during the liquid polymerization process. Wang and coworkers [151] reported a simple process to synthesis the nanocomposites of GO-doped PANI via *in situ* polymerization in the presence of GO and monomer. Although the mass ration of aniline/GO is 100/1, the obtained composite exhibited a high specific capacitance of 531 F/g in the potential range from 0 to 0.45 V at 200 mA/g by charge/discharge analysis compared with 216 F/g of individual PANI, indicating the synergistic effect between GO and PANI. The conductivity of as-prepared composite is only 10 S/cm, which is much higher than those of GO (~10^{-3} S/cm) and pure PANI (2 S/cm). This is mainly ascribed to the insulation of GO. For this reason, GO is not an appropriate choice to improve the electrochemical performance of PANI, whereas highly conductive graphene is much more favorable than GO to be doped into PANI composites. Wu and coworkers [55] prepared graphene and PANI nanofiber composites by *in situ* polymerization of aniline monomer in the presence of GO under acid conditions, and the GO was subsequently reduced by using hydrazine. Reoxidation and reprotonation of the reduced PANI followed, resulting in graphene-PANI nanocomposites (Figure 5.16). They found that the composites that contained 80% of GO showed the highest specific capacitance of 480 F/g at a current density of 0.1 A/g. In addition, when the current density was increased up to 0.5 A/g and even 1 A/g, the specific capacitances still remained at a high level above 200 F/g without a significant decrease on charge/discharge

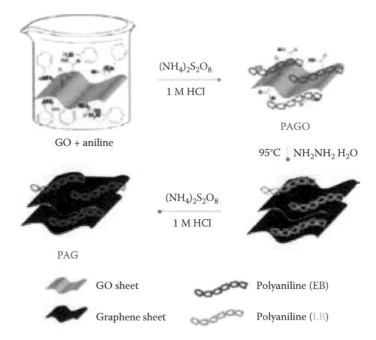

$(NH_4)_2S_2O_8$

1 M HCl

PAGO

GO + aniline

95°C $NH_2NH_2 H_2O$

$(NH_4)_2S_2O_8$

1 M HCl

PAG

GO sheet   Polyaniline (EB)

Graphene sheet   Polyaniline (LB)

FIGURE 5.16 Illustration of the process for preparation of graphene-PANI composite. (Reprinted with permission from Ref. 55, 1392–1401. Copyright 2010 The American Chemical Society.)

cycling. While the BET surface areas of all the composites were rather low (i.e., 4.3–20.2 m²/g), the much higher specific capacitance of this graphene-PANI composite compared with that of pure reduced graphene can be mainly ascribed to the pseudocapacitance from the PANI nanofibers in the composite. Furthermore, over 70% of the original capacitance was retained after over 1000 cycles, indicating that this electrode material has good cycling stability. This stability is mainly due to the small amount of PANI in the composites.

On the other hand, flexible paper with graphene sheet or GO sheet as the sole building block has already been fabricated by flow-direction assembly. Graphene paper presents excellent tensile modulus up to 35 GPa and a room temperature electrical conductivity of 7200 S/m [79]. These features make it possible to prepare flexible graphene-PANI composite films. A freestanding binder-free electrode with favorable mechanical strength and large capacitance is a vital component of a flexible supercapacitor. It is worth noting that the flexible graphene-PANI composite films combine the large pseudocapacitance of PANI and the excellent mechanical properties of graphene. Cheng and coworkers [52] prepared a graphene/polyaniline composite paper (GPCP) by *in situ* anodic electropolymerization (AEP) of aniline monomers into a PANI film on graphene paper (G-paper). The flexible black G-paper (Figure 5.17) was prepared by directional flow-guided assembly. GPCP retains the layer-by-layer structure of the G-paper and exhibits an improved mechanical tensile strength (by 43%) and electrical conductivity. As a result, the greatest gravimetric and volumetric capacitances of the GPCP electrode reach 233 F/g and 135 F/cm³, respectively, much larger than those of the G-paper (147 F/g and 64 F/cm³). Furthermore, the GPCP flexible supercapacitors also exhibit good cycling stability. These intriguing features make it quite a promising material as a freestanding electrode for flexible supercapacitors.

Apart from the use of PANI for graphene-conductive polymer composites, other graphene-polymer composites have also been explored for electrodes in supercapacitor applications. For example, polypyrrole (PPy) has also been used as the conducting polymer for supercapacitor electrodes, owing to its high energy-storage capacity,

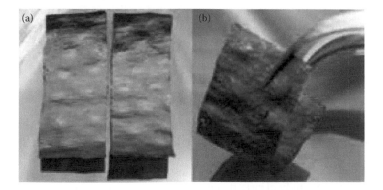

FIGURE 5.17 Digital camera images of (a) two freestanding G-papers (30 mm × 10 mm) and (b) a flexible G-paper. (Reprinted with permission from Ref. 52, 1745–1752. Copyright 2009 American Chemical Society.)

good electrical conductivity, ease of low-cost synthesis, and environmental stability. However like PANI, the achievement of high specific capacitance is only possible with effective electrolyte transport to the active sites for enhanced faradic charge transfer reactions. It is necessary to design nanoarchitecture involving PPy and graphene composites. Drzal and coworkers [57] aimed to integrate the polymerized nanostructure PPy with graphene nanosheets in a direction self-assembly approach governed by the large van der Waals force of attraction between the graphene basal plane and the π-conjugated polymer in a layered composite structure. A multilayered nanostructure of graphene nanosheets and PPy nanowires was prepared by taking advantage of capillary force driven self-assembly of the layer of PPy nanowires and graphene nanosheets. The multilayer composite electrode exhibits a high specific capacitance of ~165 F/g with a nearly ideal rectangular cyclic voltammogram at increasing voltage scanning rates and high electrochemical cyclic stability.

Graphene-conductive polymer composites offer many advantages as supercapacitors. They are flexible, highly conductive, and generally exhibit high specific capacitance, while being able to deliver energy at a relatively rapid rate. However, supercapacitors based on graphene-conductive polymer composites still have a lower cycle life than those based on carbon materials, mainly due to the volume charge of conductive polymers during anions or cations doped or undoped. Although the presence of graphene in the composite significantly improved the cyclic stability of conductive polymers, the results at present are far from satisfactory. On the other hand, similar to metal oxides, the relatively low loading of active materials in the composite is another drawback of graphene-conductive polymers composites. In the future, how to further improve the cyclic stability and active material loading is the problem that must be solved first and foremost.

5.3.3 Electrode Materials Based on Graphene with Heteroatoms in Carbon Network

Apart from utilizing the combination of graphene and pseudocapacitive materials mentioned above, including conductive polymers and metal oxide composites, another important strategy to improve graphene capacitance is doping heteroatoms, including nitrogen, boron, oxygen, and phosphorus, which can have reversible pseudocapacitance and enhance the electronic conductivity. To date, however, the exact mechanisms remain ambiguous. Recent experimental and theoretical work have shown that the total capacitance (C_T) of a doped graphene-based electrode strongly depends on the relative contributions from both the double-layer capacitance (C_D) and electrode quantum capacitance (C_Q). The C_Q of low-dimensional materials such as graphene is proportional to the electronic density of states, which can be altered from chemical doping. Very recently, Hwang and coworkers [152] investigated the influence of N-doping graphene on the interfacial capacitance (C_T) using combined density functional theory and classical molecular dynamics calculations. The computational study clearly demonstrates that nitrogen doping can lead to significant enhancement in the electrode capacitance as a result of electronic structure modification while there is virtually no charge in the double-layer capacitance.

This finding may suggest that other structural and/or chemical modification to graphene-based electrodes could significantly contribute to enhance the performance of supercapacitors.

Choi and coworkers [153] doped nitrogen into the graphene basal planes by a nitrogen plasma process. The supercapacitor performance of the nitrogen-doped graphene (NG) in organic electrolyte was thus measured and yielded a specific capacitance of 281 F/g, which was four times higher compared with pristine graphene (68 F/g). Power and energy densities were achieved up to ~8 × 10^2 kW/kg and ~48 Wh/kg, respectively, in 1 M TEABF$_4$. Moreover, the NG exhibited excellent cycling stability; after 10,000 cycles 99.8% of its initial capacitance was still retained (Figure 5.18), which suggests that the charging and discharging processes based on the electrostatic interaction are very robust over a large number of cycles and the NG supercapacitor should be able to function as a long-term energy storage device. Our group [154]

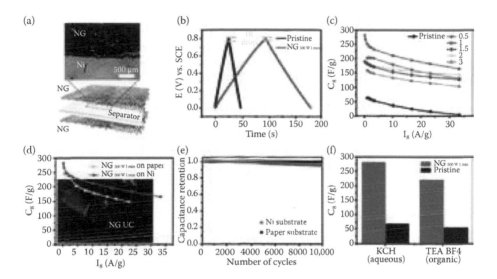

FIGURE 5.18 Ultracapacitors based on NGs and their electrochemical testing. (a) Schematic illustration of the assembled UC structure alongside a scanning electron microscopy image showing a top view of the device. (b) Charging and discharging curves measured by galvanostatic characterization. Black and red lines correspond to the pristine graphene and the NG 500 W, 1 min, respectively. The *IR* drops at the top cutoff potentials are also denoted. (c) Gravimetric capacitances of UCs based on various NGs and pristine graphene measured at a series of current densities. The numbers in the legend indicate the plasma durations in minutes. (d) Gravimetric capacitances of UGs built on Ni and paper substrates measured at a series of current densities. (inset) Photograph showing that a wearable UC wrapped around a human arm can shore the electrical energy to light up a light-emitting diode. (e) Cycling tests for the UCs based on Ni and paper substrates up to 10,000 cycles. (f) Specific capacitance measured in aqueous and organic electrolytes. For both electrolytes, the specific capacitances of NG and pristine graphene are compared. The dates in (b–e) were measured with a 6 M KOH electrolyte. All of the data presented in this figure are based on the NG UCs with mass loading of ~1 mg/cm². (Reprinted with permission from Ref. 153, 2472–2477. Copyright 2011 American Chemical Society.)

reported a simple method to prepare NG by a pressure-promoted process at relatively low temperature. NG with an atomic N higher than 10% can be obtained by heating GO and ammonium bicarbonate (NH_4HCO_3) in a sealed autoclave at a temperature as low as 150°C. As a result, NG displays a specific capacitance of 170 F/g at 0.5 A/g in 6 M KOH, and a high retention rate of 96.4% of its initial capacitance after 10,000 charge/discharge cycles at a current density of 10 A/g. In addition, this modified doping process can be easily scaled up and is more cost-effective than previous ones.

It is well known that the surface area of electrode materials is one of the critical factors that affect specific capacitance. In addition, the electronic properties of electrode materials play an important role in determining capacitor performance. Graphene doped with electron-rich N atoms has shown good performance as supercapacitor electrodes. Apart from N-doping, B-doping in carbon materials has also enhanced the performance when used as supercapacitor electrodes. Park and coworkers [69] reported novel B-doped chemically modified graphene (CMG) nanoplatelets (B-rG-O) were made by reduction of GO with a borane (BH_3)-tetrahydrofuran (THF) adduct under reflux via a one-pot synthesis using a liquid process on a large scale (Figure 5.19). During the reaction, the GO was reduced by the treatment of the BH_3 adduct and a small amount of B atoms were incorporated into the CMG nanoplatelets. The B-rG-O nanoplatelets had a high SSA of 466 m^2/g and were tested as an electrode material in supercapacitors. The

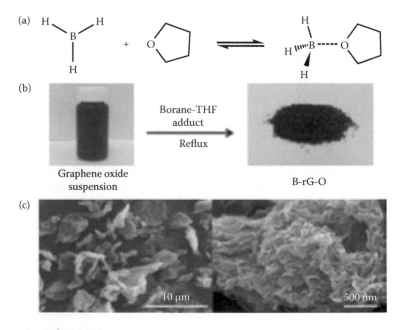

FIGURE 5.19 (a) Borane-THF adduct, (b) reaction scheme of the reduction of graphene oxide with the borane-THF adduct, and (c) SEM images of B-rG-O powder at different magnifications. (Reprinted with permission from Ref. 69, 19–26. Copyright 2013 American Chemical Society.)

specific capacitance of B-rG-O was as high as 200 F/g in 6 M KOH electrolyte. Furthermore, the B-rG-O electrode had a low resistivity for ion movement, exhibited good stability after 4500 cycles more than 95% of the original capacity was retained, and showed a good rate response.

In summary, chemical doping is one of the most effective methods to tailor the electronic properties of graphene-based electrode materials. It has demonstrated that N- or B-doped graphene-based materials exhibit high specific capacitance and excellent cycle stability. Moreover, heteroatom-doped graphene is more suitable to operate in aqueous electrolytes. However, the narrow potential window seriously affects the energy density of supercapacitors. Therefore, the combination of high-performance graphene electrode materials and stable electrolytes with high working voltage is a reasonable approach to solve the insufficient energy density issue. In Section 5.4.2, we will discuss the electrochemical performance of graphene-based electrode materials in organic electrolytes or IL electrolytes.

5.4   GRAPHENE-BASED HYBRID SUPERCAPACITOR

During the last few years, although great progress has been made in the improvement of energy density for symmetric supercapacitors, currently energy density still ranges between 5 and 10 Wh/kg, which is far lower than that of LIBs (150–200 Wh/kg). Compared to LIBs with low power density (<1000 W/kg) and poor cycle life (<1000 cycles), supercapacitors can provide much high power density (>10 kW/kg), long cycle life (>100,000 cycles), and fast charge/discharge processes (within seconds). Therefore, one of the most critical approaches in the development of hybrid supercapacitors is to enhance their energy density while retaining their intrinsic high power density. Generally, hybrid supercapacitors include asymmetric supercapacitors and supercapacitor-battery hybrids. Asymmetric supercapacitors usually consist of two different electrodes. During the charge/discharge processes, one electrode should take place redox (faradic) reactions with or without faradic reactions, and the other one mostly taking places is EDL (nonfaradic or electrostatic) adsorption/desorption. Supercapacitor-battery hybrids, also called Li-ion capacitors, are composed of a capacitor-type electrode and a Li-ion battery-type electrode in a Li salt containing organic electrolyte. The combination of two different types of electrode would be coupled with the fast-charging rate of a supercapacitor and the high-energy density of a Li-ion battery [1].

Turning to the various negative electrode materials of hybrid supercapacitors, including carbonaceous materials (porous carbons [71,74,155,156], CNTs [157–159], and graphene [72,160,161]), oxides (V_2O_5 [162,163], and Fe_3O_4 [164]) and their composites, among them, graphene has attracted increasing attention in recent years. Due to the unique 2-D structure and excellent mechanical, thermal, and electronic properties, as well as possible large SSA, graphene as an electrode material for hybrid supercapacitors has also attracted a considerable amount of interest in the field of clean energy storage systems. Hybrid supercapacitors are mainly classified into two types based on electrolytes: aqueous and nonaqueous. In Sections 5.4.1 and 5.4.2, the two types of hybrid supercapacitor will be discussed, respectively.

5.4.1 Hybrid Supercapacitor in Aqueous Electrolyte

Although most commercial supercapacitors use nonaqueous electrolytes that can provide a wider potential window (up to 3 V), aqueous electrolytes are still in favor in academia and industry due to their low cost, safety, easy assembling process, and high ionic conductivity. In aqueous hybrid supercapacitors, the most important issue is to couple appropriate electrode materials to increase the maximum operating voltage by making full use of the different potential window of the two electrodes. AC is generally used as a negative electrode and manganese oxide (MnO_2) as a positive electrode, allowing the operating voltage in the aqueous electrolyte to be extended up to 2 V. Khomenko and coworkers studied the charge storage mechanism of MnO_2 and AC in aqueous electrolyte, respectively. It was found that amorphous MnO_2 can be polarized up to potentials of 1.2 V (vs. NHE) in neutral medium and AC electrode even be stabilized in −0.8 V (vs. NHE) due to the high oxygen overpotential when MnO_2 is used as a positive electrode and high hydrogen overpotential presented by a negative AC electrode. As a result, by operating in a practical cell voltage of 2 V in aqueous medium [150], hybrid supercapacitors exhibit high energy densities that are close to the values obtained with EDLCs working in organic electrolytes.

Both AC and graphene are composed of the same element. Compared to AC, graphene can provide a large accessible surface area for fast transport of hydrate ions and superior electrical conductivity for fast electron transport. These features make graphene not only especially suitable for using as a negative electrode for hybrid supercapacitors, achieving a much higher performance operating at ~2 V, but also beneficial for a positive composite electrode. The presence of graphene in the positive composite electrode could enhance the utilization efficiency of pseudocapacitive materials and their cyclic stability. Cheng and coworkers [165] reported a high-voltage asymmetric supercapacitor with a MnO_2 nanowire/graphene (MGC) as the positive electrode and graphene as the negative electrode in an aqueous Na_2SO_4 solution. The aqueous electrolyte-based asymmetric supercapacitor could be cycled reversibly in a high-voltage region of 0–2 V and exhibited a superior energy density of 30.4 Wh/kg, which is much higher than that of symmetric supercapacitors based on graphene//graphene (2.8 Wh/kg) and MGC//MGC (5.2 Wh/kg). Moreover, it has a high power density and reasonable cycling performance. In another similar work, Fan and coworkers [72] developed an asymmetric supercapacitor device operable in neutral electrolyte using graphene/MnO_2 and activated carbon nanofibers (CAN) materials as the positive and negative electrodes, respectively. The graphene/MnO_2 composites were synthesized by self-limiting the deposition of nanoscale MnO_2 on the surface of graphene under microwave irradiation. The ACN derived from rod-shaped polyaniline was synthesized by carbonization and subsequently activated with KOH. The asymmetric supercapacitor device could be cycled reversibly between 0 and 1.8 V in 1 M Na_2SO_4 solution, and exhibited a maximum specific capacitance of 113.5 F/g with a measured energy density of 51.1 Wh/kg, which is almost twice than those of graphene/MnO_2//DWNT, MnO_2//DWNT, MnO_2//ACN asymmetric supercapacitors and CAN//ACN, graphene/MnO_2//graphene/MnO_2 symmetric supercapacitors (Figure 5.20). Additionally, the asymmetric cell exhibited excellent electrochemical stability with 97.3% specific capacitance retention

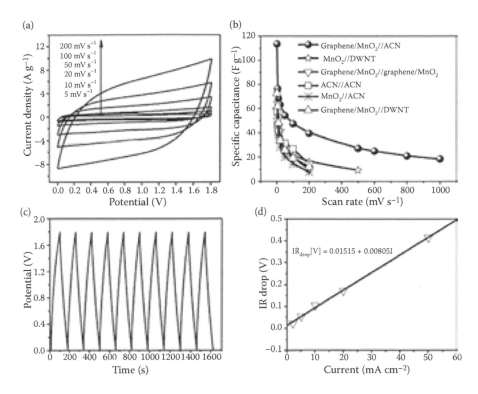

FIGURE 5.20 (a) CV curves of graphene/MnO$_2$//ACN asymmetric supercapacitor measured at different scan rates of 5, 10, 20, 50, 100, and 200 mV/s between 0 and 1.8 V in 1 M Na$_2$SO$_4$ aqueous electrolyte. (b) Comparison of specific capacitances of graphene/MnO$_2$//ACN, MnO$_2$//DWNT, MnO$_2$//ACN, graphene/MnO$_2$//ACN asymmetric cells and ACN//CAN, and graphene/MnO$_2$//graphene/MnO$_2$ symmetric supercapacitors at different scan rates. (c) Galvanostatic charge/discharge curve of graphene/MnO$_2$//ACN asymmetric supercapacitor at a current density of 1.2 A/g. (d) Potential drop associated with graphene/MnO$_2$//ACN asymmetric supercapacitor internal resistance (IR loss) vs. different discharge current densities. (Fan, Z. J., Yan, J., Wei, T. et al.: Asymmetric supercapacitors based on graphene/MnO$_2$ and activated carbon nanofiber electrodes with high power and energy density. *Adv. Funct. Matter.* 2011. 21. 2366–2375. Copyright Wiley-VHC Verlag GmbH & Co. KGaA, Weinheim. Reprinted with permission.)

after 1000 cycles. Such an asymmetric supercapacitor is a highly promising candidate for application in high-performance energy storage systems.

Ni(OH)$_2$ is another promising candidate electrode material for asymmetric supercapacitors due to its high theoretical specific capacitance (2082 F/g) [166], which is much higher than that of transition metal oxides. The redox behavior of Ni(OH)$_2$ is not very symmetric when combined with conductive materials (graphene, etc.), and their rate capability is greatly improved due to good electronic conductivity. In addition, the specific capacitance is also increased. Generally, the Ni(OH)$_2$-based composites are used as a positive electrode for asymmetric supercapacitors and its energy and power densities are much higher than those of the symmetric supercapacitors. Fan and coworkers [161] reported a fast, facile, and cost-effective microwave heating method to prepare hierarchical flowerlike $Ni(OH)_2$ decorated on graphene sheets and employed the composite material as a positive electrode coupled with porous graphene to assemble an asymmetric supercapacitor. The asymmetric supercapacitor could be operated in an aqueous solution at a voltage of 1.6 V. Due to the unique flowerlike nanostructure with short diffusion length, the $Ni(OH)_2$/graphene hybrid material showed a high specific capacitance of 1735 F/g and high rate capability compared to a pure $Ni(OH)_2$ electrode. As a result, the asymmetric supercapacitor exhibited a specific capacitance of 218.4 F/g and a maximum energy density of 77.8 Wh/kg based on the total mass of active materials. Furthermore, the device could be cycled reversibly in the voltage range of 0–1.6 V with 94.3% capacitance retention after 3000 cycles (Figure 5.21), suggesting good cycling stability.

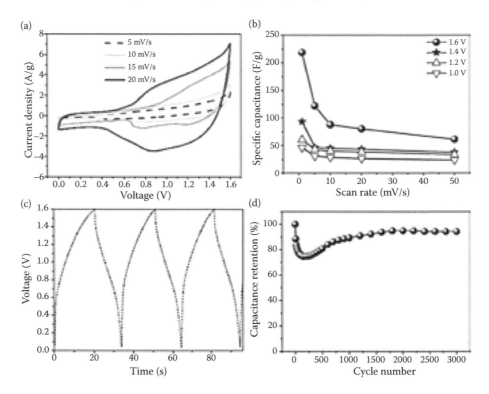

FIGURE 5.21 (a) CV curves of an optimized asymmetric supercapacitor in 6 M NaOH electrolyte at different scan rates of 5, 10, 15, and 20 mV/s. (b) Variation of specific capacitance at different scan rates for the asymmetric supercapacitor operated within different voltage windows. (c) Galvanostatic charge/discharge curves of the asymmetric supercapacitor at a current density of 0.5 A/g. (d) Cycle performance of the optimized $Ni(OH)_2$/graphene// porous graphene asymmetric supercapacitor within a voltage window of 1.6 V at a scan rate of 100 mV/s. (Yan, J., Fan, Z. J., Sun, W. et al.: Advanced asymmetric supercapacitors based on $Ni(OH)_2$/graphene and porous graphene electrodes with high energy density. *Adv. Funct. Matter.* 2012. 22. 2632–2641. Copyright Wiley-VCH Verlag GmbH & Co. KGaA, Weinheim. Reproduced with permission.)

5.4.2 ASYMMETRIC SUPERCAPACITOR IN ORGANIC ELECTROLYTE

An asymmetric supercapacitor in organic electrolyte mainly refers to the Li-ion capacitor. The main cell reactions are Li-ion intercalation/deintercalation in negative electrode materials but Li-ion adsorption/desorption on the surface of positive electrode materials. For the same reasons mentioned above, graphene can be used as a positive electrode material as well as act as an excellent conductive additive for composite electrodes. Ruoff and coworkers [167] have obtained outstanding performance using a-MEGO as electrode materials in symmetric supercapacitors. Recently, they successfully introduced a-MEGO into an asymmetric supercapacitor system. In this assembled device, a-MEGO and LTO with a highly reversible zero-strain Li insertion around 150–160 mAh/g at 1.5 V (vs. Li^+/Li) were used as positive and negative electrode materials, respectively. The hybrid cells could work at 4 V and yield specific capacitances as high as 266 F/g. Energy density was over five times that of current symmetric supercapacitors and greater than current lead acid batteries and would likely accelerate the adoption of energy storage devices based on this novel carbon. Furthermore, Chen and coworkers [1] designed a hybrid supercapacitor using graphene-enhanced negative and positive electrode materials to improve both energy density and power density performance. In this system, a well-structured Fe_3O_4 nanoparticle/graphene (Fe_3O_4/G) composite was prepared as the negative electrode material by a facile approach including solvethermal reaction and an easy heat treatment (Figure 5.22), which exhibited high reversible specific capacity exceeding 1000 mAh/g at the current density of 90 mA/g as well as excellent rate capability and cycle stability. Meanwhile, a graphene-based 3-D porous carbon material (3-D graphene) with high SSA (~3355 m^2/g) and good conductivity was used as a positive electrode material to enhance electrochemical performance compared to commercial AC. As a result, the Fe_3O_4/G//3D graphene hybrid supercapacitor with optimal mass ratio could work in the potential range of 1.0–4.0 V and exhibited an ultrahigh energy density of 204 Wh/kg (while the symmetric supercapacitor got only 45 Wh/kg) and prominent rate performance; when the power density increased to ~1000 W/kg, the energy density of the hybrid capacitor still remained higher (122 Wh/kg) than the symmetric capacitor (40 Wh/kg). Importantly, the Fe_3O_4/G//3D graphene hybrid supercapacitor also exhibited a passable cycle stability with the capacity retention of 70% after 1000 cycling numbers at a high current density of 2 A/g, with high Coulombic efficiency of ~100% during cycles.

Metal oxides, such as SnO_2 [168], Co_3O_4 [134,169], and $Fe_3O_4^1$ are considered as good candidates as negative electrode materials in Li-ion capacitors due to their large theoretical capacity (500–1000 mAh/g) and relatively low voltage plateau (~0.8 V), and their cycle stability can be improved by adding conductive carbon materials (CNTs, graphene, etc.). However, their rate capability and cycle stability still cannot match that of positive electrodes, which results in a relatively poor performance for a Li-ion capacitor. It is reasonable to expect the Li-ion capacitor to possess the high energy density of LIBs and the fast charging rate of supercapacitors. Unfortunately, the reported results indicate that obtaining a Li-ion capacitor with high rate capability and excellent cycle stability is still a great challenge. The most important issue is to

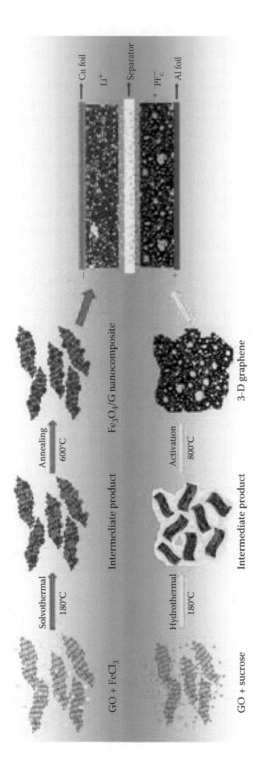

FIGURE 5.22 Schematic show of the synthesis of the negative electrode material Fe₃O₄/G nanocomposite and the positive electrode material 3-D graphene, together for the configuration of a Li-ion containing organic hybrid supercapacitor. (From Zhang, F., Zhang, T. F., Yang, X. et al. A high-performance supercapacitor-battery hybrid energy storage device based on graphene-enhanced electrode materials with ultrahigh energy density. *Energy Environ. Sci.* 6:1623–1632, 2013. Reproduced by permission of The Royal Society of Chemistry.)

select an appropriate negative electrode material, which should have high rate capability and cyclic stability. From this point of view, as the mechanism of LTO gassing is becoming more and more clearer, and the potential solutions for LTO gassing have been widely proposed, LTO can be regarded as a suitable negative electrode material. Very recently, TiO_2 [170–173] has received increasing attention as a promising Li-ion capacitor negative electrode material due to its low cost, structure stability, excellent recharge ability, and improved safety over graphite. Kang and coworkers [174] employed TiO_2 anatase nanoparticles embedded in reduced GO as the anode material, and AC was used as cathode material to assemble a high-energy hybrid supercapacitor. TiO_2 anchored on reduced graphene can substantially increase the performance of a hybrid supercapacitor, so the hybrid cell delivers 42 Wh/kg at 800 W/kg and 8.9 Wh/kg even at a 4-s charge/discharge rate (Figure 5.23). The energy and power densities could satisfy the requirements of a hybrid electric vehicle (HEV) application. It is believed that this new system with optimization could be a strong competitor for energy storage in HEVs.

Among various polymorphs of TiO_2 (anatase, rutile, and TiO_2 (B)), TiO_2(B) [175–177] shows a favorable channel structure for lithium mobility, which results in fast charge/discharge capability. In particular, the Li insertion into TiO_2(B) is a pseudocapacitive faradic process that is rather different from the solid-state diffusion process observed for anatase and rutile. Furthermore, the theoretical capacity of TiO_2(B) is 335 mAh/g, and about twice the value of commercial LTO and anatase of TiO_2. Thus, TiO_2(B) is considered as a new kind of anode material and an alternative to LTO and anatase TiO_2 for a high-power Li-ion capacitor. Zhang and coworkers [174] reported a novel sheet-belt hybrid nanostructure composed of 1-D mesoporous single-crystalline TiO_2(B) nanobelts combined with 2-D graphene. The G-TiO_2(B) hybrid was constructed *in situ* by hydrothermal treatment of TiO_2 (P25) nanoparticles in the presence of graphene. The homogeneous incorporation graphene into nanbelts not only guarantees the fast electron transport and the structural integrity of a sample, but also improves its surface-to-volume ratio. As a result, the G-TiO_2(B) hybrid exhibits an ultrahigh reversible capacity, excellent capacity retention, and superior rate performance.

In summary, development of a Li-ion capacitor is an effective approach to improve the energy density of a supercapacitor. However, the poor rate capability and cycle stability are still problems that must be solved. Carbon/TiO_2(B) hybrid may be an appropriate negative electrode material for a Li-ion capacitor but there is still much research to be done. Apart from synthesizing suitable negative electrode materials, optimization to balance the positive and negative electrodes is another important issue. Generally, the respective masses of the two electrodes must be adjusted to equalize the capacity because the specific capacity of a negative electrode is much higher than that of a positive electrode. However, this balance is only established on one selected measure condition (for example the same charge/discharge current density). It is probable that the hybrid system will become unbalanced at different charge/discharge current densities. Thus, in the future, it will be very important to develop positive and negative electrode materials with nearly the same capacity retention in different current densities.

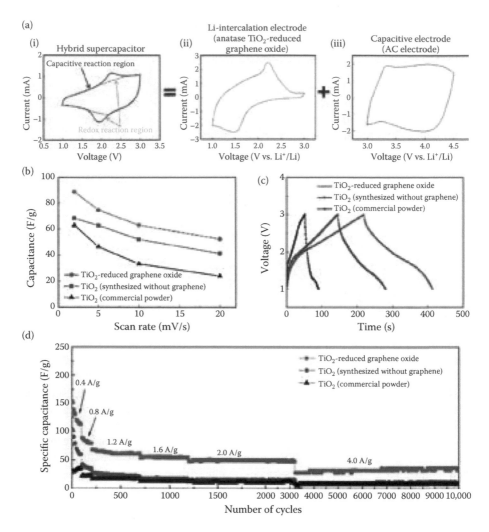

FIGURE 5.23 Electrochemical performance of TiO_2-(RGO) in a hybrid supercapacitor. (a) CV analysis of (i) a hybrid supercapacitor based on the TiO_2-(RGO) anode and AC cathode and of (ii, iii) TiO_2-(RGO) (ii) and AC electrodes (iii) in a Li half-cell. (b) Specific capacitance of TiO_2-(RGO) and TiO_2 nanoparticles in a hybrid supercapacitor. (c) Initial charge/discharge profiles of a hybrid supercapacitor. (d) Cyclability and rate capability of a hybrid supercapacitor at various current rates, from 0.4 to 4.0 A/g. (Kim, H., Cho, M. Y., Kim, M. H. et al.: A novel high-energy hybrid supercapacitor with an anatase TiO_2-reduced graphene oxide anode and an activated carbon cathode. *Adv. Energy Mater.* 2013. 3. 1500–1506. Copyright Wiley-VCH Verlag GmbH & Co. KGaA, Weinheim. Reproduced with permission.)

5.5 SUMMARY AND PERSPECTIVES

In summary, it is clear that graphene-based materials can be considered as ideal supercapacitor electrode materials due to their extremely high SSA, excellent electrical conductivity, and stable physical/chemical properties. Until now, scientists have achieved great success in the laboratory but looking ahead, there are several issues that need to be solved urgently in order to develop high-performance supercapacitors based on graphene-based materials in practice. First, large-scale production of high-quality graphene-based materials with low cost is a very challenging requirement. Numerous efforts have been made to tackle this problem, and among the various preparation methods, liquid phase exfoliation is regarded as the most likely to obtain massive preparation. First, and most recently, scientists have made breakthroughs in mass preparation and industrial experiments of graphene production are being put together. Second, graphene is very easy to restack, which seriously affects its performance in supercapacitors. Some efficient methods, including the functionalization of graphene or the addition of spacers between the graphene layers, have been presented in an attempt to solve this problem. Third, the relatively low packing density is another drawback of graphene. Graphene with chemical activation or utilization of hybrid graphene-based carbon materials can effectively improve its packing density. Subsequent mechanical compression and electrolyte capillary compression could also further enhance the packing density.

Although modifying the GO preparation and subsequent reduction methods and designing curved graphene in graphene-based EDLC applications have successfully resolved some existing problems, future efforts are still needed for the development of high-quality graphene on a large scale in order to reach enhanced performance. It is worth noting that broadening the potential window of a supercapacitor is one of the most effective approaches to improve its energy density. Graphene can be considered as good candidate electrode material for supercapacitors in high voltage. Thus, it is absolutely necessary to prepare high-purity and stable graphene for application in high-voltage supercapacitors. On the other hand, for graphene-based composites in the pesudocapacitor applications, it has been demonstrated that graphene can significantly improve the specific capacitance, rate capability, and cycle stability of pseudomaterials due to its large SSA and excellent conductivity. However, compared to EDLCs based on carbon materials, their cycle stability still cannot meet the requirements for practical applications. Therefore, the design and synthesis of a new nanostructure and architecture based on graphene will be an important task in the future. Finally, hybrid supercapacitors based on graphene-based materials are regarded as another most effective method to realize high-performance supercapacitors. It has been found that graphene can be used as a positive electrode material and can also act as skeleton or conductive material in negative electrode material. However, poor rate capability and cycle stability are still pressing problems that need to be solved and that will need a great deal of research to overcome.

ACKNOWLEDGMENTS

This work was supported by the National Nature Science Foundation of China (Grant No. 21201173), the Key Research program of the Chinese Academy of Sciences (Grant No. KGZD-EW-202-4), the Ningbo Science and Technology Innovation Team (Grant No. 2012B82001), the 973 program (Grant No. 2011CB935900), the Natural Science Foundation of Ningbo (Grant No. 2013A610029), and the Zhejiang Province Preferential Postdoctoral Found Project (Grant No. Bsh1302054).

REFERENCES

1. Zhang, F., Zhang, T., Yang, X. et al. 2013. A high-performance supercapacitor-battery hybrid energy storage device based on graphene-enhanced electrode materials with ultrahigh energy density. *Energy Environ. Sci.* 6:1623–1632.
2. Kim, Y. J., Yang, C.-M., Park, K. C. et al. 2012. Edge-enriched, porous carbon-based, high energy density supercapacitors for hybrid electric vehicles. *Chemsuschem* 5:535–541.
3. Mastragostino, M. and Soavi, F. 2007. Strategies for high-performance supercapacitors for HEV. *J. Power Sources* 174:89–93.
4. Naoi, K., Ishimoto, S., Miyamoto, J.-I. et al. 2012. Second generation "nanohybrid supercapacitor": Evolution of capacitive energy storage devices. *Energy Environ. Sci.* 5:9363–9373.
5. Balducci, A., Dugas, R., Taberna, P. L. et al. 2007. High temperature carbon–carbon supercapacitor using ionic liquid as electrolyte. *J. Power Sources* 165:922–927.
6. Lota, G., Grzyb, B. and Machnikowska, H. et al. 2005. Effect of nitrogen in carbon electrode on the supercapacitor performance. *Chem. Phys. Lett.* 404:53–58.
7. Qu, D. Y. 2002. Studies of the activated carbons used in double-layer supercapacitors. *J. Power Sources* 109:403–411.
8. Raymundo-Pinero, E., Leroux, F. and Beguin, F. 2006. A high-performance carbon for supercapacitors obtained by carbonization of a seaweed biopolymer. *Adv. Mater.* 18:1877–1882.
9. Xu, G., Zheng, C., Zhang, Q. et al. 2011. Binder-free activated carbon/carbon nanotube paper electrodes for use in supercapacitors. *Nano Res.* 4:870–881.
10. Fischer, U., Saliger, R., Bock, V. et al. 1997. Carbon aerogels as electrode material in supercapacitors. *J. Porous. Mat.* 4:281–285.
11. Meng, Q. H., Liu, L., Song, H. H. et al. 2004. Electrochemical properties of carbon aerogels electrode for super-capacitor. *J. Inorg. Mater.* 19:593–598.
12. Probstle, H., Wiener, M. and Fricke, J. 2003. Carbon aerogels for electrochemical double layer capacitors. *J. Porous. Mat.* 10:213–222.
13. Liu, D., Shen, J., Liu, N. et al. 2013. Preparation of activated carbon aerogels with hierarchically porous structures for electrical double layer capacitors. *Electrochim. Acta* 89:571–576.
14. Mezzavilla, S., Zanella, C., Aravind, P. R. et al. 2012. Carbon xerogels as electrodes for supercapacitors. The influence of the catalyst concentration on the microstructure and on the electrochemical properties. *J. Mater. Sci.* 47:7175–7180.
15. Frackowiak, E. and Beguin, F. 2002. Electrochemical storage of energy in carbon nanotubes and nanostructured carbons. *Carbon* 40:1775–1787.
16. Frackowiak, E., Metenier, K., Bertagna, V. et al. 2000. Supercapacitor electrodes from multiwalled carbon nanotubes. *Appl. Phys. Lett.* 77:2421–2423.
17. Lu, W., Qu, L., Henry, K. et al. 2009. High performance electrochemical capacitors from aligned carbon nanotube electrodes and ionic liquid electrolytes. *J. Power Sources* 189:1270–1277.
18. Yu, C., Masarapu, C., Rong, J. et al. 2009. Stretchable supercapacitors based on buckled single-walled carbon nanotube macrofilms. *Adv. Mater.* 21:4793–4797.
19. Futaba, D. N., Hata, K., Yamada, T. et al. 2006. Shape-engineerable and highly densely packed single-walled carbon nanotubes and their application as super-capacitor electrodes. *Nat. Mater.* 5:987–994.
20. Hiraoka, T., Izadi-Najafabadi, A., Yamada, T. et al. 2010. Compact and light supercapacitor electrodes from a surface-only solid by opened carbon nanotubes with 2 200 m(2) g(-1) surface area. *Adv. Funct. Mater.* 20:422–428.
21. Izadi-Najafabadi, A., Yamada, T., Futaba, D. N. et al. 2011. High-power supercapacitor electrodes from single-walled carbon nanohorn/nanotube composite. *ACS Nano* 5:811–819.

22. Izadi-Najafabadi, A., Yasuda, S., Kobashi, K. et al. 2010. Extracting the full potential of single-walled carbon nanotubes as durable supercapacitor electrodes operable at 4 V with high power and energy density. *Adv. Mater.* 22:E235–E241.

23. Fuertes, A. B., Lota, G., Centeno, T. A. et al. 2005. Templated mesoporous carbons for supercapacitor application. *Electrochim. Acta* 50:2799–2805.

24. Li, H.-Q., Luo, J.-Y., Zhou, X.-F. et al. 2007. An ordered mesoporous carbon with short pore length and its electrochemical performances in supercapacitor applications. *J. Electrochem. Soc.* 154:A731–A736.

25. Lei, Z., Christov, N., Zhang, L, L. et al. 2011. Mesoporous carbon nanospheres with an excellent electrocapacitive performance. *J. Mater. Chem.* 21:2274–2281.

26. Zhao, J., Lai, C., Dai, Y. et al. 2007. Pore structure control of mesoporous carbon as supercapacitor material. *Mater. Lett.* 61:4639–4642.

27. Cai, T., Zhou, M., Ren, D. et al. 2013. Highly ordered mesoporous phenol-formaldehyde carbon as supercapacitor electrode material. *J. Power Sources* 231:197–202.

28. Wang, Q., Yan, J., Wei, T. et al. 2013. Two-dimensional mesoporous carbon sheet-like framework material for high-rate supercapacitors. *Carbon* 60:481–487.

29. Arulepp, M., Leis, J., Latt, M. et al. 2006. The advanced carbide-derived carbon based supercapacitor. *J. Power Sources* 162:1460–1466.

30. Korenblit, Y., Rose, M., Kockrick, E. et al. 2010. High-rate electrochemical capacitors based on ordered mesoporous silicon carbide-derived carbon. *ACS Nano* 4:1337–1344.

31. Perez, C. R., Yeon, S.-H., Segalini, J. et al. 2013. Structure and electrochemical performance of carbide-derived carbon nanopowders. *Adv. Funct. Mater.* 23:1081–1089.

32. Portet, C., Yang, Z., Korenblit, Y. et al. 2009. Electrical double-layer capacitance of zeolite-templated carbon in organic electrolyte. *J. Electrochem. Soc.* 156:A1–A6.

33. Presser, V., Zhang, L., Niu, J. J. et al. 2011. Flexible nano-felts of carbide-derived carbon with ultra-high power handling capability. *Adv. Energy Mater.* 1:423–430.

34. Frackowiak, E., Jurewicz, K., Delpeux, S. et al. 2001. Nanotubular materials for supercapacitors. *J. Power Sources* 97–98:822–825.

35. Novoselov, K. S., Geim, A. K., Morozov, S. V. et al. 2004. Electric field effect in atomically thin carbon films. *Science* 306:666–669.

36. Balandin, A. A., Ghosh, S., Bao, W. et al. 2008. Superior thermal conductivity of single-layer graphene. *Nano Lett.* 8.902–907.

37. Lee, C., Wei, X., Kysar, J. W. et al. 2008. Measurement of the elastic properties and intrinsic strength of monolayer graphene. *Science* 321:385–388.

38. Jang, B. Z. and Zhamu, A. 2008. Processing of nanographene platelets (NGPs) and NGP nanocomposites: A review. *J. Mater. Sci.* 43:5092–5101.

39. Liu, C., Yu, Z., Neff, D. et al. 2010. Graphene-based supercapacitor with an ultrahigh energy density. *Nano Lett.* 10:4863–4868.

40. Huang, Y., Liang, J. and Chen, Y. 2012. An overview of the applications of graphene-based materials in supercapacitors. *Small* 8:1805–1834.

41. Gujar, T. P., Shinde, V. R., Lokhande, C. D. et al. 2007. Spray deposited amorphous RuO_2 for an effective use in electrochemical supercapacitor. *Electrochem. Commun.* 9:504–510.

42. Lee, B. J., Sivakkumar, S. R., Ko, J. M. et al. 2007. Carbon nanofibre/hydrous RuO_2 nanocomposite electrodes for supercapacitors. *J. Power Sources* 168:546–552.

43. Wu, Z.-S., Wang, D.-W., Ren, W. et al. 2010. Anchoring hydrous RuO_2 on graphene sheets for high-performance electrochemical capacitors. *Adv. Funct. Mater.* 20:3595–3602.

44. Dubal, D. P., Gund, G. S., Holze, R. et al. 2013. Solution-based binder-free synthetic approach of RuO_2 thin films for all solid state supercapacitors. *Electrochim. Acta* 103:103–109.

45. Vellacheri, R., Pillai, V. K. and Kurungot, S. 2012. Hydrous RuO_2-carbon nanofiber electrodes with high mass and electrode-specific capacitance for efficient energy storage. *Nanoscale* 4:890–896.

46. Subramanian, V., Zhu, H. W., Vajtai, R. et al. 2005. Hydrothermal synthesis and pseudo-capacitance properties of MnO_2 nanostructures. *J. Phys. Chem. B* 109:20207–20214.
47. Xu, M., Kong, L., Zhou, W. et al. 2007. Hydrothermal synthesis and pseudocapaci-tance properties of alpha-MnO_2 hollow spheres and hollow urchins. *J. Phys. Chem. C* 111:19141–19147.
48. Yan, J., Fan, Z., Wei, T. et al. 2010. Fast and reversible surface redox reaction of graphene-MnO_2 composites as supercapacitor electrodes. *Carbon* 48:3825–3833.
49. Yan, J., Khoo, E., Sumboja, A. et al. 2010. Facile coating of manganese oxide on tin oxide nanowires with high-performance capacitive behavior. *ACS Nano* 4:4247–4255.
50. Yu, G., Hu, L., Liu, N. et al. 2011. Enhancing the supercapacitor performance of graphene/MnO_2 nanostructured electrodes by conductive wrapping. *Nano Lett.* 11:4438–4442.
51. Snook, G. A., Kao, P. and Best, A. S. 2011. Conducting-polymer-based supercapacitor devices and electrodes. *J. Power Sources* 196:1–12.
52. Wang, D.-W., Li, F., Zhao, J. et al. 2009. Fabrication of graphene/polyaniline composite paper via in situ anodic electropolymerization for high-performance flexible electrode. *ACS Nano* 3:1745–1752.
53. Wang, Y.-G., Li, H.-Q. and Xia, Y.-Y. 2006. Ordered whiskerlike polyaniline grown on the surface of mesoporous carbon and its electrochemical capacitance performance. *Adv. Mater.* 18:2619–2623.
54. Wu, Q., Xu, Y., Yao, Z. et al. 2010. Supercapacitors based on flexible graphene/polyaniline nanofiber composite films. *ACS Nano* 4:1963–1970.
55. Zhang, K., Zhang, L. L., Zhao, X. S. et al. 2010. Graphene/polyaniline nanofiber com-posites as supercapacitor electrodes. *Chem. Mater.* 22:1392–1401.
56. An, K. H., Jeon, K. K., Heo, J. K. et al. 2002. High-capacitance supercapacitor using a nanocomposite electrode of single-walled carbon nanotube and polypyrrole. *J. Electrochem. Soc.* 149:A1058–A1062.
57. Biswas, S. and Drzal, L. T. 2010. Multi layered nanoarchitecture of graphene nanosheets and polypyrrole nanowires for high performance supercapacitor electrodes. *Chem. Mater.* 22:5667–5671.
58. Fan, L. Z. and Maier, J. 2006. High-performance polypyrrole electrode materials for redox supercapacitors. *Electrochem. Commun.* 8:937–940.
59. Fang, Y., Liu, J., Yu, D. J. et al. 2010. Self-supported supercapacitor membranes: Polypyrrole-coated carbon nanotube networks enabled by pulsed electrodeposition. *J. Power Sources* 195:674–679.
60. Mi, H., Zhang, X., Ye, X. et al. 2008. Preparation and enhanced capacitance of core-shell polypyrrole/polyaniline composite electrode for supercapacitors. *J. Power Sources* 176:403–409.
61. Park, J. H., Ko, J. M., Park, O. O. et al. 2002. Capacitance properties of graphite/polypyrrole composite electrode prepared by chemical polymerization of pyrrole on graphite fiber. *J. Power Sources* 105:20–25.
62. Alvi, F., Basnayaka, P. A., Ram, M. K. et al. 2012. Graphene-polythiophene nanocom-posite as novel supercapacitor electrode material. *J. New Mat. Electr. Sys.* 15:89–95.
63. Ambade, R. B., Ambade, S. B., Shrestha, N. K. et al. 2013. Polythiophene infiltrated TiO_2 nanotubes as high-performance supercapacitor electrodes. *Chem. Commun.* 49: 2308–2310.
64. Frackowiak, E. and Beguin, F. 2001. Carbon materials for the electrochemical storage of energy in capacitors. *Carbon* 39:937–950.
65. An, B., Xu, S., Li, L. et al. 2013. Carbon nanotubes coated with a nitrogen-doped carbon layer and its enhanced electrochemical capacitance. *J. Mater. Chem. A* 1:7222–7228.
66. Chen, L.-F., Zhang, X.-D., Liang, H.-W. et al. 2012. Synthesis of nitrogen-doped porous carbon nanofibers as an efficient electrode material for supercapacitors. *ACS Nano* 6:7092–7102.

67. Su, P., Guo, H.-L., Peng, S. et al. 2012. Preparation of nitrogen-doped graphene and its supercapacitive properties. *Acta Phys.-Chim. Sin.* 28:2745–2753.
68. Wang, D.-W., Li, F., Chen, Z.-G. et al. 2008. Synthesis and electrochemical property of boron-doped mesoporous carbon in supercapacitor. *Chem. Mater.* 20:7195–7200.
69. Han, J., Zhang, L. L., Lee, S. et al. 2013. Generation of B-doped graphene nanoplatelets using a solution process and their supercapacitor applications. *ACS Nano* 7:19–26.
70. Hulicova-Jurcakova, D., Fiset, E., Lu, G. Q. M. et al. 2012. Changes in surface chemistry of carbon materials upon electrochemical measurements and their effects on capacitance in acidic and neutral electrolytes. *Chemsuschem* 5:2188–2199.
71. Brousse, T., Taberna, P.-L., Crosnier, O. et al. 2007. Long-term cycling behavior of asymmetric activated carbon/MnO₂ aqueous electrochemical supercapacitor. *J. Power Sources* 173:633–641.
72. Fan, Z., Yan, J., Wei, T. et al. 2011. Asymmetric supercapacitors based on graphene/MnO₂ and activated carbon nanofiber electrodes with high power and energy density. *Adv. Funct. Mater.* 21:2366–2375.
73. Park, J. H. and Park, O. O. 2002. Hybrid electrochemical capacitors based on polyaniline and activated carbon electrodes. *J. Power Sources* 111:185–190.
74. Khomenko, V., Raymundo-Pinero, E. and Beguin, F. 2008. High-energy density graphite/AC capacitor in organic electrolyte. *J. Power Sources* 177:643–651.
75. Leng, K., Zhang, F., Zhang, L. et al. 2013. Graphene-based Li-ion hybrid supercapacitors with ultrahigh performance. *Nano Res.* 6:581–592.
76. Simon, P. and Gogotsi, Y. 2008. Materials for electrochemical capacitors. *Nat. Mater.* 7:845–854.
77. Stoller, M. D., Park, S., Zhu, Y. et al. 2008. Graphene-based ultracapacitors. *Nano Lett.* 8:3498–3502.
78. Stankovich, S., Dikin, D. A., Piner, R. D. et al. 2007. Synthesis of graphene-based nanosheets via chemical reduction of exfoliated graphite oxide. *Carbon* 45:1558–1565.
79. Li, D., Mueller, M. B., Gilje, S. et al. 2008. Processable aqueous dispersions of graphene nanosheets. *Nat. Nanotechnol.* 3:101–105.
80. Stankovich, S., Piner, R. D., Chen, X. Q. et al. 2006. Stable aqueous dispersions of graphitic nanoplatelets via the reduction of exfoliated graphite oxide in the presence of poly(sodium 4 styrenesulfonate). *J. Mater. Chem.* 16:155–158.
81. Gao, W., Alemany, L. B., Ci, L. et al. 2009. New insights into the structure and reduction of graphite oxide. *Nat. Chem.* 1:403–408.
82. Che, J., Shen, L. and Xiao, Y. 2010. A new approach to fabricate graphene nanosheets in organic medium: Combination of reduction and dispersion. *J. Mater. Chem.* 20:1722–1727.
83. Wang, G., Yang, J., Park, J. et al. 2008. Facile synthesis and characterization of graphene nanosheets. *J. Phys. Chem. C* 112:8192–8195.
84. Fan, X., Peng, W., Li, Y. et al. 2008. Deoxygenation of exfoliated graphite oxide under alkaline conditions: A green route to graphene preparation. *Adv. Mater.* 20:4490–4493.
85. Lei, Z., Lu, L. and Zhao, X. S. 2012. The electrocapacitive properties of graphene oxide reduced by urea. *Energy Environ. Sci.* 5:6391–6399.
86. Zhu, Y., Stoller, M. D., Cai, W. et al. 2010. Exfoliation of graphite oxide in propylene carbonate and thermal reduction of the resulting graphene oxide platelets. *ACS Nano* 4:1227–1233.
87. Jung, I., Dikin, D., Park, S. et al. 2008. Effect of water vapor on electrical properties of individual reduced graphene oxide sheets. *J. Phys. Chem. C* 112:20264–20268.
88. Lin, Z., Yao, Y., Li, Z. et al. 2010. Solvent-assisted thermal reduction of graphite oxide. *J. Phys. Chem. C* 114:14819–14825.
89. Pei, S. and Cheng, H.-M. 2012. The reduction of graphene oxide. *Carbon* 50:3210–3228.
90. Ye, J., Zhang, H., Chen, Y. et al. 2012. Supercapacitors based on low-temperature partially exfoliated and reduced graphite oxide. *J. Power Sources* 212:105–110.

91. Zhu, Y., Murali, S., Stoller, M. D. et al. 2010. Microwave assisted exfoliation and reduction of graphite oxide for ultracapacitors. *Carbon* 48:2118–2122.

92. Yan, J., Liu, J., Fan, Z. et al. 2012. High-performance supercapacitor electrodes based on highly corrugated graphene sheets. *Carbon* 50:2179–2188.

93. Cheng, Y., Lu, S., Zhang, H. et al. 2012. Synergistic effects from graphene and carbon nanotubes enable flexible and robust electrodes for high-performance supercapacitors. *Nano Lett.* 12:4206–4211.

94. Jiang, H., Lee, P. S. and Li, C. 2013. 3D carbon based nanostructures for advanced supercapacitors. *Energy Environ. Sci.* 6:41–53.

95. Xu, J., Wang, K., Zu, S.-Z. et al. 2010. Hierarchical nanocomposites of polyaniline nanowire arrays on graphene oxide sheets with synergistic effect for energy storage. *ACS Nano* 4:5019–5026.

96. Fan, Z., Yan, J., Zhi, L. et al. 2010. A three-dimensional carbon nanotube/graphene sandwich and its application as electrode in supercapacitors. *Adv. Mater.* 22:3723–3728.

97. Jha, N., Ramesh, P., Bekyarova, E. et al. 2012. High energy density supercapacitor based on a hybrid carbon nanotube-reduced graphite oxide architecture. *Adv. Energy Mater.* 2:438–444.

98. Zhang, L., Zhang, F., Yang, X. et al. 2013. Porous 3D graphene-based bulk materials with exceptional high surface area and excellent conductivity for supercapacitors. *Sci. Rep.* 3:1408.

99. Fischbein, M. D. and Drndic, M. 2008. Electron beam nanosculpting of suspended graphene sheets. *Appl. Phys. Lett.* 93:113107.

100. Ramasse, Q. M., Zan, R., Bangert, U. et al. 2012. Direct experimental evidence of metal-mediated etching of suspended graphene. *ACS Nano* 6:4063–4071.

101. Ning, G., Fan, Z., Wang, G. et al. 2011. Gram-scale synthesis of nanomesh graphene with high surface area and its application in supercapacitor electrodes. *Chem. Commun.* 47:5976–5978.

102. Fang, Y., Lv, Y., Che, R. et al. 2013. Two-dimensional mesoporous carbon nanosheets and their derived graphene nanosheets: Synthesis and efficient lithium ion storage. *J. Am. Chem. Soc.* 135:1524–1530.

103. Kim, T., Jung, G., Yoo, S. et al. 2013. Activated graphene-based carbons as supercapacitor electrodes with macro- and mesopores. *ACS Nano* 7:6899–6905.

104. Wang, J. and Kaskel, S. 2012. KOH activation of carbon-based materials for energy storage. *J. Mater. Chem.* 22:23710–23725.

105. Zhang, L. L., Zhao, X., Stoller, M. D. et al. 2012. Highly conductive and porous activated reduced graphene oxide films for high-power supercapacitors. *Nano Lett.* 12:1806–1812.

106. Zhu, Y., Murali, S., Stoller, M. D. et al. 2011. Carbon-based supercapacitors produced by activation of graphene. *Science* 332:1537–1541.

107. Raymundo-Pinero, E., Azais, P., Cacciaguerra, T. et al. 2005. KOH and NaOH activation mechanisms of multiwalled carbon nanotubes with different structural organisation. *Carbon* 43:786–795.

108. Lillo-Rodenas, M. A., Cazorla-Amoros, D. and Linares-Solano, A. 2003. Understanding chemical reactions between carbons and NaOH and KOH: An insight into the chemical activation mechanism. *Carbon* 41:267–275.

109. Kierzek, K., Frackowiak, E., Lota, G. et al. 2004. Electrochemical capacitors based on highly porous carbons prepared by KOH activation. *Electrochim. Acta* 49:515–523.

110. Wang, T., Tan, S. and Liang, C. 2009. Preparation and characterization of activated carbon from wood via microwave-induced $ZnCl_2$ activation. *Carbon* 47:1880–1883.

111. Yue, Z. R., Mangun, C. L. and Economy, J. 2002. Preparation of fibrous porous materials by chemical activation 1. $ZnCl_2$ activation of polymer-coated fibers. *Carbon* 40:1181–1191.

112. Guo, Y. P. and Rockstraw, D. A. 2006. Physical and chemical properties of carbons synthesized from xylan, cellulose, and Kraft lignin by H_3PO_4 activation. *Carbon* 44:1464–1475.
113. Toles, C., Rimmer, S. and Hower, J. C. 1996. Production of activated carbons from a Washington lignite using phosphoric acid activation. *Carbon* 34:1419–1426.
114. Murali, S., Quarles, N., Zhang, L. L. et al. 2013. Volumetric capacitance of compressed activated microwave-expanded graphite oxide (a-MEGO) electrodes. *Nano Energy* 2:764–768.
115. Yang, X., Cheng, C., Wang, Y. et al. 2013. Liquid-mediated dense integration of graphene materials for compact capacitive energy storage. *Science* 341:534–537.
116. Kim, Y. A., Hayashi, T., Kim, J. H. et al. 2013. Important roles of graphene edges in carbon-based energy storage devices. *J. Energy Chem.* 22:183–194.
117. Yuan, W., Zhou, Y., Li, Y. et al. 2013. The edge- and basal-plane-specific electrochemistry of a single-layer graphene sheet. *Sci. Rep.* 3:2248.
118. Son, Y.-W., Cohen, M. L. and Louie, S. G. 2006. Energy gaps in graphene nanoribbons. *Phys. Rev. Lett.* 97:216803.
119. Jiao, L., Zhang, L., Wang, X. et al. 2009. Narrow graphene nanoribbons from carbon nanotubes. *Nature* 458:877–880.
120. Kosynkin, D. V., Higginbotham, A. L., Sinitskii, A. et al. 2009. Longitudinal unzipping of carbon nanotubes to form graphene nanoribbons. *Nature* 458:872–876.
121. Kumar, P., Panchakarla, L. S. and Rao, C. N. R. 2011. Laser-induced unzipping of carbon nanotubes to yield graphene nanoribbons. *Nanoscale* 3:2127–2129.
122. Shinde, D. B., Debgupta, J., Kushwaha, A. et al. 2011. Electrochemical unzipping of multi-walled carbon nanotubes for facile synthesis of high-quality graphene nanoribbons. *J. Am. Chem. Soc.* 133:4168–4171.
123. Parashar, U. K., Bhandari, S., Srivastava, R. K. et al. 2011. Single step synthesis of graphene nanoribbons by catalyst particle size dependent cutting of multiwalled carbon nanotubes. *Nanoscale* 3:3876–3882.
124. Elias, A. L., Botello-Mendez, A. R., Meneses-Rodriguez, D. et al. 2010. Longitudinal cutting of pure and doped carbon nanotubes to form graphitic nanoribbons using metal clusters as nanoscalpels. *Nano Lett.* 10:366–372.
125. Wang, H., Wang, Y., Hu, Z. et al. 2012. Cutting and unzipping multiwalled carbon nanotubes into curved graphene nanosheets and their enhanced supercapacitor performance. *ACS Appl. Mater. Interfaces* 4:6826–6833.
126. Zheng, C., Zhou, X. F., Cao, H. L. et al. 2014. Edge-enriched porous graphene nanoribbons for high energy density supercapacitors. *J. Mater. Chem. A.* 2:7484–7490. doi: 10.1039/C4TA00727A.
127. Izadi-Najafabadi, A., Yamada, T., Futaba, D. N. et al. 2010. Impact of cell-voltage on energy and power performance of supercapacitors with single-walled carbon nanotube electrodes. *Electrochem. Commun.* 12:1678–1681.
128. Arbizzani, C., Biso, M., Cericola, D. et al. 2008. Safe, high-energy supercapacitors based on solvent-free ionic liquid electrolytes. *J. Power Sources* 185:1575–1579.
129. Balducci, A., Bardi, U., Caporali, S. et al. 2004. Ionic liquids for hybrid supercapacitors. *Electrochem. Commun.* 6:566–570.
130. Frackowiak, E. 2006. Supercapacitors based on carbon materials and ionic liquids. *J. Brazil. Chem. Soc.* 17:1074–1082.
131. Vivekchand, S. R. C., Rout, C. S., Subrahmanyam, K. S. et al. 2008. Graphene-based electrochemical supercapacitors. *J. Chem. Sci.* 120:9–13.
132. Kim, T. Y., Lee, H. W., Stoller, M. et al. 2011. High-performance supercapacitors based on poly(ionic liquid)-modified graphene electrodes. *ACS Nano* 5:436–442.
133. Zhang, L. L. and Zhao, X. S. 2009. Carbon-based materials as supercapacitor electrodes. *Chem. Soc. Rev.* 38:2520–2531.

134. Yan, J., Wei, T., Qiao, W. et al. 2010. Rapid microwave-assisted synthesis of graphene nanosheet/Co$_3$O$_4$ composite for supercapacitors. *Electrochim. Acta* 55:6973–6978.

135. Xia, X.-H., Tu, J.-P., Mai, Y.-J. et al. 2011. Self-supported hydrothermal synthesized hollow Co$_3$O$_4$ nanowire arrays with high supercapacitor capacitance. *J. Mater. Chem.* 21:9319–9325.

136. Liu, J., Jiang, J., Cheng, C. et al. 2011. Co$_3$O$_4$ nanowire@MnO$_2$ ultrathin nanosheet core/shell arrays: A new class of high-performance pseudocapacitive materials. *Adv. Mater.* 23:2076–2081.

137. Wee, G., Soh, H. Z., Cheah, Y. L. et al. 2010. Synthesis and electrochemical properties of electrospun V$_2$O$_5$ nanofibers as supercapacitor electrodes. *J. Mater. Chem.* 20:6720–6725.

138. Sathiya, M., Prakash, A. S., Ramesha, K. et al. 2011. V$_2$O$_5$-anchored carbon nanotubes for enhanced electrochemical energy storage. *J. Am. Chem. Soc.* 133:16291–16299.

139. Chen, Z., Augustyn, V., Wen, J. et al. 2011. High-performance supercapacitors based on intertwined CNT/V$_2$O$_5$ nanowire nanocomposites. *Adv. Mater.* 23:791–795.

140. Liu, H., Lu, B., Wei, S. et al. 2012. Electrodeposited highly-ordered manganese oxide nanowire arrays for supercapacitors. *Solid State Sci.* 14:789–793.

141. Wang, X. Y., Wang, X. Y., Huang, W. G. et al. 2005. Sol-gel template synthesis of highly ordered MnO$_2$ nanowire arrays. *J. Power Sources* 140:211–215.

142. Feng, Z.-P., Li, G.-R., Zhong, J.-H. et al. 2009. MnO$_2$ multilayer nanosheet clusters evolved from monolayer nanosheets and their predominant electrochemical properties. *Electrochem. Commun.* 11:706–710.

143. Chu, Q., Du, J., Lu, W. et al. 2012. Synthesis of a MnO$_2$ nanosheet/graphene flake composite and its application as a supercapacitor having high rate capability. *Chempluschem* 77:872–876.

144. Nam, H.-S., Yoon, J.-K., Ko, J. M. et al. 2010. Electrochemical capacitors of flower-like and nanowire structured MnO$_2$ by a sonochemical method. *Mater. Chem. Phys.* 123:331–336.

145. Yuan, C., Hou, L., Yang, L. et al. 2011. Facile interfacial synthesis of flower-like hierarchical a-MnO$_2$ sub-microspherical superstructures constructed by two-dimension mesoporous nanosheets and their application in electrochemical capacitors. *J. Mater. Chem.* 21:16035–16041.

146. Wu, M.-S., Guo, Z.-S. and Jow, J.-J. 2010. Highly regulated electrodeposition of needle-like manganese oxide nanofibers on carbon fiber fabric for electrochemical capacitors. *J. Phys. Chem. C* 114:21861–21867.

147. Ma, J., Cheng, Q., Pavlinek, V. et al. 2013. Morphology-controllable synthesis of MnO$_2$ hollow nanospheres and their supercapacitive performance. *New J. Chem.* 37:722–728.

148. Munaiah, Y., Raj, B. G. S., Kumar, T. P. et al. 2013. Facile synthesis of hollow sphere amorphous MnO$_2$: The formation mechanism, morphology and effect of a bivalent cation-containing electrolyte on its supercapacitive behavior. *J. Mater. Chem. A* 1:4300–4306.

149. Chen, H., Zhou, S., Chen, M. et al. 2012. Reduced graphene Oxide-MnO$_2$ hollow sphere hybrid nanostructures as high-performance electrochemical capacitors. *J. Mater. Chem.* 22:25207–25216.

150. Khomenko, V., Raymundo-Pinero, E. and Beguin, F. 2006. Optimisation of an asymmetric manganese oxide/activated carbon capacitor working at 2 V in aqueous medium. *J. Power Sources* 153:183–190.

151. Wang, H., Hao, Q., Yang, X. et al. 2009. Graphene oxide doped polyaniline for supercapacitors. *Electrochem. Commun.* 11:1158–1161.

152. Paek, E., Pak, A. J., Kweon, K. E. et al. 2013. On the origin of the enhanced supercapacitor performance of nitrogen-doped graphene. *J. Phys. Chem. C* 117:5610–5616.

153. Jeong, H. M., Lee, J. W., Shin, W. H. et al. 2011. Nitrogen-doped graphene for high-performance ultracapacitors and the importance of nitrogen-doped sites at basal planes. *Nano Lett.* 11:2472–2477.

154. Cao, H., Zhou, X., Qin, Z. et al. 2013. Low-temperature preparation of nitrogen-doped graphene for supercapacitors. *Carbon* 56:218–223.

155. Wang, D.-W., Li, F. and Cheng, H.-M. 2008. Hierarchical porous nickel oxide and carbon as electrode materials for asymmetric supercapacitor. *J. Power Sources* 185:1563–1568.

156. Xue, Y., Chen, Y., Zhang, M.-L. et al. 2008. A new asymmetric supercapacitor based on lambda-MnO$_2$ and activated carbon electrodes. *Mater. Lett.* 62:3884–3886.

157. Wang, Q., Wen, Z. and Li, J. 2007. Carbon nanotubes/TiO$_2$ nanotubes hybrid supercapacitor. *J. Nanosci. Nanotechnol.* 7:3328–3331.

158. Tang, Z., Tang, C.-H. and Gong, H. 2012. A high energy density asymmetric supercapacitor from nano-architectured Ni(OH)$_2$/carbon nanotube electrodes. *Adv. Funct. Mater.* 22:1272–1278.

159. Amitha, F. E., Reddy, A. L. M. and Ramaprabhu, S. 2009. A non-aqueous electrolyte-based asymmetric supercapacitor with polymer and metal oxide/multiwalled carbon nanotube electrodes. *J. Nanopart. Res.* 11:725–729.

160. Zhang, J., Jiang, J., Li, H. et al. 2011. A high-performance asymmetric supercapacitor fabricated with graphene-based electrodes. *Energy Environ. Sci.* 4:4009–4015.

161. Yan, J., Fan, Z., Sun, W. et al. 2012. Advanced asymmetric supercapacitors based on Ni(OH)$_2$/graphene and porous graphene electrodes with high energy density. *Adv. Funct. Mater.* 22:2632–2641.

162. Lu, X., Yu, M., Zhai, T. et al. 2013. High energy density asymmetric quasi-solid-state supercapacitor based on porous vanadium nitride nanowire anode. *Nano Lett.* 13:2628–2633.

163. Qu, Q. T., Shi, Y., Li, L. L. et al. 2009. V$_2$O$_5$ center dot 0.6H(2)O nanoribbons as cathode material for asymmetric supercapacitor in K$_2$SO$_4$ solution. *Electrochem. Commun.* 11:1325–1328.

164. Du, X., Wang, C., Chen, M. et al. 2009. Electrochemical performances of nanoparticle Fe$_3$O$_4$/activated carbon supercapacitor using koh electrolyte solution. *J. Phys. Chem. C* 113:2643–2646.

165. Wu, Z.-S., Ren, W., Wang, D.-W. et al. 2010. High-energy MnO$_2$ nanowire/graphene and graphene asymmetric electrochemical capacitors. *ACS Nano* 4:5835–5842.

166. Wang, H., Casalongue, H. S., Liang, Y. et al. 2010. Ni(OH)(2) nanoplates grown on graphene as advanced electrochemical pseudocapacitor materials. *J. Am. Chem. Soc.* 132:7472–7477.

167. Stoller, M. D., Murali, S., Quarles, N. et al. 2012. Activated graphene as a cathode material for Li-ion hybrid supercapacitors. *Phys. Chem. Chem. Phys.* 14:3388–3391.

168. Li, R., Ren, X., Zhang, F. et al. 2012. Synthesis of Fe$_3$O$_4$@SnO$_2$ core-shell nanorod film and its application as a thin-film supercapacitor electrode. *Chem. Commun.* 48:5010–5012.

169. Vijayanand, S., Kannan, R., Potdar, H. S. et al. 2013. Porous Co$_3$O$_4$ nanorods as superior electrode material for supercapacitors and rechargeable Li-ion batteries. *J. Appl. Electrochem.* 43:995–1003.

170. Wang, G., Liu, Z. Y., Wu, J. N. et al. 2012. Preparation and electrochemical capacitance behavior of TiO$_2$-B nanotubes for hybrid supercapacitor. *Mater. Lett.* 71:120–122.

171. Wang, Q., Wen, Z. and Li, J. 2006. A hybrid supercapacitor fabricated with a carbon nanotube cathode and a TiO$_2$-B nanowire anode. *Adv. Funct. Mater.* 16:2141–2146.

172. Aravindan, V., Shubha, N., Ling, W. C. et al. 2013. Constructing high energy density non-aqueous Li-ion capacitors using monoclinic TiO$_2$-B nanorods as insertion host. *J. Mater. Chem. A* 1:6145–6151.

173. Hosono, E., Matsuda, H., Honma, I. et al. 2007. High-rate lithium ion batteries with flat plateau based on self-nanoporous structure of tin electrode. *J. Electrochem. Soc.* 154:A146–A149.
174. Kim, H., Cho, M. Y., Kim, M. H. et al. 2013. A novel high-energy hybrid supercapacitor with an anatase TiO_2-reduced graphene oxide anode and an activated carbon cathode. *Adv. Energy Mater.* 3:1500–1506.
175. Armstrong, A. R., Armstrong, G., Canales, J. et al. 2005. TiO_2-B nanowires as negative electrodes for rechargeable lithium batteries. *J. Power Sources* 146:501–506.
176. Armstrong, G., Armstrong, A. R., Canales, J. et al. 2006. TiO_2(B) nanotubes as negative electrodes for rechargeable lithium batteries. *Electrochem. Solid State* 9:A139–A143.
177. Armstrong, A. R., Armstrong, G., Canales, J. et al. 2005. Lithium-ion intercalation into TiO_2-B nanowires. *Adv. Mater.* 17:862–865.

6 Applications of Graphene in Solar Cells

Fuqiang Huang, Dongyun Wan,
Hui Bi, and Tianquan Lin

CONTENTS

6.1 Introduction 217
6.2 Graphene Materials as Transparent Conducting Front Contact 218
    6.2.1 Low-Defect Graphene Films via a Smart Janus Substrate 218
    6.2.2 Few-Layer Graphene Films on Dielectric Substrates 221
    6.2.3 Vertically-Erected Graphene Walls on Dielectric Substrates 222
    6.2.4 Cu/Ni Sphere-Wrapped Graphene on Highly Conductive
        Graphene Films 222
    6.2.5 Transparent Conducting Front Contact Applications in PV
        Solar Cells 224
6.3 Graphene Materials as Back Electrodes 228
    6.3.1 Pristine Graphene and B-Doped Graphene Powders 228
    6.3.2 Highly Conductive 3-D Graphene 230
    6.3.3 Back Electrode Applications in PV Solar Cells 231
6.4 Photoanode Applications in New-Concept Solar Cells 234
6.5 Summary and Outlook 237
Acknowledgments 237
References 238

6.1 INTRODUCTION

The technology of solar cells converts light to electricity and harvests solar energy. Current solar cells suffer from high cost compared with fossil fuel technologies. In order to seek more effective costs, we need to develop third-generation solar cells [1–5], including quantum dot cells [6–8], hot carrier cells [9,10], intermediate band cells [11,12], and tandem cells [13,14]. Increasing the energy conversion efficiency can provide cost-effective solar cells by maximizing the incident spectral range to allow as much sunlight to reach the absorption layer as possible, promoting adequate absorption of broad solar spectrum to realize perfect matching between the photovoltaic (PV) absorption spectrum and solar spectrum, and establishing reasonable band offset structure to enforce electron collection.

Transparent conducting oxide films are used as transparent electrodes in solar cells, which cannot promise the transmittance of near IR bands in the solar spectrum.

Conventional transparent electrode materials are indium tin oxide (ITO), fluorine tin oxide (FTO), and aluminum zinc oxide (AZO). Inflexibility, poor raw resources, and an exorbitant price make ITO uncompetitive for future PV use [15]. Furthermore, the transmittance of transparent conducting glasses is only 80%–85% in the visible light band and especially poor in the IR band. Based on its merits and capacity of contributing to broad spectrum absorption, graphene is regarded as an ideal alternative material to ITO. For a semiconductor material, its absorption spectrum can never cover the entire solar spectrum. To overcome this difficulty, multijunction and quantum dot solar cells were put forward. These PV cells urgently demand back electrode with a suitable work function [16,17], which restricts the promotion of the separation of an electron-hole pair and curbing charge recombination.

The unique advantages of graphene potentially satisfy the above PV requirements, including high optical transmittance (97.7% for single layer graphene) [18,19], high electron mobility (about 200,000 cm^2/Vs) [20], and high work function (4.5 eV) [21]. Graphene is also highly flexible, environmentally friendly, and rich in raw resources [22–33]. Many efforts have been made to apply graphene as electrodes in Schottky cells [34–36], CdTe cells [37], dye-sensitized cells [38–40], organic cells [41–43], and hybrid solar cells [44,45] aiming at increasing PV efficiency and reducing manufacturing cost. Single-layer graphene is a typical semimetal with zero band gap, its density of states (DOS) at the Dirac point is equal to zero [46,47], and its work function is about 4.5 eV [21]. N- or p-type doping enables the increase of carrier concentration and regulates work function in a rather wide span [48–50] to fit the requirements of back electrodes according to the band structure for high-efficiency solar cells. Pristine graphene, B-doped graphene [51], 3-D graphene [52], and Cu-nanowire-doped 3-D graphene [53] were used accordingly as back electrodes of CdTe cells, respectively.

Graphene-based PV applications have become an important direction that is attracting more and more attention. New-concept solar cells require a new charge collector with high electronic conductivity, light capture ability, and high surface area to load light absorbers. Due to the extremely high electron mobility, continuous graphene skeleton was used as a high-speed transport path for charge transfer [54,55]. The graphene network supported the load light absorber in dye sensitized solar cell (DSSC) [56] and quantum dot sensitized solar cell (QDSSC) [57]. Doping and tailoring of graphene are very powerful for tuning electronic structure and work function, which are the key approaches for graphene to be assembled into PV cells. It is worthy of investigating the photogenerated charge migration in a continuous 3-D graphene network. This chapter focuses mainly on the recent synthesis of high-performance graphene-based materials in terms of their practical usage as back contacts, transparent conductive electrodes, and photoanodes in PV cells. The challenges and prospects of graphene-based new-concept solar cells are also discussed.

6.2 GRAPHENE MATERIALS AS TRANSPARENT CONDUCTING FRONT CONTACT

6.2.1 Low-Defect Graphene Films via a Smart Janus Substrate

Large-area graphene films were synthesized widely on copper and nickel metal substrates by chemical vapor deposition [57–61]. A Ni metal layer can provide an efficient way of producing a multilayer graphene film due to Ni with a moderate C solubility of 0.4%–2.7 at.% over 700°C–1300°C [58,62–64]. Controlling the layer number has not been realized and the graphene layers consisting of small grains are not uniform [57–59]. In contrast, Cu foil with little C solubility (<0.001 at.% at 1000°C) has been used to produce monolayer graphene [65,66] by self-limiting growth. Unfortunately, the latter's stringent process conditions (temperature, pressure, gas composition, flow rate, etc.) [60,67] requiring a low C source content and a high temperature close to the melting point of Cu limit its utility for practical application, and controlling the number of graphene layers has never been accessible [61].

Controlling the number of graphene layers with high uniformity is a prerequisite for numerous applications [68]. Because the various physicochemical properties of graphene are sensitive to its thickness [69–71], the capability of synthesizing uniform graphene with well-controlled layer numbers has been one of the major challenges in graphene research [72], and lacking control of the number of graphene layers generates deteriorated transport properties. For instance, the band gap is open in bilayer graphene, which is useful for transistors. Reference [73] reported a rational design of a binary metal alloy that effectively suppresses the carbon precipitation process and activates a self-limited growth mechanism for homogeneous monolayer graphene. As demonstrated by a Ni-Mo alloy, the use of a binary alloy to suppress the carbon precipitation in the CVD process allowed us to grow strict single-layer graphene with 100% surface coverage. Except for this study, little work has been done in the area of strict layer controlled research.

A new method for large-area growth of graphene films was reported to realize the strict layer control and overcome the narrow processing window of the CVD method [74]. A composite substrate made of a C-dissolving top metal layer (e.g., Ni) and a C-rejecting base metal layer (e.g., Cu) was designed (Figure 6.1). Over time, the two mutually miscible metals form an alloy (e.g., 0.9Cu:0.1Ni having a C solubility of 0.07 at.% at 1000°C), which is too C-rejecting [75]. At the start of CVD, the top layer can initially dissolve adsorbed C and serve as a C sink, lessening the surface supersaturation of C and building up a pool of C supply. Next, after interdiffusion of the two metals, an alloy forms that acts as a C source, rejecting the stored C and supplying it to the surface. In this way, the self-evolving, smart Janus substrate can

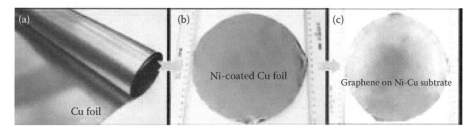

FIGURE 6.1 Cu foil (a) coated with 600 nm thick Ni film was used as a smart substrate (b) to grow Wafer-sized (ϕ18 cm) graphene (c). (Wan, D. Y., Lin, T. Q., Bi, H. et al., Autonomously controlled homogenous growth of wafer-sized high-quality graphene via a smart Janus substrate. *Adv. Funct. Mater.* 2012. 22. 1033–1039. Copyright Wiley-VCH Verlag GmbH & Co. KGaA, Weinheim. Reproduced with permission.)

dynamically and autonomously regulate the C content at and near the surface, maintaining it at an elevated yet stable level for an extended time. This helps stabilize the nucleation and growth kinetics for graphene, thus lessening the sensitivity to the processing conditions.

The above mechanism involving counteracting thermodynamic (C solubility and Cu-Ni mixing) and kinetic (C and Cu/Ni diffusion) forces that drive the self-evolution of the smart Janus substrate to maintain an elevated C content and to control graphene synthesis is fundamentally different from previously proposed ideas for large-area graphene synthesis, such as the surface-catalysis or precipitation mechanism for CVD growth on Cu and Ni [58,60,76–78] and the segregation growth mechanism for non-CVD growth on Cu-Ni [79,80]. The smart Janus substrate makes CVD much more robust, having a processing temperature spanning over 350°C, a CH_4 (carbon source) pressure ranging over one order of magnitude, and a carbon layer number tunable from one to five (Figure 6.2). This in turn yields much better film quality (transmittance, resistance, and mobility) over a coverage area that has no apparent size limit. As the concept of a "smart" substrate along with the theoretical idea for autonomous control can also be adopted to other supports—metals, semiconductors, or dielectrics, it will hopefully help steer the future development of large-scale fabrication of device-quality, defined layer-thickness graphene films for practical applications.

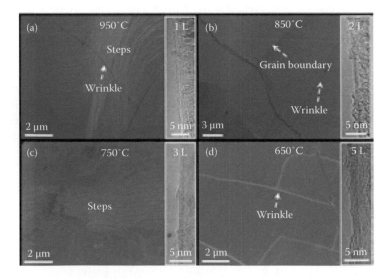

FIGURE 6.2 (a–d) SEM images of graphene grown on Ni-Cu substrate at different temperatures. Insets of (a–d): HRTEM images of folded edges showing graphene layer thickness. (Wan, D. Y., Lin, T. Q., Bi, H. et al., Autonomously controlled homogenous growth of wafer-sized high-quality graphene via a smart Janus substrate. *Adv. Funct. Mater.* 2012. 22. 1033–1039. Copyright Wiley-VCH Verlag GmbH & Co. KGaA, Weinheim. Reproduced with permission.)

6.2.2 Few-Layer Graphene Films on Dielectric Substrates

Generally, CVD graphene grown on a metal substrate [58,60] needs to be transferred on a desired substrate for the replacement of conventional transparent electrodes of solar cells, including In_2O_3: Sn (ITO), and SnO_2: F (FTO) [37,81,82]. The efficiencies of these cells cannot compete with conventional solar cells due to the wet-transfer process, which makes graphene defective and high-resistance [83–86]. Therefore, it is desirable to develop the approach of a direct synthesis of highly conductive graphene films on dielectric substrates (BN, Si, SiO_2, Al_2O_3, GaN, MgO, Si_3N_4, etc.) to replace the conventional transparent conductive films (TCFs) for PV applications. Considerable efforts were made to directly grow graphene films on these substrates [87–92] but so far continuous and highly conductive films are hard to obtain.

A new growth approach for the direct deposition of graphene films on SiO_2 substrates by ambient pressure CVD (APCVD) at 1100°C–1200°C was developed [93], the layer number and in-plane crystal size of the graphene films can be controlled by varying the growth conditions. High growth temperature, long growth time, and large flow rate ratios of CH_4 to H_2 is beneficial for obtaining thicker and large-size graphene sheets for the formation of continuous graphene films on SiO_2 substrates (Figure 6.3). Moreover, nearly the same quality graphene films can also grow on BN, SiO_2/Si, AlN, and Si substrates. The graphene films directly grown on SiO_2 substrates exhibit outstanding photoelectric properties, having a transparency from 98.1% (30 min growth) to 39.9% (120 min growth) at 550 nm, and the respective sheet resistances vary from 16540 to 158 Ω/sq. In addition, graphene films with a sheet resistance of ~63.0 Ω/sq and electron mobility of ~201.4 cm^2/Vs can be obtained by further increasing the growth time and the flow rate of CH_4 (Figure 6.4).

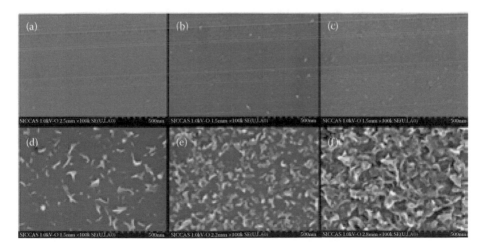

FIGURE 6.3 SEM images of graphene films directly grown on SiO_2 substrates with growth times of (a) 10, (b) 20, (c) 30, (d) 60, (e) 90, and (f) 120 min, respectively, at 1200°C. (From Bi, H., Sun, S. R., Huang, F. Q. et al., Direct growth of few-layer graphene films on SiO_2 substrates and their photovoltaic applications. *J. Mater. Chem.* 22:411–416, 2012. Reproduced by permission of The Royal Society of Chemistry.)

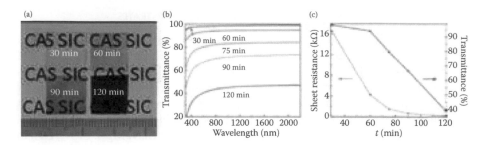

FIGURE 6.4 (a) Optical images of graphene directly grown on SiO_2 substrates at 1200°C with growth times of 30, 60, 90, and 120 min, respectively. (b) Optical transmittance spectra of the transferred graphene films. (c) Transmittance (at 550 nm) and sheet resistance as a function of growth time. (From Bi, H., Sun, S. R., Huang, F. Q. et al., Direct growth of few-layer graphene films on SiO_2 substrates and their photovoltaic applications. *J. Mater. Chem.* 22:411–416, 2012. Reproduced by permission of The Royal Society of Chemistry.)

6.2.3 VERTICALLY-ERECTED GRAPHENE WALLS ON DIELECTRIC SUBSTRATES

Methane needs high temperature (>1200°C) to be dissociated to the activated C atoms without a metal catalyst. Most of the transparent substrates cannot stand this high temperature process. PE-CVD has the advantage of achieving deposition at a low temperature and low pressure by use of a reactive species generated in the plasma. Plasma shows an excellent motivation for the reactions of gas predecessors and is a promising way to combine plasma with conventional thermal CVD conditions to directly form graphene films without any metal catalyst.

PE-CVD was developed to grow uniform vertically erected graphene walls at 900°C directly on SiO_2 substrates without using any catalyst, and it possesses a unique two-level structure composed of continuous vertically erected graphene sheets (the second level) on a nanocrystalline graphene film (the first level) [94]. Qualities and properties of graphene walls are strongly dependent on plasma power and growth time. Meanwhile, the 3-D microstructure of graphene walls can be modified by plasma power, growth temperature, growth time, and seed layer coating on the substrates. The graphene walls with well-connected 3-D structure (Figure 6.5) possess high hydrophobicity (contact angle: 141°), outstanding electron conductivity (sheet resistance: 198 Ω/sq), and tunable transparency (91.9%~38.0% at 550 nm). Hence, PE-CVD as a simple, low-cost, and scalable approach to fabricate excellent multifunction electrodes has very promising potential in future applications.

6.2.4 CU/NI SPHERE-WRAPPED GRAPHENE ON HIGHLY CONDUCTIVE GRAPHENE FILMS

Graphene structure with various dimensions have been reported by conventional CVD method, including 1-D GNR, 1-D graphene nanotube, and 3-D graphene network. A multidimensional graphene structure has controlled periodicity, large surface area, and extremely high conductivity. Thus, compared with 2-D graphene electrodes, a unique structure would show more superior properties and many more applications.

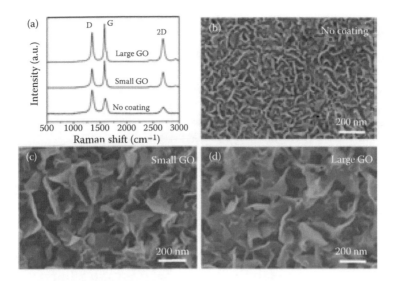

FIGURE 6.5 (a) Raman spectra and (b–d) SEM images of graphene walls grown on different preprocessed substrates. (From Yang, C. Y., Bi, H., Wan, D. Y. et al., Direct PECVD growth of vertically erected graphene walls on dielectric substrates as excellent multifunctional electrodes. *J. Mater. Chem. A* 1:770–775, 2013. Reproduced by permission of The Royal Society of Chemistry.)

DSSCs as next-generation PV devices have attracted much attention due to high energy conversion efficiency (PCE), simple device fabrication process, and relatively low cost. They composed of a dye, an electrolyte, a working electrode, and a counterelectrode (CE). Among them, the counterelectrode material in DSSCs has usually been Pt- or Pt-deposited on TCOs such as FTO and ITO. However, noble metal Pt and expensive TCO are not desirable for the low-cost fabrication of DSSC devices. Thus, it is imperative to develop a CE with high stability, conductivity, and outstanding electrochemical activity.

Accordingly, 3-D architecture of graphene-wrapped Cu-Ni sphere/graphene was synthesized by CVD method utilizing a Cu/Ni alloy sphere as both catalyst and template on highly conductive graphene films [95]. A whole flowchart for graphene sphere growth is illustrated in Figure 6.6. Two processes are included, including dip coating of Cu/Ni alloy catalyst and subsequent chemical vapor deposition of graphene. The layer numbers of graphene were controlled by the volume ratio of CH_4/H_2 and the diameters of the graphene nano-/microsphere were changed from 50 to 200 nm via tuning the Cu/Ni catalyst precursor solution concentration from 0.01 M to 0.5 M. The 3-D graphene-based CE showed superior electrocatalytic activity compared with 2-D graphene film, and a power conversion efficiency of 5.46% was achieved based on graphene-wrapped Cu-Ni sphere/graphene film CE in DSSC devices, which was comparable to that of a Pt/FTO electrode (6.19%). The novel 3-D architecture of graphene-wrapped Cu-Ni sphere/graphene film CE may be used as Pt-free and TCO-free CEs for low-cost, high-performance DSSCs.

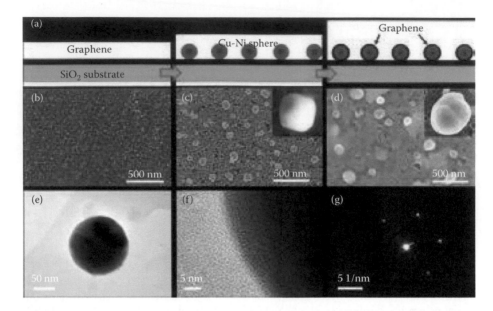

FIGURE 6.6 (a) Schematic illustration of the synthesis process of graphene/Cu-Ni sphere/ graphene film. (b–d) FESEM images of graphene film, Cu-Ni sphere/graphene, graphene/ Cu-Ni sphere/graphene film. The insets of Figure 6.1c and d are the corresponding enlarged FESEM images. (e, f) HRTEM images and (g) SAED pattern of graphene-wrapped Cu-Ni sphere. (From Bi, H. and Huang, F. Q., Novel architecture counter electrode of copper/nickel sphere wrapped graphene on highly conductive graphene films for energy conversion device. *J. Phys. Chem. C*, unpublished observations, 2013.)

6.2.5 TRANSPARENT CONDUCTING FRONT CONTACT APPLICATIONS IN PV SOLAR CELLS

The high cost of commercial solar cells is in no small part due to the high cost of the TCFs: ~40% of the fabrication cost for a-Si thin-film cells and >50% for TiO_2 dye-sensitized solar cells is attributed to the TCF front electrodes. Graphene films may substantially lower the cost of these cells if they are competitive in their perfor-mance. Graphene TCFs have been employed as electrodes for thin-film solar cells in several reports. Graphene films reduced from exfoliated graphite oxide were used as window electrodes of a DSC with a power conversion efficiency (PCE) of 0.26% [38]. Similarly, graphene films were used as transparent electrodes in organic PV (OPV) cells with an initial PCE of 0.13% [94], which was recently improved to 1.18% [95] and 3% [81]. Graphene TCFs have also been introduced to solar cells based on CdS nanowires and n-Si wafer Schottky cells, with efficiencies of 1.5% and 1.65%, respec-tively [35,96,97]. However, there have been few reports on using graphene TCFs to replace ITO or FTO in commercial a-Si, cadmium telluride (CdTe), and copper indium gallium selenide (CIGS) cells. B-doped graphene powders and 3-D graphene networks were reported as being used as back electrodes in CdTe cells [51,98,99]. Accordingly, the use of low-cost atmospheric pressure chemical vapor deposition

(AP-CVD) graphene films as front electrodes for CdTe solar cells was also studied [37], achieving an efficiency of 4.17% in two new CdTe cell structures (Figure 6.7).

As mentioned above, graphene thin films exhibit high transmittance in a wide optical range, which is ideal for use as the front contact in a PV solar cell. In this regard it shows the optic transmittance spectra of the as-prepared four-layer graphene, ITO, AZO, and commercial FTO films in the range of 350–2500 nm. Overall, graphene films exhibit a very good transparency in the whole range, and their values are superior to other TCFs, especially at long wavelengths, as shown in Figure 6.7a. Also, graphene films exhibit an extremely high mobility (up to 600 cm²/Vs) compared with commercial TCFs. Such high and wide transparency is advantageous for IR absorption for PV conversion. Meanwhile, the much higher mobility may allow light-induced electrons to be more promptly collected.

The low-cost, ambient-pressure fabrication of high-performance graphene films were incorporated as transparent front electrodes in CdTe solar cells. Graphene/cadmium sulfide (CdS) and graphene/ZnO composite films were fabricated and used in two prototype devices of glass/graphene/CdS/CdTe/(graphite paste) and

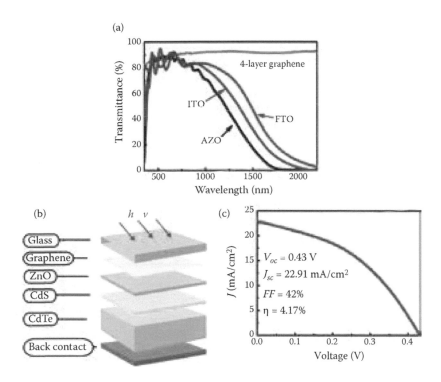

FIGURE 6.7 (a) Transmittance and transparent conductive properties of the four-layer graphene, ITO, AZO, and commercial FTO films. (b) Schematic diagram and (c) *J-V* characteristics of a glass/graphene/ZnO/CdS/CdTe/(graphite paste) solar cell. (Bi, H., Huang, F. Q., Liang, L. et al., Transparent conductive graphene films synthesized by ambient pressure chemical vapor deposition used as front electrode of CdTe solar cells. *Adv. Mater.* 2011. 23. 3202–3206. Copyright Wiley-VCH Verlag GmbH & Co. KGaA, Weinheim. Reproduced with permission.)

glass/graphene/ZnO/CdS/CdTe/(graphite paste) configurations (Figure 6.7b). The latter configuration has a PV efficiency of 4.17% as shown in Figure 6.7c, providing encouraging evidence that the graphene TCFs may be used as a new low-cost front electrode material for thin-film PV devices.

Graphene films have been employed as electrodes in CdTe solar cells [37] for the replacement of conventional TCFs. The as-used CVD graphene films were transferred from those on metal substrates. The inevitable transfer step of large-area graphene grown on a metal layer created defects, impurities, wrinkles, and cracks, which may greatly decrease the conductivity and performance of the films and is unfavorable in current PV device fabrication. Hence, the approach of direct synthesis of highly conductive graphene films on dielectric substrates was developed.

To demonstrate the applications of graphene films directly grown on SiO_2 substrates in thin-film solar cells [93], graphene films are used as CEs to fabricate DSSC prototype devices with a SiO_2/graphene-Pt/electrolyte/dye-sensitized TiO_2/FTO/ glass configuration, as shown in Figure 6.8a. The best PV efficiency reaches up to 4.25%, quite close to the performance (4.32%) of FTO electrode devices (Figure 6.8b). Such high efficiency is mainly attributed to the fact that graphene films directly grown on SiO_2 substrates exhibit a relatively low sheet resistance (~63.0 Ω/sq). More importantly, graphene films also supply an extremely high mobility (~201.4 cm²/Vs), which is conducive to separate carriers (holes/electrons) and allows light-induced carriers to be more promptly collected. In addition, the surface of graphene films is not smooth, and many graphene sheets can form a 3-D porous network structure, which can supply a large surface area for the support of Pt nanoparticles as much as possible. The unusual 3-D network of graphene films is responsible for the high electrochemical activity of graphene CEs, resulting in the excellent PV performance of DSSCs.

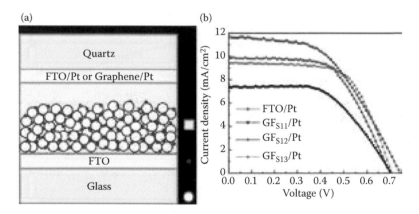

FIGURE 6.8 (a) Schematic diagram and (b) *J-V* characteristics of dye-sensitized solar cell based on FTO/Pt and graphene/Pt counterelectrodes. (From Bi, H., Sun, S. R., Huang, F. Q. et al., Direct growth of few-layer graphene films on SiO_2 substrates and their photovoltaic applications. *J. Mater. Chem.* 22:411–416, 2012. Reproduced by permission of The Royal Society of Chemistry.)

To demonstrate the applications as excellent electrode and catalyst supports, graphene films on SiO_2 substrates were also assembled as CEs in DSSC prototype devices, shown in Figure 6.9a. The DSSC, based on the graphene film with a sheet resistance of 198 Ω/sq, shows a short-circuit photocurrent density (J_{sc}) of 6.7 mA/cm², an open-circuit voltage (V_{oc}) of 0.74 V, a fill factor (FF) of 44.1%, and an overall PCE of 2.19% (Figure 6.9b). This demonstrates that graphene film can be used as the CE in DSSCs. To further optimize the cell efficiency, graphene film with a small amount of Pt by sputtering (30 s) is used as CEs, and its efficiencies reach up to 6.01% (V_{oc} = 0.77 V, J_{sc} = 12.1 mA/cm², FF = 64.4%). This property is quite close to the reference sample with Pt/FTO as a CE (6.10%), which is widely used as a CE in DSSCs [20,21]. These two works provide encouraging evidence that highly conductive graphene films are very promising CEs for DSSCs, especially for the substitution of expensive FTO.

In order to demonstrate the applications of highly conductive and flexible graphene papers in DSSCs, a prototype device containing a graphene-based CE, a dye-sensitized porous TiO_2 photoanode, and electrolyte between the two electrodes was fabricated. The reference device was prepared by using a widely used Pt/FTO CE in DSSC. This study was mainly based on the substitution of expensive FTO and noble metal Pt (30%–50% cost in DSSCs) with highly conductive papers [100]. Figure 6.10 illustrates the J-V curves of the DSSCs based on graphene, Cu-Ni alloy sphere/graphene, graphene/Cu-Ni alloy sphere/graphene, and CuSe-NiSc/graphene CEs, respectively. The Pt/graphene CE exhibited a high efficiency of 6.09%, and the corresponding J_{sc}, V_{oc}, and FF are 13.3 mA/cm², 0.70 V, and 65.9%, respectively, as shown in Figure 6.10. The cell efficiency of Pt/graphene is slightly higher than that of Pt/FTO (V_{oc} = 0.71 V, J_{sc} = 13.0 mA/cm², FF = 62.0%, η = 5.76%).

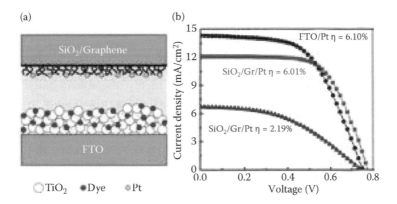

FIGURE 6.9 (a) Schematic diagram and (b) J-V characteristics of dye-sensitized solar cells based on FTO/Pt, graphene walls/Pt, and graphene wall counterelectrodes. (From Yang, C. Y., Bi, H., Wan, D. Y. et al., Direct PECVD growth of vertically erected graphene walls on dielectric substrates as excellent multifunctional electrodes. *J. Mater. Chem. A* 1:770–775, 2013. Reproduced by permission of The Royal Society of Chemistry.)

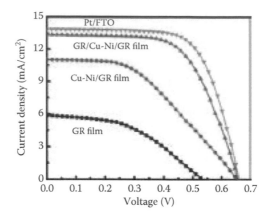

FIGURE 6.10 Current density-voltage of DSSC curves with Pt/FTO, graphene film, Cu-Ni sphere/graphene film, and graphene wrapped Cu-Ni spheres/graphene film CEs. (From Bi, H. and Huang, F. Q., Novel architecture counter electrode of copper/nickel sphere wrapped graphene on highly conductive graphene films for energy conversion device. *J. Phys. Chem. C*, unpublished observations, 2013.)

6.3 GRAPHENE MATERIALS AS BACK ELECTRODES

6.3.1 PRISTINE GRAPHENE AND B-DOPED GRAPHENE POWDERS

Great progress has been made in graphene preparation, but highly conductive graphene is hard to synthesize. Pristine graphene prepared by a conventional chemical exfoliation method is full of defects whose transport properties are not suitable for PV applications [101,102]. It is also necessary to tune by electron or hole doping to further induce novel physicochemical properties and greatly expand the application range [48].

Solvothermal reaction is a promising approach to prepare high-quality and freestanding graphene in a supercritical reaction environment with the presence of a reductant of alkali metal and an organic sovelnt (e.g., sodium and ethanol [57] and potassium and tetrachloromethane [97]). However, such an approach does not always ensure the formation of graphene, as evidenced by the formation of various carbon products reported [51,103]. Therefore, it is highly desirable to develop a rapid and cost-effective approach to the mass production of high-quality and freestanding graphene. On the other hand, the work function of graphene was calculated to be ~4.5 eV [21], and the Fermi levels of the 2 at.% B- and N-doped graphene systems can be shifted by −0.65 and 0.59 eV, respectively [48]. As an electron-deficient dopant, B can increase the work function of graphene, allowing it to serve as an ohmic back contact for CdTe solar cells that until now lacked a suitable metal for such a function.

Accordingly, a new Wurtz-type reductive coupling (WRC) reaction except solvothermal reaction was proposed as a new bottom-up method for the rapid preparation of freestanding and high-quality pristine graphene (PG) and boron-doped graphene (BG) nanosheets [51,104] (Figure 6.11). The WRC reaction is different from the solvothermal reaction, which needs an excessive amount of CCl$_4$ to form a supercritical

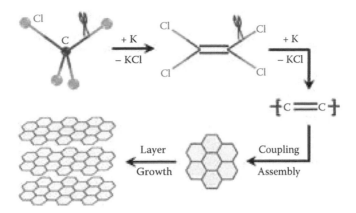

FIGURE 6.11 Schematic illustration of the formation mechanism of graphene. (From Lü, X. J., Wu, J. J., Lin, T. Q. et al., Low-temperature rapid synthesis of high-quality pristine or boron-doped graphene via Wurtz-type reductive coupling reaction. *J. Mater. Chem.* 21:10685–10689, 2011. Reproduced by permission of The Royal Society of Chemistry.)

reaction environment. The present WRC process can be a nearly stoichiometric reaction of CCl_4 and K. During the reaction, CCl_4 is first converted to dichlorocarbene via stripping off the chlorines by metallic K, and is further coupled to form tetrachloroethylene ($Cl_2C=CCl_2$). $Cl_2C=CCl_2$ species can continuously be dechlorinated to turn into $-C=C-$. Subsequently, the freshly formed $-C=C-$ readily couples together and assembles into two-dimensional hexagonal carbon clusters, and then they grow to be sp^2-structural graphene nanosheets. Graphene formed rapidly from nascent C and B in the reduction reactions of tetrachloromethane (CCl_4) and boron tribromide (BBr_3) using an alkali metal (K) reductant (Figure 6.12). As-prepared graphene with fewer defects has better electrical transport performance than chemically reduced

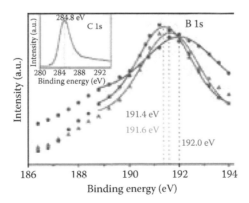

FIGURE 6.12 B 1s XPS spectra of B-doped graphene with B contents of 1.02 at.% (●), 1.90 at.% (▲), and 2.56 at.% (■). Inset shows C 1s XPS spectrum. (From Lü, X. J., Wu, J. J., and Lin, T. Q., et al., Low-temperature rapid synthesis of high-quality pristine or boron-doped graphene via Wurtz-type reductive coupling reaction. *J. Mater. Chem.* 21:10685–10689, 2011. Reproduced by permission of The Royal Society of Chemistry.)

exfoliated GO sheets using hydrazine hydrate as a reductant. As discussed in the report, B doping introduces more holes to the valence band of the graphene sheet, and a larger carrier concentration of the sample is achieved. The B-doped graphene sample has an even better electrical conductivity and a higher work function than the PG due to a larger DOS generated near the Fermi level.

6.3.2 HIGHLY CONDUCTIVE 3-D GRAPHENE

Many approaches (chemical exfoliation of graphite oxide [105,106], liquid ultrasonic exfoliation of bulk graphite [107], etc.) introduce severe structural defects in graphene sheets during preparation, and the graphene sizes are normally small (~50–500 nm). The assembled graphene films possess very high resistance (10–100 kΩ/sq), which is mainly attributed to the small size of defective graphene sheets and large intersheet contact resistance. CVD graphene films exhibit extremely high carrier mobility (up to 5100 cm^2/Vs) and low sheet resistance (~100 Ω/sq) and satisfy the requirements of PV applications [67,81,82,96] while the graphene yield is still low due to the limited growth area of the 2-D substrate.

In order to overcome the large resistance from structural defects and interfacial contacts of small-size graphene sheets to obtain highly conductive graphene films, a new growth strategy was proposed to take advantage of the CVD method to grow graphene on 3-D skeletons (porous substrates) with large surface area, and from that, the entire network of interconnected graphene sheets can be formed. Three-dimensional graphene networks have been used as the backbone of polymer composites to significantly enhance the mechanical and electrical properties [108]. Therefore, the large-size 3-D network of CVD graphene is expected to possess excellent electrical properties and promising PV applications.

Ni foam was used to grow a 3-D graphene network by APCVD [98]. The Ni foam framework, used as a template for the growth of graphene network, is a 3-D porous structure with a tortoiseshell-like morphology with pore sizes ranging from 60 to 130 μm, as shown in Figure 6.13a. The optimal APCVD was performed at 1000°C with the CH$_4$: H$_2$: Ar gas rates equal to 2: 50: 300 in standard cubic centimeters per minute (sccm). After 10-min CVD growth, the entire Ni foam surface was observed to be fully covered by continuous graphene film and graphene with many wrinkles was adhered to the Ni foam surface (Figure 6.13b). Much suspended graphene bridging these gaps between the interfacial Ni grains was found to form continuous large-size graphene sheets (Figure 6.13c), consistent with graphene growth on metal foils or films [109]. After etching away the Ni foam, a freestanding graphene network was obtained that exhibited an interconnected 3-D network structure, and all the graphene sheets (grains) in the graphene network were connected with each other without any breakage. The graphene sheets had lateral sizes of 10–50 μm and the 3-D graphene yield was 10–40 mg per gram of Ni foam.

The high-yield large-size graphene network shows high crystalline quality and excellent electrical conductivity. The electrical conductivities of our graphene films (550–600 S·cm^{-1}) are much higher than those of the graphene films from chemical or thermal reduction of GO (2–200 S·cm^{-1}) [105,110,111], B-/N-doped graphene films (1.6–3.1 S·cm^{-1}) [48], and CNT papers (~200 S·cm^{-1}) [112]. The higher electrical conductivities of our graphene films are attributed to not only the large size and

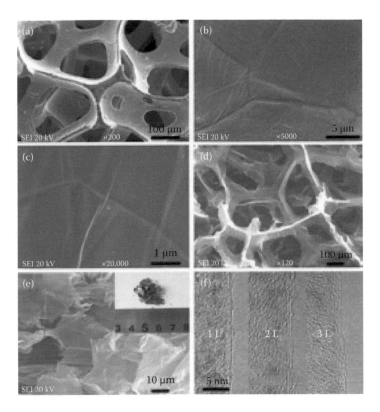

FIGURE 6.13 (a) SEM image of a Ni foam template for graphene CVD growth. (b) Low- and (c) high-magnification SEM images of as-grown graphene on a Ni foam. (d) SEM image of a 3-D graphene network without an Ni skeleton. (e) SEM image of graphene sheets after ultrasonication and an optical image of graphene sheets in the inset. (f) TEM images of graphene sheets with one, two, and three layers. (From Bi, H., Huang, F. Q., Liang, J. et al., Large-scale preparation of highly conductive three dimensional graphene and its applications in CdTe solar cells. *J. Mater. Chem.* 21:17366–17370, 2011. Reproduced by permission of The Royal Society of Chemistry.)

high crystalline CVD graphene sheets but also the 3-D interconnected graphene. The 3-D continuous network structure of the CVD graphene formed by the Ni foam template greatly eliminated the contact resistance from grain boundaries to further improve the electrical conductivity. Therefore, a 3-D graphene network with unique electrical characteristics may provide various applications in polymer composites, Li-ion batteries, supercapacitors, PV devices, and so forth [113,114].

6.3.3 Back Electrode Applications in PV Solar Cells

Conductive graphite paste is a good electrode (back contact) material to be used in CdTe solar cells. The work function of graphite nearly matches the CdTe absorber, and the highly polarized valence orbitals of tellurium (Te) (6s and 6p) have a rather intense chemical interaction with the delocalized C $3p_z$ orbitals. The explored BG sample was applied as back contact in CdTe solar cells [97], which was found to have an even better electrical conductivity and a higher work function than PG due to a larger DOS generated near the Fermi level. Accordingly, the BG had a favorable effect on improving the hole-collection ability and the PV efficiency of CdTe solar cells.

By adopting these graphenes as back electrodes, the PV efficiency of CdTe cells was increased from 6.50% for chemically (hydrazine hydrate) reduced GO to 7.41% for PG to 7.86% for BG, indicating that the highly conductive BG can improve the hole-collecting ability and reduce the barrier height to enhance the overall power conversion efficiency (Figure 6.14). The J_{sc} value of the BG-based cell, 21.96 mA cm^{-2}, is also higher than those of the rGO- and PG-based cells. The improved overall power conversion efficiency of the BG-based cell also reflects its higher work function helping to form an ohmic contact with the p-CdTe.

The as-reported large-size, continuous (Figure 6.15a), and highly conductive 3-D graphene sheets were successfully demonstrated as back electrodes in CdTe solar cells [98]. The solar cells consisted of FTO transparent electrode, CdS window layer, CdTe absorber layer, and 3-D graphene back electrode, as illustrated in Figure 6.15b, and the cell band structure is shown in Figure 6.15c. The fabricated CdTe solar cell (area: 1.0 cm^2) with a CVD graphene-based back electrode had a short-circuit photocurrent density (J_{sc}) of 20.1 mA·cm^{-2}, an open-circuit voltage (V_{oc}) of 0.78 V, a calculated fill factor (FF) of 58.0%, and an overall PCE (η) of 9.1%, which is superior to those of RGO, BG, and CNT back electrodes in this type of solar cell. The higher efficiency due to the larger V_{oc} and FF indicates that the highly conductive graphene back electrode improves the hole-collecting ability and reduces the barrier height to form a better back contact with the p-CdTe compared with other C-based back electrodes. The excellent graphene-based conducting materials may be used as new low-cost back electrodes for PV devices in the future.

The BG and highly conductive 3-D graphene networks were applied in CdTe solar cells as the back electrode [97,98]. However, the integrated graphene-based back contact still shows poor electrical characteristics (7.86% and 9.1% of cell efficiency).

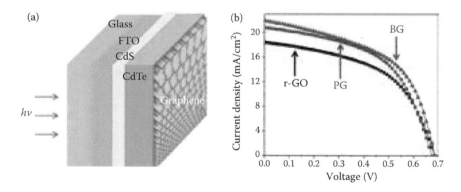

FIGURE 6.14 (a) Schematic of a CdTe solar cell with a graphene back electrode, and (b) J-V characteristics of CdTe solar cells with rGO-modified, PG-modified, and BG-modified electrodes. (From Lin, T. Q., Huang, F. Q., Liang, J. et al., A facile preparation route for boron-doped graphene, and its CdTe solar cell application. *Energ. Environ. Sci.* 4:862–865, 2011. Reproduced by permission of The Royal Society of Chemistry.)

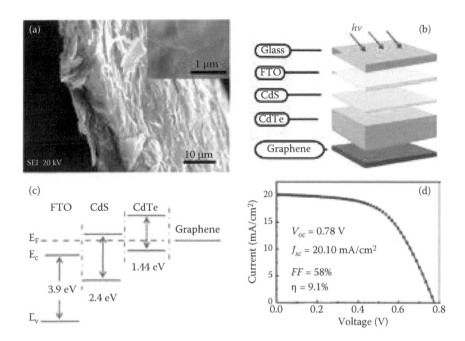

FIGURE 6.15 (a) SEM cross-section image of a graphene film and top-view of the graphene film in the inset. (b) Schematic CdTe solar cell with a 3D graphene back electrode. (c) Band structure of the graphene-based CdTe solar cell. (d) *J-V* characteristics of CdTe solar cell with graphene back electrode. (From Bi, H., Huang, F. Q., Liang, J. et al., Large-scale preparation of highly conductive three dimensional graphene and its applications in CdTe solar cells, *J. Mater. Chem.* 21:17366–17370, 2011. Reproduced by permission of The Royal Society of Chemistry.)

In addition, Cu dopant also plays a significant role in the performance improvement of CdTe solar cells. The conventional method is to dope Cu particles (Cu Ps) with a diameter of ~75 μm in graphite as a diffusion source to the CdTe layer. The large-size Cu Ps are not completely active under the condition of cell fabrication. A large amount of free Cu cannot be captured and diffuse into the CdS/CdTe PN junction, which causes electrical shorts and greatly affects the cell performance.

A novel back contact structure of 1-D Cu nanowires (NWs) doped graphene (Cu NWs/graphene) was proposed and applied in CdTe solar cells [99]. Large-size, high-quality graphene was obtained using a 3-D Ni foam catalyst by APCVD. One-dimensional single-crystal Cu NWs were prepared via hydrothermal method. The CdTe solar cells based on Cu NWs/graphene back contact showed superior improved performance of efficiency and thermal stability to those of Cu-doped graphite and Cu/Ni back contact. The electrical conductivity of Cu NWs/graphene was 16.7 S/cm and carrier mobility was 16.2 cm²/Vs. The Cu NWs/graphene as back contact exhibited high hole-collection and the short current density of the cell reached 22.4 mA/cm². Eventually the efficiency was up to 12.1% (Figure 6.16). The thermal stability of cells also benefitted from the 1-D Cu NWs/graphene, which is attributed to forming high-quality Cu$_x$Te at the surface of the CdTe layer. One-dimensional Cu NWs/graphene can be seen to provide a potential application for highly efficient and improved stable solar cells.

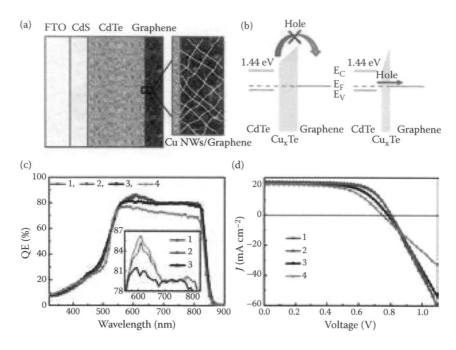

FIGURE 6.16 (a) Schematic solar cell and (b) band structure of FTO/CdS/CdTe/Cu NWs/ graphene. (c) QE of CdTe solar cells with (1) Cu NWs/graphene, (2) Cu Ps/graphene, (3) Cu Ps/ graphite, and (4) Cu/Ni. The inset is magnified QE spectra in the visible range. (d) Corresponding *J-V* characteristics of CdTe solar cells. (Liang, J., Bi, H., Wan, D. Y. et al., Novel Cu nanowires/ graphene as back contact for CdTe solar cells. *Adv. Funct. Mater.* 2012. 22. 1267–1271. Copyright Wiley-VCH Verlag GmbH & Co. KGaA, Weinheim. Reproduced with permission.)

6.4 PHOTOANODE APPLICATIONS IN NEW-CONCEPT SOLAR CELLS

The robust but flexible structure of graphene with high carrier mobility and the high specific surface area of graphene (2630 m²/g) [115–119] becomes an ideal support material with improved interfacial contact to have a profound effect in new-concept solar cells. In dye-sensitized solar cells, it should be particularly interesting to incorporate graphene into TiO_2 photoanodes due to efficient electron transfer paths to improve cell performance [54].

A simple method of heterogeneous coagulation was utilized to prepare graphene/ P25 nanocomposites [54]. P25 particles were adsorbed onto the surface of Nafion®- functionalized graphene because of the strong electrostatic attractive force between each other. The composites were used to prepare thin films acting as the electrodes of DSSCs and it was found that a small amount of graphene could effectively enhance the performance of DSSCs by providing rapid electron transport paths and extending electron lifetime. The short-circuit photocurrent density and overall energy conversion efficiency of DSSCs with P25-graphene electrodes increased by 66% and 59%, respectively, compared with P25 electrodes (Figure 6.17).

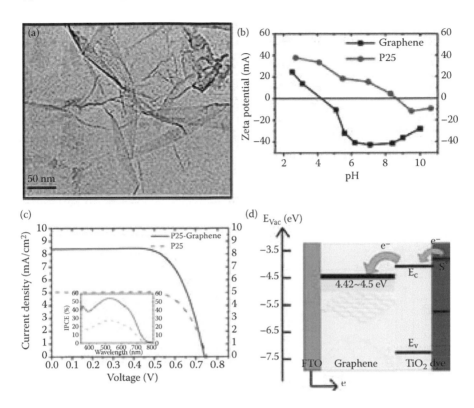

FIGURE 6.17 (a) TEM image of graphene dispersed by Nafion. (b) Zeta potential as a function of pH for Nafion-functionalized graphene and P25 (see color version online). Photocurrent *J-V* characteristics of DSSCs with P25 and P25-graphene electrodes. The inset shows the IPCE curves of two films (see color version online). (d) Schematic diagram for energy band matching. (Reprinted with permission from Sun, S. R., Gao, L., Liu, Y. Q. et al., Enhanced dye-sensitized solar cell using graphene-TiO$_2$ photoanode prepared by heterogeneous coagulation. *Appl. Phys. Lett.* 96:083113–083115. Copyright 2010, American Institute of Physics.)

Classical QDSSCs constructed by narrow band gap semiconductor QDs, nanocrystalline metal oxides, electrolytes, and CEs have attracted the interest of many researchers because of QDs' tunable absorption spectra, high molar extinction coefficient, and chemical stability [120–122]. Although the theoretical efficiency of QDSSCs is as high as 44% [123], practical performance still lags behind that of DSSCs at present. One reason is the existence of a surface state in QDs, which can capture photoinduced electrons and lead to recombination of carriers. Moreover, if the surface state is located below the conduction band (CB) of photoanode material, injection of photoinduced electrons into photoanode will be inhibited, therefore limiting the amount of photoinduced electrons collected by conducting substrate and reducing the photocurrent density of QDSSCs.

Compared with CNTs and C$_{60}$, graphene with high surface areas, low manufacturing cost, and excellent electron mobility have been proved to be more suitable for application such as conductive networks [42,124,125]. Also, the work function of graphene is in the range of 4.42–4.5 eV, which is much lower than the CB of many metal oxide photoanode material such as TiO$_2$, ZnO, and so forth. Therefore, more effective electron injections are expected to occur into graphene than into traditional nanocrystalline metal oxides. In the above work, graphene was incorporated into TiO$_2$ as a photoanode of DSSCs and was found to effectively improve the photocurrent density of DSSCs by providing fast electron transfer paths and prolonging electron lifetime [54]. Graphene employed as conducting scaffolds to anchor CdSe QDs for light harvesting and constructing QDSSCs was further reported [58]. Considering the excellent exciton generation properties of CdSe and the electron capture and transport characteristics of graphene, graphene-CdSe composite is believed to have the potential to show unique photoelectrochemical performance in QDSSCs.

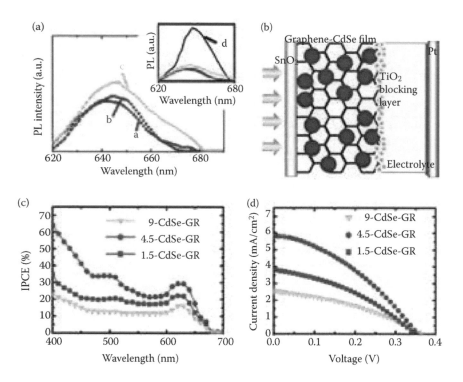

FIGURE 6.18 (a) PL spectra of graphene-CdSe composites with a CdSe-to-graphene ratio of 1.5:1 (curve a), 4.5:1 (curve b), and 9:1 (curve c) (excitation wavelength 480 nm). The inset is the PL spectra of CdSe (curve d) and the above composites, (b) structure illustration for QDSSC, (c) IPCE spectrum of, and (d) *J-V* characteristics of the QDSSCs based on different graphene-CdSe composites. (Reprinted with permission from Sun, S. R., Gao, L., Liu, Y. Q. et al., Assembly of CdSe nanoparticles on graphene for low-temperature fabrication of quantum dot sensitized solar cell. *Appl. Phys. Lett.* 98:093112–093114. Copyright 2011 American Institute of Physics.)

Graphene-CdSe composite has been prepared by chemical bath deposition and used to fabricate graphene-CdSe films at low temperature for assembly of QDSSC without traditional wide band gap semiconductor material. Through a study of TEM images and PL spectra, it was found that increasing the amount of graphene could decrease the aggregation of QDs CdSe and inhibit charge recombination. The above results pave the way for graphene-CdSe composites used in PV applications. In a QDSSC system, a graphene network could work as an excellent conducting scaffold to capture and transport photoinduced charge carriers, which has been proven by IPCE and J-V measurements (Figure 6.18). This study also optimized the weight ratio of GO to CdSe and found that J_{sc} and η increased to 5.80 mA/cm^2 and 0.76%, respectively, for the QDSSCs with 4.5-CdSe-GR. Compared with previous reports, the way the films were prepared in this work allowed control of the amount of sensitizers conveniently and increased the surface coverage of QDs effectively. Meanwhile, the low-temperature fabrication process might greatly benefit the preparation of flexible solar cells.

6.5 SUMMARY AND OUTLOOK

As a novel and unique member of the carbon nanomaterials group, graphene manifests attractive application potentials in PV cell applications [126]. Some recent developments of advanced graphene synthesis and their related devices applications are reviewed in this chapter, and these works constitute significant advances toward the production of high-performance graphene-based materials and their practical usage as back contacts, transparent conductive electrodes, and photoanodes in PV cells.

Solar cells are one of the most practical solutions for our energy sources in the future and are one of the most promising devices that could satisfy global energy requirements. All efforts related to solar cells need to be focused on improving efficiency and lowering their cost. Graphene is an attractive candidate for these purposes, although the applications of graphene in solar cells are still in the initial stages. To achieve these goals, further efforts should be directed at fabricating scalable high-quality graphene films efficiently, confining electron-band gap engineering, synthesizing graphene-based light absorbers with suitable band gaps (~0.5–3 eV), and assembling them into devices as all-C-based full solar spectrum PV cells in novel or effective ways. It is foreseeable that much more effort will be focused on exploring the full potential of graphene and will push the mature solar cell technologies into the PV market.

ACKNOWLEDGMENTS

Financial support from National 973 & 863 Program of China Grant Nos. 2009CB939903 and 2011AA050505, NSF of China Grant Nos. 11274328, 91122034, 51125006, 51102263, and 21101164, and NSF of Shanghai Grant No.11ZR1441900. The authors thank Dr. Jun Liang, Dr. Hui Bi, Dr. Xujie Lv, Dr. Shengrui Sun, Mr. Tianquan Lin, and Mr. Chongyin Yang for their helpful suggestions.

REFERENCES

1. Werner, J. H. 2004. Second and third generation photovoltaics—dreams and reality. *Adv. Solid State Phys.* 44:51–67.
2. Green, M. A. Editor. 2003. Third generation concepts for photovoltaics. *Proceedings of 3rd World Conference on Photovoltaic Energy Conversion*, Vols. A–C, Osaka, Japan.
3. Green, M. A. 2002. Third generation photovoltaics: Solar cells for 2020 and beyond. *Physica E* 14:65–70.
4. Conibeer, G. 2007. Third-generation photovoltaics. *Mater. Today* 10:42–50.
5. Brown, G. F. and Wu, J. Q. 2009. Third-generation photovoltaics. *Laser Photon. Rev.* 3:394–405.
6. Aroutiounian, V., Petrosyan, S., Khachatryan, A. et al. 2001. Quantum dot solar cells. *J. Appl. Phys.* 89:2268–2271.
7. Nozik, A. J. 2002. Quantum dot solar cells. *Physica E* 14:115–120.
8. Beard, M. C., Luther, J. M., Midgett, A. G. et al. 2010. Third generation photovoltaics: Multiple exciton generation in colloidal quantum dots, quantum dot arrays, and quantum dot solar cells. *35th IEEE Photovoltaic Specialists Conference*, Honolulu, HI.
9. Ross, R. T. and Nozik, A. J. 1982. Efficiency of hot-carrier solar energy converters. *J. Appl. Phys.* 53:3813–3818.
10. Konig, D., Casalenuovo, K., Takeda, Y. et al. 2010. Hot carrier solar cells: Principles, materials and design. *Physica E* 42:2862–2866.
11. Luque, A. and Marti, A. 1997. Increasing the efficiency of ideal solar cells by photon induced transitions at intermediate levels. *Phys. Rev. Lett.* 78:5014–5017.
12. Marti, A. and Luque, A. Editors. 2009. Intermediate band solar cells. *14th OptoElectronics and Communications Conference*, Hong Kong, Peoples Republic of China.
13. Dharmadasa, I. M. 2005. Third generation multi-layer tandem solar cells for achieving high conversion efficiencies. *Sol. Energy Mat. Sol. C.* 85:293–300.
14. King, R. R., Law, D. C., Edmondson, K. M. et al. 2007. 40% efficient metamorphic GaInP/GaInAs/Ge multijunction solar cells. *Appl. Phys. Lett.* 90:183516–183518.
15. Wang, L., Yang, Y., Marks, T. J. et al. 2005. Near-infrared transparent electrodes for precision Teng–Man electro-optic measurements: In_2O_3 thin-film electrodes with tunable near-infrared transparency. *Appl. Phys. Lett.* 87:161107–161109.
16. Ponpon, J. P. 1985. A review of ohmic and rectifying contacts on cadmium telluride. *Solid State Electron.* 28:689–706.
17. Fritsche, J., Kraft, D., Thissen, A. et al. 2002. Band energy diagram of CdTe thin film solar cells. *Thin Solid Films* 403:252–257.
18. Nair, R. R., Blake, P. and Grigorenko, A. N. 2008. Fine structure constant defines visual transparency of graphene. *Science* 320:1308.
19. Stauber, T., Peres, N. M. R. and Geim, A. K. 2008. Optical conductivity of graphene in the visible region of the spectrum. *Phys. Rev. B* 78:085432–085439.
20. Bolotin, K. I., Sikes, K. J. and Jiang, Z. 2008. Ultrahigh electron mobility in suspended graphene. *Solid State Commun.* 146:351–355.
21. Sque, S. J., Jones, R. and Briddon, P. R. 2007. The transfer doping of graphite and graphene. *Phys. Status Solidi Appl. Res.* 204:3078–3084.
22. Vollmer, A., Feng, X. L. and Wang, X. 2009. Electronic and structural properties of graphene-based transparent and conductive thin film electrodes. *Appl. Phys. A Mater.* 94:1–4.
23. Wang, J. L., Ma, L., Yuan, Q. et al. 2011. Transition-metal-catalyzed unzipping of single-walled carbon nanotubes into narrow graphene nanoribbons at low temperature. *Angew. Chem. Int. Ed.* 50:8041–8045.
24. Jiao, L. Y., Zhang, L. and Wang, X. R. G. 2009. Narrow graphene nanoribbons from carbon nanotubes. *Nature* 458:877–880.

25. Kosynkin, D. V., Higginbotham, A. L., Sinitskii, A. et al. 2009. Longitudinal unzipping of carbon nanotubes to form graphene nanoribbons. *Nature* 458:872–876.
26. Ma, L., Zhu, L. Y. and Wang, J. L. 2011. Boron and nitrogen doping induced half-metallicity in zigzag triwing graphene nanoribbons. *J. Phys. Chem. C* 115:6195–6199.
27. Cai, J. M., Ruffieux, P., Jaafar, R. et al. Atomically precise bottom-up fabrication of graphene nanoribbons. *Nature* 466:470–473.
28. Ma, L., Wang, J. L. and Ding, F. 2012. Strain-induced orientation-selective cutting of graphene into graphene nanoribbons on oxidation. *Angew. Chem. Int. Ed.* 51:1161–1164.
29. Sprinkle, M., Ruan, M., Hu, Y. et al. 2010. Scalable templated growth of graphene nanoribbons on SiC. *Nat. Nanotechnol.* 5:727–731.
30. Wei, D. C., Liu, Y. Q., Zhang, H. L. et al. 2009. Scalable synthesis of few-layer graphene ribbons with controlled morphologies by template method and their applications in nanoelectromechanical switches. *J. Am. Chem. Soc.* 131:11147–11154.
31. Cano-Márquez, A. G., Rodríguez-Macías, F. J., Campos-Delgado, J. et al. 2009. Graphene sheets and ribbons produced by lithium intercalation and exfoliation of carbon nanotubes. *Nano Lett.* 9:1527–1533.
32. Campos-Delgado, J., Romo-Herrera, J. M., Jia, X. T. et al. 2008. Bulk production of a new form of sp(2) carbon: Crystalline graphene nanoribbons. *Nano Lett.* 8:2773–2778.
33. Yang, X. Y., Dou, X., Rouhanipour, A. et al. 2008. Two-dimensional graphene nanoribbons. *J. Am. Chem. Soc.* 130:4216–4217.
34. Feng, T. T., Xie, D., Lin, Y. X. et al. 2011. A ZnO cross-bar array resistive random access memory stacked with heterostructure diodes for eliminating the sneak current effect. *Appl. Phys. Lett.* 99:233505–233507.
35. Li, X. M., Zhu, H. W., Wang, K. L. et al. 2010. Graphene-on-silicon Schottky junction solar cells. *Adv. Mater.* 22:2743–2748.
36. Li, C. Y., Li, Z., Zhu, H. W. et al. 2010. Graphene nano-"patches" on carbon nanotube network for highly transparent/conductive thin film applications. *J. Phys. Chem. C* 114:14008–14012.
37. Bi, H., Huang, F. Q., Liang, L. et al. 2011. Transparent conductive graphene films synthesized by ambient pressure chemical vapor deposition used as front electrode of CdTe solar cells. *Adv. Mater.* 23:3202–3206.
38. Wang, X., Zhi, L. J. and Mullen, K. 2008. Transparent, conductive graphene electrodes for dye-sensitized solar cells. *Nano Lett.* 8:323–327.
39. Zhang, D. W., Li, X. D., Li, H. B. et al. 2011. Graphene-based counter electrode for dye-sensitized solar cells. *Carbon* 49:5382–5388.
40. Cruz, R., Tanaka, D. A. P. and Mendes, A. 2012. Reduced graphene oxide films as transparent counter-electrodes for dye-sensitized solar cells. *Sol. Energy* 86:716–724.
41. Park, H., Rowehl, J. A., Kim, K. K. et al. 2010. Doped graphene electrodes for organic solar cells. *Nanotechnology* 21:505204–505209.
42. Wu, J. B., Becerril, H. A., Bao, Z. N. et al. 2008. Organic solar cells with solution-processed graphene transparent electrodes. *Appl. Phys. Lett.* 92:263302–263304.
43. Kalita, G., Matsushima, M., Uchida, H. et al. 2010. Graphene constructed carbon thin films as transparent electrodes for solar cell applications. *J. Mater. Chem.* 20:9713–9717.
44. Yang, K. K., Xu, C. K., Huang, L. W. et al. 2011. Hybrid nanostructure heterojunction solar cells fabricated using vertically aligned ZnO nanotubes grown on reduced graphene oxide. *Nanotechnology* 22:405401–405408.
45. Wang, Z. B., Puls, C. P., Staley, N. E. et al. 2011. Technology ready use of single layer graphene as a transparent electrode for hybrid photovoltaic devices. *Physica E* 44:521–524.
46. Uchoa, B. and Neto, A. H. C. 2007. Superconducting states of pure and doped graphene. *Phys. Rev. Lett.* 98:146801–146804.

47. Wang, X. R., Li, X. L., Zhang, L. et al. 2009. N-doping of graphene through electrothermal reactions with ammonia. *Science* 324:768–771.
48. Panchokarla, L. S., Subrahmanyam, K. S., Saha, S. K. et al. 2009. Synthesis, structure, and properties of boron- and nitrogen-doped graphene. *Adv. Mater.* 21:4726–4730.
49. Shi, Y. M., Kim, K. K., Reina, A. et al. 2010. Work function engineering of graphene electrode via chemical doping. *ACS Nano* 4:2689–2694.
50. Matis, B. R., Burgess, J. S., Bulat, F. A. et al. 2012. Surface doping and band gap tunability in hydrogenated graphene. *ACS Nano* 6:17–22.
51. Lin, T. Q., Huang, F. Q., Liang, J. et al. 2011. A facile preparation route for boron-doped graphene, and its CdTe solar cell application. *Energy Environ. Sci.* 4:862–865.
52. Yang, N. L., Zhai, J., Wang, D. et al. 2010. Two-dimensional graphene bridges enhanced photoinduced charge transport in dye-sensitized solar cells. *ACS Nano* 4:887–894.
53. Ng, Y. H., Lightcap, I. V., Goodwin, K. et al. 2010. To what extent do graphene scaffolds improve the photovoltaic and photocatalytic response of TiO_2 nanostructured films? *J. Phys. Chem. Lett.* 1:2222–2227.
54. Sun, S. R., Gao, L., Liu, Y. Q. et al. 2010. Enhanced dye-sensitized solar cell using graphene-TiO_2 photoanode prepared by heterogeneous coagulation. *Appl. Phys. Lett.* 96:083113–083115.
55. Sun, S. R., Gao, L., Liu, Y. Q. et al. 2011. Assembly of CdSe nanoparticles on graphene for low-temperature fabrication of quantum dot sensitized solar cell. *Appl. Phys. Lett.* 98:093112–093114.
56. Reina, A., Jia, X., Ho, J. et al. 2008. Large area, few-layer graphene films on arbitrary substrates by chemical vapor deposition. *Nano Lett.* 9:30–35.
57. Kim, K. S., Zhao, Y., Jang, H. et al. 2009. Large-scale pattern growth of graphene films for stretchable transparent electrodes. *Nature* 457:706–710.
58. Gunes, F., Han, G. H., Kim, K. K. et al. 2009. Graphene-based flexible transparent conducting films. *NANO* 4:83–90.
59. Li, X., Cai, W., An, J. et al. 2009. Large-area synthesis of high-quality and uniform graphene films on copper foils. *Science* 324:1312–1314.
60. Han, G. H., Shin, H. J., Kim, E. S. et al. 2011. One-step transfer of ultra-large graphene. *NANO* 6:59–65.
61. Lander, J. J., Kern, H. E. and Beach, A. L. 1952. Solubility and diffusion coefficient of carbon in nickel: reaction rates of nickel-carbon alloys with barium oxide. *J. Appl. Phys.* 23:1305–1309.
62. Eizenberg, M. and Blakely, J. M. 1979. Carbon monolayer phase condensation on Ni(111). *Surf. Sci.* 82:228–236.
63. Shelton, J. C., Patil, H. R. and Blakely, J. M. 1974. Equilibrium segregation of carbon to a nickel (111) surface: a surface phase transition. *Surf. Sci.* 43:493–520.
64. Bhaviripudi, S., Jia, X., Dresselhaus, M. S. et al. 2010. Role of kinetic factors in chemical vapor deposition synthesis of uniform large area graphene using copper catalyst. *Nano Lett.* 10:4128–4133.
65. López, G. A. and Mittemeijer, E. J. 2004. The solubility of C in solid Cu. *Scripta Mater.* 51:1–5.
66. Bae, S., Kim, H., Lee, Y. et al. 2010. Roll-to-roll production of 30-inch graphene films for transparent electrodes. *Nat. Nanotechnol.* 5:574–578.
67. Shin, H. J., Choi, W. M., Yoon, S. M. et al. 2011. Transfer-free growth of few-layer graphene by self-assembled monolayers. *Adv. Mater.* 23:4392–4397.
68. Zhang, Y. B., Tang, T. T., Girit, C. et al. 2009. Direct observation of a widely tunable bandgap in bilayer graphene. *Nature* 459:820–823.
69. Craciun, M. F., Russo, S., Yamamoto, M. et al. 2009. Trilayer graphene is a semimetal with a gate-tunable band overlap. *Nat. Nanotechnol.* 14:383–388.

70. Partoens, B. and Peeters, F. M. 2006. From graphene to graphite: Electronic structure around the K point. *Phys. Rev. B* 74:075404–075414.
71. Wei, D. C. and Liu, Y. Q. 2010. Controllable synthesis of graphene and its applications. *Adv. Mater.* 22:3225–3241.
72. Dai, B. Y., Fu, L., Zou, Z. Y. et al. 2011. Rational design of a binary metal alloy for chemical vapour deposition growth of uniform single-layer graphene. *Nat. Commun.* 2:522–525.
73. Wan, D. Y., Lin, T. Q., Bi, H. et al. 2012. Autonomously controlled homogenous growth of wafer-sized high quality graphene via a smart Janus substrate. *Adv. Funct. Mater.* 22:1033–1039.
74. Nicholson, M. E. 1962. Molecular processes on solid surfaces. *Trans. Metall. Soc. AIME* 224:533–535.
75. Li, X., Cai, W., Colombo, L. et al. 2009. Evolution of graphene growth on Ni and Cu by carbon isotope labeling. *Nano Lett.* 9:4268–4272.
76. Chae, S. J., Güneş, F., Kim, K. K. et al. 2009. Synthesis of large-area graphene layers on poly-nickel substrate by chemical vapor deposition: Wrinkle formation. *Adv. Mater.* 21:2328–2333.
77. Li, X., Magnuson, C. W., Venugopal, A. et al. 2010. Graphene films with large domain size by a two-step chemical vapor deposition process. *Nano Lett.* 10:4328–4334.
78. Liu, N., Fu, L., Dai, B. et al. 2010. Universal segregation growth approach to wafer-size graphene from non-noble metals. *Nano Lett.* 11:297–303.
79. Wang, Y., Tong, S. W., Xu, X. F. et al. 2011. Interface engineering of layer-by-layer stacked graphene anodes for high-performance organic solar cells. *Adv. Mater.* 23:1514–1518.
80. Li, X., Zhu, H., Wang, K. et al. 2010. Graphene-on-silicon Schottky junction solar cells. *Adv. Mater.* 22:2743–2748.
81. Li, X., Zhu, Y., Cai, W. et al. 2009. Transfer of large-area graphene films for high-performance transparent conductive electrodes. *Nano Lett.* 9:4359–4363.
82. Levendorf, M. P., Ruiz-Vargas, C. S., Garg, S. et al. 2009. Transfer-free batch fabrication of single layer graphene transistors. *Nano Lett.* 9:4479–4483.
83. Hofrichter, J., Szafranek, B. N., Otto, M. et al. 2010. Synthesis of graphene on silicon dioxide by a solid carbon source. *Nano Lett.* 10:36–42.
84. Ismach, A., Druzgalski, C., Penwell, S. et al. 2010. Direct chemical vapor deposition of graphene on dielectric surfaces. *Nano Lett.* 10:1542–1548.
85. Jerng, S. K., Yu, D. S., Kim, Y. S. et al. 2011. Nanocrystalline graphite growth on sapphire by carbon molecular beam epitaxy. *J. Phys. Chem. C* 115:4491–4494.
86. Kim, K. B., Lee, C. M. and Choi, J. 2011. Catalyst-free direct growth of triangular nanographene on all substrates. *J. Phys. Chem. C* 115:14488–14493.
87. Lee, C. M. and Choi, J. 2011. Direct growth of nanographene on glass and postdeposition size control. *Appl. Phys. Lett.* 98:183106–183108.
88. Zhang, L., Shi, Z., Wang, Y. et al. 2011. Catalyst-free growth of nanographene films on various substrates. *Nano Res.* 4:315–321.
89. Ding, X., Ding, G., Xie, X. et al. 2011. Direct growth of few layer graphene on hexagonal boron nitride by chemical vapor deposition. *Carbon* 49:2522–2525.
90. Rümmeli, M. H., Bachmatiuk, A., Scott, A. et al. 2010. Direct low temperature nanographene synthesis over a dielectric insulator. *ACS Nano* 4:4206–4210.
91. Bi, H., Sun, S. R., Huang, F. Q. et al. 2012. Direct growth of few-layer graphene films on SiO_2 substrates and their photovoltaic applications. *J. Mater. Chem.* 22:411–416.
92. Yang, C. Y., Bi, H., Wan, D. Y. et al. 2013. Direct PECVD growth of vertically erected graphene walls on dielectric substrates as excellent multifunctional electrodes. *J. Mater. Chem. A* 1:770–775.

93. Bi, H. and Huang, F. Q. 2014. Novel architecture counter electrode of copper/nickel sphere wrapped graphene on highly conductive graphene films for energy conversion device. *Adv. Funct. Mater.* (Submitted).

94. Xu, Y. F., Long, G. K., Huang, L. et al. 2010. Polymer photovoltaic devices with transparent graphene electrodes. *Carbon* 48:3293–3311.

95. Arco, L. G. D., Zhang, Y., Schlenker, C. W. K. et al. 2010. Continuous, highly flexible, and transparent graphene films by chemical vapor deposition for organic photovoltaics. *ACS Nano* 25:2865–2873.

96. Ye, Y., Dai, Y., Dai, L. et al. 2010. High-performance single CdS nanowire (Nanobelt) Schottky junction solar cells with au/graphene Schottky electrodes. *ACS Appl. Mater. Interfaces* 2:3406–3410.

97. Kholmanov, I. N., Stoller, M. D., Edgeworth, J. et al. 2012. Nanostructured hybrid transparent conductive films with antibacterial properties. *ACS Nano* 6:5157–5163.

98. Bi, H., Huang, F. Q., Liang, J. et al. Large-scale preparation of highly conductive three dimensional graphene and its applications in CdTe solar cells. *J. Mater. Chem.* 21:17366–17370.

99. Liang, J., Bi, H., Wan, D. Y. et al. Novel Cu nanowires/graphene as back contact for CdTe solar cells. *Adv. Funct. Mater.* 22:1267–1271.

100. Wang, Y., Wu, M., Lin, X. et al. 2012. Several highly efficient catalysts for Pt-free and FTO-free counter electrodes of dye-sensitized solar cells. *J. Mater. Chem.* 22: 4009–4014.

101. Uchoa, B. and Castro Neto, A. H. 2007. Superconducting states of pure and doped graphene. *Phys. Rev. Lett.* 98:146801–146804.

102. Wang, X., Li, X., Zhang, L. et al. 2009. N-doping of graphene through electrothermal reactions with ammonia. *Science* 324:768–771.

103. Kuang, Q., Xie, S. Y., Jiang, Z. Y. et al. 2004. Low temperature solvothermal synthesis of crumpled carbon nanosheets. *Carbon* 42:1737–1741.

104. Lü, X. J., Wu, J. J., Lin, T. Q. et al. 2011. Low-temperature rapid synthesis of high-quality pristine or boron-doped graphene via Wurtz-type reductive coupling reaction. *J. Mater. Chem.* 21:10685–10689.

105. Eda, G., Fanchini, G., Chhowalla, M. et al. 2008. Large-area ultrathin films of reduced graphene oxide as a transparent and flexible electronic material. *Nat. Nanotechnol.* 3:270–274.

106. Xu, Y., Bai, H., Lu, G. et al. 2008. Flexible graphene films via the filtration of water-soluble noncovalent functionalized graphene sheets. *J. Am. Chem. Soc.* 130:5856–5857.

107. Hernandez, Y., Nicolosi, V. and Lotya, M. et al. 2008. High-yield production of graphene by liquid-phase exfoliation of graphite. *Nat. Nanotechnol.* 3:563–568.

108. Chen, Z. P., Ren, W. C., Gao, L. B. et al. 2011. Researchers in Shenyang have made graphene–polymer composites that are not only light and flexible, but also conduct electricity. *Nat. Mater.* 10:424–428.

109. Zhao, W., Kozlov, S. M., Höfert, O. et al. 2011. Graphene on Ni(111): Coexistence of different surface structures. *J. Phys. Chem. Lett.* 2:759–764.

110. Gao, W., Alemany, L. B., Ci, L. et al. 2009. New insights into the structure and reduction of graphite oxide. *Nat. Chem.* 1:403–408.

111. Gwon, H., Kim, H. S., Lee, K. et al. 2011. Flexible energy storage devices based on graphene paper. *Energy Environ. Sci.* 4:1277–1283.

112. Wang, D., Song, P., Liu, C. et al. 2008. Highly oriented carbon nanotube papers made of aligned carbon nanotubes. *Nanotechnology* 19:075609–075614.

113. Pumera, M. 2011. Graphene-based nanomaterials for energy storage. *Energy Environ. Sci.* 4:668–674.

114. Sun, Y. Q., Wu, Q., Shi, G. Q. et al. 2011. Graphene based new energy materials. *Energy Environ. Sci.* 4:1113–1132.

115. Novoselov, K. S., Geim, A. K., Morozov, S. V. et al. 2004. Electric field effect in atomically thin carbon films. *Science* 306:666–669.
116. Peigney, A., Laurent, C., Flahaut, E. et al. 2001. Specific surface area of carbon nanotubes and bundles of carbon nanotubes. *Carbon* 39:507–514.
117. Shi, Y. M., Fang, W. J., Zhang, K. K. et al. 2009. Photoelectrical response in single-layer graphene transistors. *Small* 5:2005–2011.
118. Vivekchand, S. R., Rout, C. C. S., Subrahmanyam, K. S. et al. 2008. Graphene-based electrochemical supercapacitors. *J. Chem. Sci.* 120:9–13.
119. Wang, Y., Chen, X. H., Zhong, Y. L. et al. 2009. Large area, continuous, few-layered graphene as anodes in organic photovoltaic devices. *Appl. Phys. Lett.* 95:063302–063304.
120. Alivisatos, A. P. 1996. Semiconductor clusters, nanocrystals, and quantum dots. *Science* 271:933–937.
121. Kamat, P. V. 2008. Quantum dot solar cells. Semiconductor nanocrystals as light harvesters. *J. Phys. Chem. C* 112:18737–18753.
122. Yu, W. W., Qu, L. H., Guo, W. Z. et al. 2003. Experimental determination of the extinction coefficient of CdTe, CdSe, and CdS nanocrystals. *Chem. Mater.* 15:2854–2860.
123. Klimov, V. I. 2006. Mechanisms for photogeneration and recombination of multiexcitons in semiconductor nanocrystals: Implications for lasing and solar energy conversion. *J. Phys. Chem. B* 110:16827–16845.
124. Rao, C. N. R., Sood, A. K., Subrahmanyam, K. S. et al. 2009. Graphene: The new two-dimensional nanomaterial. *Angew. Chem. Int. Ed.* 48:7752–7777.
125. Wei, T., Luo, G. L., Fan, Z. J. et al. 2009. Preparation of graphene nanosheet/polymer composites using in-situ reduction-extractive dispersion. *Carbon* 47:2296–2299.
126. Wan, D., Cui, H., Huang, F. et al. 2013. Graphene fabrication and its photovoltaic applications. *Rev. Adv. Sci. Eng. (RASE)* 2:1–16.

7 Applications of Graphene in Fuel Cells

Xuejun Zhou, Jinli Qiao, and Yuyu Liu

CONTENTS

7.1   Introduction to Polymer Electrolyte Fuel Cells 246
    7.1.1   History of Fuel Cells...246
    7.1.2   Operating Principles of Fuel Cells and Their Components..............247
    7.1.3   Challenges of Fuel Cells and Mitigation Strategies........................250
7.2   Graphene as Support Material for Pt/Pt-Alloy Catalysts 252
    7.2.1   Pristine Single-Layer Graphene as Support Material......................252
    7.2.2   Conductive Polymer-Functionalized GNSs as Support Material 256
    7.2.3   Hybrid Supporting Material by Intercalating Carbon Black and CNTs 258
    7.2.4   Nitrogen-Doped Graphene as Support Material..............................261
7.3   Graphene/Nitrogen-Doped Graphene as Support Material for Nonprecious Metal Catalysts 264
    7.3.1   Metal Oxides...265
    7.3.2   Pyrolyzed Metal Macrocyclic Compounds 267
    7.3.3   Transition Metal-Nitrogen-Containing Complexes 267
7.4   Graphene as a Metal-Free Catalyst for ORR and MOR..............................270
    7.4.1   NG as a Metal-Free Catalyst 270
        7.4.1.1   Synthesis Validation of NG...271
        7.4.1.2   Morphology Control of NG 273
        7.4.1.3   Active Sites of NG for ORR...275
    7.4.2   Other Heteroatom-Doped Graphene as a Catalyst for ORR and MOR 276
7.5   Graphene for MEA Assembly and Fuel Cell Performance 278
7.6   Graphene as Additives for Membrane Preparation and Modifications 281
7.7   Conclusions and Outlook..286
Acknowledgments...287
References...287

7.1 INTRODUCTION TO POLYMER ELECTROLYTE FUEL CELLS

The need for increased energy supply and continuous and rapid depletion of fossil fuels with alarming increase in the concentration of greenhouse gases have stimulated intense research on the development of alternative sustainable energy technologies. In this way, electrochemical energy storage and conversion technologies such as fuel cells, batteries, supercapacitors, and water electrolysis have been recognized as the most feasible and efficient technologies for portable, stationary, and transportation applications. Among these technologies, fuel cells converted the energy of a fuel into electricity through a direct electrochemical reaction without any harmful emissions with efficiencies in excess of 80%, in contrast to the limiting efficiency of the Carnot cycle in principle. Thus, fuel cell technologies have received a great deal of interest among the scientific and engineering communities because they are extremely environmentally friendly and have a high energy power density.

According to the type of electrolyte employed, fuel cells can be mainly divided into five categories: (1) polymer electrolyte fuel cells (PEFCs) including proton-exchange membrane fuel cell (PEMFC) and direct methanol fuel cell (DMFC) depending on their different fuels used (H_2 is used for the former and methanol is used for the latter), (2) alkaline fuel cell (AFC), (3) phosphoric acid fuel cell (PAFC), (4) molten carbonate fuel cell (MCFC), and (5) solid oxide fuel cell (SOFC). Among various kinds of fuel cells, PEMFCs and DMFCs have been extensively studied, since both are able to efficiently generate high power densities at relatively low temperature (50°C to 100°C), making them particularly attractive for transportation and certain mobile and portable applications.

7.1.1 HISTORY OF FUEL CELLS

Despite the recent considerable amount of attention given to it, the fuel cell has a long history. As early as 1838, the principle of the fuel cell was discovered by German scientist Christian Friedrich Schönbein and published in one of the scientific magazines of the times. Based on this work, William Robert Grove, known as the father of the fuel cell, constructed the first fuel cell in 1839 (Figure 7.1). Grove conducted a series of experiments with what he termed a gas voltaic battery, which ultimately proved that an electric current could be produced from an electrochemical reaction between hydrogen and oxygen over a platinum catalyst. Charles Langer and Ludwig Mond, who researched fuel cells using coal gas as a fuel, first used the term "fuel cell" in 1889. Further attempts to convert coal directly into electricity were made in the early twentieth century but the technology generally remained obscure until 1932, when Francis Thomas Bacon, the Cambridge engineering professor, developed the first AFC. However, it was not until 1959 that he demonstrated a practical 5-kW fuel cell system. At around the same time, a team led by Harry Ihrig fitted a modified 15-kW Bacon cell to an Allis-Chalmers agricultural tractor. This system used aqueous KOH liquid as the electrolyte and compressed hydrogen and oxygen as the reactants. In the late 1950s and early 1960s, the National Aeronautics and Space Administration (NASA), in collaboration with industrial partners, began to develop

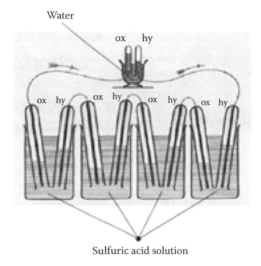

Water

ox hy

ox hy   ox hy   ox hy   ox hy

Sulfuric acid solution

FIGURE 7.1 Sketch of William Grove's 1839 fuel cell.

fuel cell generators for manned space missions. The first PEMFC unit was one result of this, for which Willard Thomas Grubb at General Electric (GE) was credited with the invention. Three years later another GE chemist, Leonard Niedrach, refined Grubb's PEMFC by depositing platinum as a catalyst onto the membranes. Prompted by concerns over energy shortages and higher oil prices, many national governments and large companies began to initiate research projects to develop more efficient forms of energy generation in the 1970s, which helped to push along the research effort of the fuel cell. The substantial technical and commercial development continued in the 1980s. In the 1990s, government policies to promote clean transport also helped drive the development of fuel cells. Today, more and more universities and institutes all over the world are participating in the development of fuel cells and fuel cell technologies.

7.1.2 OPERATING PRINCIPLES OF FUEL CELLS AND THEIR COMPONENTS

To illustrate their working principle, Figure 7.2 shows the schematic of a PEFC including fuel cell components. A key component of PEFC is a membrane electrode assembly (MEA) [1]. In MEA, electrocatalysts and a polymer electrolyte membrane (PEM) are believed to be two key fuel cell components. The electrocatalysts are used to promote the cathode oxygen reduction reaction and the anode fuel (H_2 or methanol) oxidation reaction, while the PEM is used to separate the anode from cathode as well as to conduct reactants depending on the PEM used: the acidic PEFC in which the PEM conducts protons (H^+), and the alkaline PEFC in which the PEM conducts hydroxide ions (OH^-). For acidic PEFC, fuel is processed at the anode where electrons are separated from H^+ on the surface of a catalyst (mainly Pt-based one). The H^+ passes through the membrane to the cathode side of

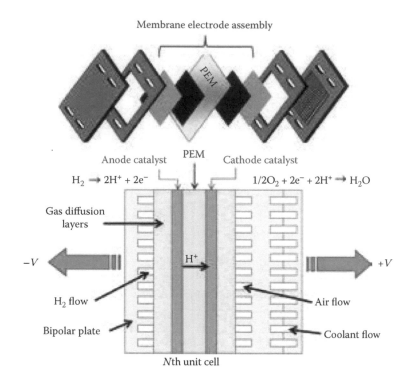

FIGURE 7.2 Schematic of a PEFC including fuel-cell components. (Reprinted by permission from Macmillan Publishers Ltd. *Nature*, Ref. 1, copyright 2012.)

the cell while the electrons travel in an external circuit, generating the electrical output of the cell. On the cathode side, the same Pt-based precious metal electrode combines the H^+ and electrons with oxygen (O_2) (from the air) to produce water (H_2O), which is expelled as the only waste product. For alkaline PEFC, on the other hand, the catalyzed cathode oxygen reduction reaction can produce OH^- ion which then passes through the alkaline PEM to the anode, where it reacts with hydrogen (H_2) to produce H_2O. If methanol is used as a fuel, for example, for DMFC, methanol is then oxidized at the anode. Table 7.1 shows the electrode reactions for both acidic and alkaline PEFC using H_2 and methanol as anode fuel and an O_2 cathode. Whether it is for acid PEFCs or for alkaline PEFCs, and for PEMFC or DMFC, the catalyst with high catalytic activity and the membrane with high ionic conductivity (both for H^+ and OH^-) are the two key factors, since it allows for higher performances to be achieved. The theoretical open circuit voltage (OCV) of a fuel cell is 1.23 V at ambient conditions.

Similar to batteries, fuel cells share the electrochemical nature of the power generation process. However, in contrast with conventional batteries, fuel cells do not need recharging. They consume reactants, which are supplied to the devices from external sources, and can continue to work as long as the fuels are maintained.

TABLE 7.1

Electrode Reactions for Acidic and Alkaline PEFC Using H_2 and Methanol as Anode Fuel and an O_2 Cathode

Fuel	Anode Reaction (Fuel Oxidation)	E_a^0 (V/SHE)	Cathode Reaction (Oxidant)	E_a^0 (V/SHE)	E (V/SHE)
		For Acidic PEFC			
Hydrogen	$H_2 \rightarrow 2H^+ + 2e, E = 0.0\ V$	0.00	$2H^+ + 2e + 1/2O_2 \rightarrow H_2O$	1.23	1.23
Methanol	$CH_3OH + H_2O \rightarrow CO_2 + 6H^+ + 6e$	0.046	$3/2O_2 + 6H^+ + 6e \rightarrow 3H_2O$	1.23	1.18
		For Alkaline PEFC			
Hydrogen	$H_2 + 2OH^- \rightarrow 2H_2O + 2e^-$	−0.83	$1/2O_2 + H_2O + 2e^- \rightarrow OH^-$	0.40	1.23
Methanol	$CH_3OH + 6OH^- \rightarrow CO_2 + 5H_2O + 6e^-$	−0.81	$3/2O_2 + 3H_2O + 6e^- \rightarrow 6OH^-$	0.40	1.21

Source: Jiang, Z., Zhao, X., Fu, Y. and Manthiram, A., Composite membranes based on sulfonated poly(ether ether ketone) and SDBS-adsorbed graphene oxide for direct methanol fuel cells. *J. Mater. Chem.* 22(47):24862–24869, 2012. Reproduced by permission of The Royal Society of Chemistry.

Note: SHE, standard hydrogen electrode; V, voltage.

7.1.3 CHALLENGES OF FUEL CELLS AND MITIGATION STRATEGIES

In fact, due to their high energy and power densities, high efficiency, and low/zero emissions, PEFCs have been developed for almost five decades for reducing environmental pollution and improving energy efficiency as strong drivers. However, for real fuel cell technology commercialization, several challenges have to be identified including their high cost and insufficient durability caused by the two key fuel cell components: electrocatalysts and the polymer electrolyte membrane. Therefore, the main objective in fuel cell technologies is to develop low-cost, high-performance durable materials. To move toward a genuinely practical technique that can be mass-produced cost-effectively, we should take a critical look at how we need to develop key components determining fuel cell cost, durability, and performance.

Regarding the catalysts, for PEMFC, ORR at the cathode has sluggish kinetics compared with an anodic hydrogen oxygen reaction (HOR). This is because the strong $O = O$ bond (498 kJ mol^{-1}) with very slow ORR rate suffers from a much larger overpotential (or polarization) of nearly ~300 mV than that of HOR, thus dominating the overall electrochemical processes in fuel cells. DMFCs, on the other hand, employing liquid methanol as a fuel, offer an attractive option due to their high energy density, simplicity in the system structure (easy storage and supply), and with no need for fuel reforming or humidification. This system simplification gives DMFC an advantage over PEMFC with reformed H_2 fuel. It also gives perspective to replace even the most advanced rechargeable batteries (e.g., nickel–metal hydride and Li-ion) currently used for such applications. However, in DMFCs, the fuel methanol is oxidized via 6-electron transfer process compared to a 2-electron transfer hydrogen oxidation in PEMFC. It, similarly, results in a slow kinetics of the methanol oxidation reaction (MOR) on the anode catalyst surface. Therefore, to make PEFCs technically feasible and practical, Pt has to be used as the most efficient catalyst either for ORR or for MOR. Unfortunately, the high cost and some unsolved technological problems related to Pt-based electrode materials, such as their susceptibility to time-dependent drift and CO deactivation, which greatly decrease the cathode/anode potential and reduce fuel efficiency, continues to delay commercial production. In this context, the development of both cathode catalysts suitable for the highly efficient catalysis of ORR and anode catalysts for MOR has therefore been the predominant focus in the past three decades.

In fact, metal catalysts are generally not used alone but are supported on conductive materials, since metal catalysts frequently suffer from dissolution, sintering, and agglomeration during operation of the fuel cell, which can result in catalyst degradation. Therefore, for improvement of the catalyst activity, stability, and utilization as well as the production cost for very large applications, in particular for Pt-based catalysts, the choice of conductive materials as catalyst supports is crucial and has been identified as the major contributor to the catalyst success or failure for fuel cell technologies. In fact, the supported materials can change or modify the electronic character of the catalyst particles, and at the same time also affect the shape/morphologies of the catalyst particles. These effects improve the catalyst efficiency and decrease the catalyst loss [2]. By this approach, both the activity and stability of the catalysts are highly improved compared to unsupported bulk metal catalysts. Generally, surface area, porosity, electrical conductivity, electrochemical stability, and surface functional groups characterize support materials. The ideal characteristics of catalyst support materials should possess the following advantages: (1) large surface area for dispersion of catalyst, (2) chemical stability at relevant temperatures, (3) high electrochemical stability under operating conditions to maintain a stable catalyst structure, and (4) high electrical conductivity to promote fast electron transfer in redox reactions [3].

Carbon nanomaterials including carbon blacks, CNTs, and graphene have been used as catalysts supports. Such support materials not only can maximize the availability of nano-sized electrocatalyst surface area for electron transfer but also provide better mass transport of reactants to the electrocatalysts. In addition, carbon nanomaterials as the conductive support facilitate efficient collection and transfer of electrons to the collecting electrode surface. Among the carbon materials described, carbon blacks (especially Vulcan XC-72) are the most commonly used carbon supports for Pt and Pt-alloy catalysts due to its high availability, high conductivity, and low cost. However, they suffer from some significant issues, such as thermochemically unstable and corrosion caused by electrochemical oxidation under fuel cell operating conditions. These easily lead to the aggregation of catalyst nanoparticles and thus the shortening of the catalyst durability and reliability by a dissolution and a precipitation, as follows [4]:

1. Pt^0 (smaller particle) $\rightarrow Pt^{2+}$ (liquid and/or ionomer) $+ 2e^-$ (carbon)
2. Pt^{2+} (liquid and/or ionomer) $+ 2e^-$ (carbon) $\rightarrow Pt^0$ (larger particle)
3. $C + 2H_2O$ $\rightarrow CO_2 + 4H^+ + 4e^-$

CNTs were also investigated as carbon black alternative catalysts supports for low-temperature fuel cells. Comparatively, the CNTs have shown better stability than that of carbon black. A problem for the commercialization of CNTs is their higher cost compared to that of carbon blacks [5]. To further improve catalytic activity and stability, developing novel support materials is necessary in the design and synthesis of new electrocatalysts.

The recent emergence of graphene (G), a 1-atom thick planar sheet of hexagonally arrayed 2-D sp^2 carbon atoms, due to its unique physicochemical properties of excellent electrical and thermal conductivities, charge-transport mobility, good transparency, great mechanical flexibility, and huge specific surface area (theoretically 2630 $m^2\,g^{-1}$ for single-layer) [6–9], has ushered in a new era for both the experimental and theoretical scientific communities since its discovery in 2004 [10]. Intensive research efforts, therefore, have been undertaken to develop G-based nanomaterials and explore their potential applications in energy devices including Li-ion batteries [11,12], supercapacitors [13–15], solar cells [16,17], and fuel cells [18,19].

As a promising supporting material, platinum nanoparticles supported on 2-D-structure G when applied as an electrocatalyst for electrochemical ORRs and methanol oxidation in fuel cells, exhibited a higher catalytic activity and less susceptibility to carbon monoxide poisoning when compared to Pt supported on carbon black. Moreover, G has high chemical, thermal, and electrochemical stabilities, which can possibly improve the lifetime of catalysts [20]. What is more, G can be obtained at relatively low cost in large quantities by using graphite, graphite oxide, and their derivatives as the starting materials [21,22]. The unique 2-D planar structure of the carbon sheet allows the edge planes to interact with the catalyst nanoparticles. The rippled but planar sheet structure also provides a very high surface area for the attaching catalyst nanoparticles. In this chapter, progress in recent significant research in the development of G and G-based materials in the field of fuel cells are introduced. Attention is focused on catalyst material selection, design, synthesis, and characterization for both catalytic performance and catalytic mechanism, and also theoretical approach with emphasis on applications in ORR and MOR. Their fabrication into MEAs and the corresponding performance/durability are also discussed. Considering the potential wide application of G, the latest developments using G and G-based materials as additives for membrane preparation and modifications are also discussed.

7.2 GRAPHENE AS SUPPORT MATERIAL FOR PT/PT-ALLOY CATALYSTS

7.2.1 PRISTINE SINGLE-LAYER GRAPHENE AS SUPPORT MATERIAL

Pt-based electrocatalysts has long been regarded as the most efficient catalyst for ORR. One of the most widely used techniques for maximizing the activity and efficiency of Pt-based electrocatalysts is loading these catalysts on the supporting nanomaterials with low cost, a large specific surface area, good conductivity, and electrochemical stability. As expected, G has been proven to be the most promising material for carbonaceous support and has been extensively explored in this field. G used as a carbon support for noble metals can effectively reduce the Pt loading and thus reduce the cost. In fact, there are many methods to prepare G/metal nanoparticles composites, such as the ethylene glycol reduction method [23–25], microwave polyol synthesis [26–28], electrochemical deposition [29,30], thermal treatment [31,32], and self-assembly methods [33–35]. Among them, the most popular way to prepare G-noble metal catalysts is based on the simultaneous reduction of metal precursor solutions and graphite oxide (GO). It has been reported that metallic nanoparticles not only play an essential role in catalytic reduction of GO, but also prevent the aggregation and restacking of G nanosheets (GNSs) during the chemical reduction process by the formation of G-based nanocomposites [23]. For example, Ha et al. synthesized a G-based catalyst by means of borohydride reduction of H_2PtCl_6 in a GO suspension [36]. In this case, GO/H_2PtCl_6 composite was prepared by adding H_2PtCl_6 to the GO nanosheet colloid suspension under stirring. The composites of graphene decorated by Pt nanoparticles were obtained via chemical reduction of GO and Pt ion using ethylene glycol (EG) and $NaBH_4$ as a reducing agent. TEM analysis showed that Pt nanoparticles with diameter of about 3 nm were uniformly loaded on the surface of G. In electrochemical measurements, the resulting catalyst Pt/G exhibited much better ORR catalytic performance than that of Pt/C catalyst. At a given polarized potential of 0.6 V where ORR is under kinetic control, the reduction current of ORR on Pt/G and JM Pt/C catalyst are 4.94 and 3.17 A g^{-1} (Pt). The authors argued that this enhanced catalytic activity can be attributed to the large electrochemical surface area (ECSA) of Pt/G, which facilitates the diffusion of O$_2$ on the surface of G to the Pt metal sites [36].

By using similar experimental techniques, a Pt/G composite has also been synthesized by Xin et al. [37]. In their work, lyophilization was introduced to avoid irreversible aggregation of GNSs during liquid water departure in the conventional drying process. The as-prepared Pt/G catalysts exhibited higher catalytic activity than that of Pt/C for the ORR in terms of both onset potential and the half-wave potential. Not only that, the performance could be further improved after the catalyst was heat-treated (Pt/G-A catalysts), for example, at 300°C for 2 h. However, both Pt/G and Pt/G-A were found to show a smaller limiting current in ORR than that of Pt/C. It was believed that the diffusion-limiting currents were strongly affected by the structure of the catalyst supporting material. The sheet structure of G might block oxygen diffusion a little compared with spherical carbon black particles. In the same work, the authors found that the resulting Pt/G catalysts were also highly active for methanol oxidation. The peak current density of MOR on Pt/G was 3.5 times higher than Pt/C in 0.5 M H$_2$SO$_4$ containing 0.5 M CH$_3$OH. The enhancement of the activity and stability could be ascribed to four effects induced by heat treatment: (1) the enhanced interaction between Pt and G, (2) the additional Pt active sites were exposed by the rolling of G sheets, (3) less defects on G by decomposing partially surface functional groups, which improve the stability of G, and (4) more active catalytic sites as a result of Pt surface morphology from amorphous to more ordered states.

In another study, Park and coworkers prepared a Pt nanoparticle-reduced graphene oxide (Pt-rGO) nanohybrid with 40% Pt/GO through the chemical reduction of H$_2$PtCl$_6$ with GO nanosheets in a mixed solution of EG and deionized water [38] The size of the Pt nanoparticles in the resulting Pt-RGO showed a narrow distribution in the range of 3–5 nm and homogenously dispersed on the RGO nanosheets. The ECSA value (32.64 m^2 g^{-1}) of Pt-RGO showed an even higher number of active sites in the catalyst after activation (33.26 m^2 g^{-1}).

In addition to these reports, other similar work has also demonstrated that using graphene as a support for noble metal can effectively improve the catalytic activity [39–41].

Electrochemical deposition was also successfully developed as an alternative approach to synthesize G-supported noble metal catalysts. By taking advantage of the redox potential differences between substrate and the metal ions (M^{m+}/M) and the excellent conductivity of GNSs, Liu et al. fabricated metal (M)-G hybrid materials (M = Pt, Pd, Ag, Au, and Cu) by immersing GNSs coated on metal (Cu or Zn) foils into solutions containing the corresponding precursors using electrochemical deposition [42]. The size and density of the metal nanoparticles on GNSs surface could be controlled by adjusting the reaction conditions. Stacks of M-decorated G were obtained by repeating the processes of G-coating and M-depositing. Electrochemical tests showed that the Pt/G hybrid materials exhibited excellent electrocatalytic activity toward methanol oxidation in 1 M H$_2$SO$_4$. RGO film-coated ITO as the substrate was proposed by Liu et al. for electrodepositing Pt nanoparticles from an aqueous solution of H_2PtCl_6 [43]. Confirmed by TEM images, the Pt nanoparticles with an average diameter of about 15 nm were well dispersed on the expandable graphene sheet (EGS) surface. Further evaluated by cyclic voltammetry (CV) measurements on these Pt/ EGS films, the peak current obtained for methanol oxidation at the Pt/EGS catalysts was 7.41 mA cm^{-2}, which was much higher than obtained at the Pt/GC with a peak current value of 4.31 mA cm^{-2}. The onset potential for methanol oxidation on Pt/EGS catalyst was about 0.92 V (vs. Ag/AgCl), which is also more positive than that of Pt/GC. The authors attributed this improvement to the residual oxygen-containing functional groups on the surface of EGS. Furthermore, the stability of the synthesized catalysts was better than that of Pt/GC for methanol oxidation. The higher activity of the Pt/EGS catalysts was ascribed to the peculiar microstructure and surface topography of EGS and a strong metal-support interaction.

To further lower the mass of precious metal, the development of Pt alloy nanoelectrocatalysts of higher ORR activity has been pursued. Typically, a self-assembly method was utilized as a versatile platform for the preparation of G-noble metal composite. In this regard, a facile solution-phase self-assembly method to deposit PtFe nanoparticles on the G-surface (PtFe/G) was demonstrated by Guo et al. [34]. By this method, the catalysts have a narrow size distribution with an average diameter of 7 ± 0.5 nm. As the CV curves show in Figure 7.3, the double-layer capacitance of the FePt/G nanoparticles is much larger than that of FePt/C and the commercial Pt/C catalysts, indicating that G-support has a much larger surface area than the carbon black support. This is important for mass activity enhancement of Pt-based catalysts. In addition, ORR measurements in 0.1 M $HClO_4$ demonstrated that the half-wave potential of FePt/G (0.557 V) is higher than those of FePt/C (0.532 V) and commercial Pt/C catalyst (0.512 V), indicating that FePt/G nanoparticles indeed have better electrocatalytic activity toward ORR than FePt/C nanoparticles at the same Pt loading. In addition, the FePt/G nanoparticles are very stable under the ORR operating conditions and show nearly no activity loss after accelerated durability testing (ADT), where 10,000 potential sweeps were proceeded between 0.4 and 0.8 V in O_2-saturated 0.1 M $HClO_4$. As a comparison, under the same reaction conditions, the ORR polarization curve of commercial Pt/C NPs shows an obvious negative shift after a stability test.

Binary materials composed of noble metals including Au, Ag, Pt, and Pd with enhanced electrocatalytic performance have also been reported for ORR [44–48]. Graphene has been linked to bimetallic catalysts to enhance their catalytic activity, decrease the cost, and increase the resistance to CO poisoning. To date, previously reported G-Pt alloy nanoparticle hybrids all exhibited high electrochemical activity and low overpotential for the ORR. Rao et al. also succeeded in synthesizing Pt_3M/G (M = Co, Cr) hybrids by the ethylene glycol reduction method [49]. They demonstrated that the Pt_3M/G hybrids have remarkably increased activity toward the ORR, which is 3–4 times higher than that of Pt/G. Also, Pt_3M/G electrodes exhibited overpotential 40–70 mV lower than that of Pt/G. The enhanced ORR activity

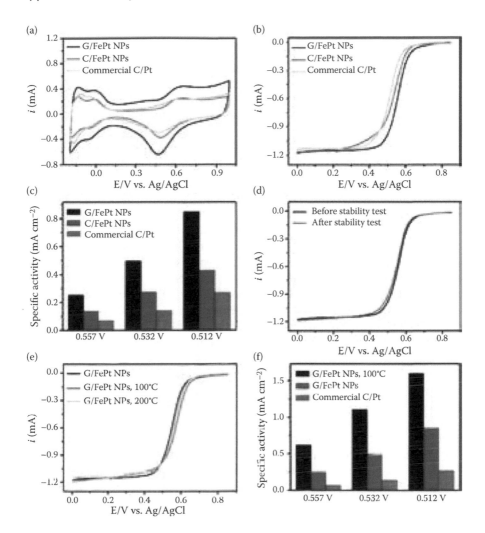

FIGURE 7.3   (a) CVs in N_2-saturated 0.1 M $HClO_4$ solution at a scan rate of 50 mV/s. (b) Polarization curves for ORR in O_2-saturated 0.1 M $HClO_4$ solution at 295 K. The potential scan rate was 10 mV s^{-1} and the electrode rotation speed was 1600 rpm. (c) ORR specific activities of the FePt/G, FePt/C and commercial Pt/C catalysts. (d) ORR polarization curves of the FePt/G nanaoparticles before and after 10,000 potential sweeps between 0.4 and 0.8 V and (e) FePt/G nanaoparticles annealed at different temperatures. (f) Comparison of ORR-specific activities. (Reprinted with permission from Ref. 34, 2492–2495. Copyright 2012 American Chemical Society.)

was ascribed to the inhibition of formation of (hydr) oxy species on Pt surface by the presence of alloying elements. In addition to these studies, a number of other bimetallic and ternary catalysts, such as PtAu [50], PdAg [47], PtNi [51], and PtPdAu [24] have also been reported in combination with graphene with increased electrocatalytic activity and improved stability.

7.2.2 Conductive Polymer-Functionalized GNSs as Support Material

It is a well-accepted fact that the performance of a catalyst in fuel cells strongly depends on the distributing state and stability of catalyst nanoparticles on the surface of support material. The structures and properties of the catalyst support have a significant effect on the activity and durability of the catalysts. In fact, the efficient immobilization of metal nanoparticles uniformly on GNSs remains a big challenge because of the inert nature of the graphitized surface of graphene. Moreover, due to the nature of 2-D material, GNSs tend to form irreversible agglomerates through the strong van der Waals interactions when they are dried, even when they are loaded with metal catalysts. In this case, the contribution of the G support to the catalytic reaction is remarkably decreased. Because the stacking of G would block a substantial amount of catalytic sites and induced higher resistance for the diffusion of the reactant molecules. Therefore, great efforts have been made to address this issue. One strategy is introducing a third component to control the layer stacking and distribution of graphene. Positively charged PDDA is a water-soluble quaternary ammonium and strong polyelectrolyte, which would be a good choice for this purpose. As an example, Luo et al. prepared functionalized GNSs by PDDA and employed them as support materials for *in situ* deposition of Pt nanoparticles [52]. During the functionalization process, NaCl was added to allow the PDDA chain to adopt a random configuration leading to a highly functionalized or covered PDDA chains on GNSs. The PDDA-GNSs were then mixed with H_2PtCl_6 solution and subsequent reduction by ethylene glycol. Pt nanoparticles were successfully deposited and well distributed on the PDDA-GNSs. Compared with pristine GNSs, PDDA functionalization provided considerably higher density and homogeneity of surface functional groups.

With the aid of PDDA, the G-supported Pd-Pt nanoparticles by reducing the mixture of GO and Pt and Pd ions was also successfully synthesized [25]. First, PDDA was introduced to functionalize GO. Second, PDDA-functionalized GO was used to load the Pt and Pd ions through the electrostatic interaction between the positive-charged PDDA and negative-charged $[PtCl_6]^{2-}$ and $[PdCl_4]^{2-}$. Finally, the GO and the attached metal ions were reduced simultaneously in an ethylene glycol solution. The obtained Pt-Pd/RGO composites showed an increased catalytic activity toward the ORR. Also, the ORR activity of this catalyst was enhanced even after ADT (i.e., 3000 potential sweeps between 0.6 to 1.1 V at a scan rate of 50 mV s^{-1} in an oxygen-saturated 0.1 M $HClO_4$). This increased activity was explained by an enrichment of Pt on the surface of the nanoparticles that occurred after repetitive potential cycling, as confirmed by XPS measurements. It seems that repetitive potential cycling used in the study can dramatically accelerate the surface dealloying of a Pd-Pt/RGO catalyst, leading to the formation of Pd-rich core and Pt-rich shell nanoparticles. The smaller lattice parameter of Pd-rich core would induce compressive strain in the outer shell, which tends to downshift the d-band center of the Pt-rich shell, weakening the adsorption energy of the oxygenated species and thus enhancing the ORR kinetics of the dealloyed Pd–Pt/RGO catalysts.

Functionalized RGO with N-(trimethoxysilylpropyl) ethylenediamine triacetic acid (EDTA-silane) was also reported to be favorable for improving nanoparticle dispersion and suppressing the agglomeration of the nanoparticles [53]. Figure 7.4a

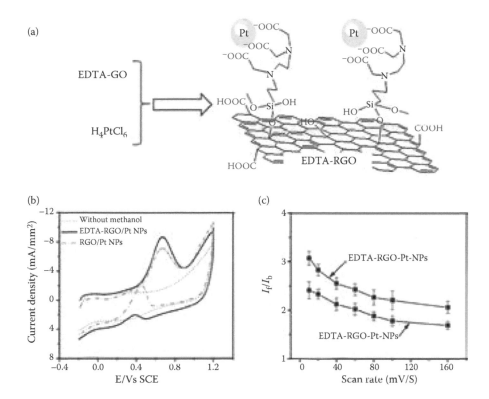

FIGURE 7.4 (a) Scheme to load Pt nanoparticles on EDTA-GO surface. (b) CVs of EDTA-RGO/Pt-NPs/GC electrode in 0.5 M bulk H_2SO_4 solution and in 0.5 M methanol + 0.5 M H_2SO_4 and GO/Pt-NPs/GC in the same methanol solution. Scan rate 60 mV s^{-1}. (c) The relationship of I_f/I_b ratio vs. scan rate of RGO/Pt-NPs/GC and EDTA-RGO/Pt-NPs electrodes. (Reprinted from *J. Power Sources*, 198, Wietecha, M. S., Zhu, J., Gao, G. et al., Platinum nanoparticles anchored on chelating group-modified graphene for methanol oxidation, 30–35, Copyright 2012, with permission from Elsevier.)

shows the postulated structure of load Pt nanoparticles on EDTA-GO surface. According to a study by Wietecha and coworkers, EDTA played a significant role in enhancing catalytic activity of Pt nanoparticle for methanol oxidation. The observed I_f (defined as the forward anodic peak current density) of methanol oxidation on the EDTA-RGO/Pt-NPs/GC electrode is ~65% higher than that on the RGO/Pt-NPs/GC electrode, whereas the observed I_b (designated as the backward anodic peak current density) of CO_{ads}-like species oxidation is ~50% lower on the EDTA-RGO/Pt-NPs/GC electrode than on the RGO/Pt-NPs/GC electrode (Figure 7.4b). Further investigation indicated that EDTA-RGO/Pt-NPs displayed much higher I_f/I_b ratios (used to explain the catalyst tolerance to CO and other carbonaceous species) compared to RGO/Pt-NPs as the scan rate increases (Figure 7.4c). Compared with carbon-black-supported Pt-NPs, EDTA-RGO/Pt-NPs have a ratio of 2.0 to 2.6 (such as at a scan rate 60 mV s^{-1}), which is much higher than that of carbon black (1.39) and graphite (1.03). This superior antipoisoning behavior EDTA-RGO/Pt-NPs can be explained by the following facts: (1) the EDTA groups contribute nitrogen atoms, –OH and –COOH groups, which provide a complex interaction between EDTA and Pt nanoparticles and prevent the Pt nanoparticles from aggregation. This strong interaction can induce modulation in the electronic structure of Pt nanoparticles, control the structure and shape of the nanoparticles, lower the Pt–CO binding energy, and thus reduce the CO adsorption on Pt; (2) the EDTA groups enhance the hydrophilic properties of RGO and thus promote water activation and as a result, the adsorbed OH^- species at the Pt nanoparticles promote the oxidation of CO; (3) the EDTA groups can provide additional reaction sites to bind and stabilize Pt nanoparticles, and then to enhance the stability and efficiency of Pt nanoparticles; and (4) the EDTA groups can also enhance charge transfer between the reactants and electrodes. Therefore, the EDTA-RGO/Pt-NPs catalysts showed higher catalytic activity, longer stability, and excellent tolerance to CO poisoning in DMFCs. In addition to PDDA and EDTA, other conducting polymers such as PPy [54], PANI [55], and poly(pyrogallol) (PPG) [56] are also found to have the same functions by other researchers.

7.2.3 HYBRID SUPPORTING MATERIAL BY INTERCALATING CARBON BLACK AND CNTS

An effective approach to prevent agglomeration can also be realized by intercalating carbon black (CB) particles between Pt-loaded G sheets. This may disrupt the preferred horizontal stacking of G sheets and make them randomly distributed in the catalyst layer. As a result, more Pt nanoparticles are effective for electrochemical fuel cell reactions in the presence of a spacer between G sheets. Based on this, Li et al. demonstrated that the Pt/RGO/CB composite (prepared just by mixing Pt/RGO with CB in IPA and stirring overnight) shows greatly enhanced ORR activity compared to the simple Pt/RGO structure [35]. They prepared two types of Pt/RGO/CB catalysts with RGO/CB mass ratios of 1:1 (termed Pt/RGO/CB-1) and 2:1 (Pt/RGO/CB-2). Given a comparison of the polarization curves, the RGO-supported Pt nanocrystals (NCs) show lower ORR activity than Pt NCs on CB. This is partly because when Pt/RGO catalyst is dried, RGO sheets tend to form a closely packed film, inhibiting diffusion of O_2 through the film to approach the Pt surface, thus lowering the reduction rate on Pt NCs. However, with the addition of CB into Pt-loaded RGO sheets as a secondary support, currents can be dramatically promoted. The inserted CB particles can enlarge the gaps between RGO sheets, thus providing enough space for fast oxygen diffusion through the film and enhancing the oxygen supply in the film, and finally, accelerating the reduction rate.

In contrast with the previous catalysts mentioned, perfect stability of these Pt/RGO/CB composites were demonstrated after ADT, which was carried out by repetitive potential cycling between 0.6 and 1.1 V in 0.1 M $HClO_4$ exposed to the atmosphere. CV curves between 0 and 1.1 V were recorded every 4000 cycles to compare the ECSAs during potential cycling. After 20,000 cycles of ADT, the Pt/RGO/CB-2 was still 87% of the initial ECSA remaining and the Pt/RGO/CB-1 retained >95% of its initial ECSA with higher additions of CB. Comparatively, the final ECSA of JM

Pt/C after 20,000 cycles dropped to ~51% of the initial value. ORR activities after 20,000 cycles of ADT in O_2-saturated $HClO_4$ electrolyte demonstrated that the specific activities of Pt/RGO/CB at 0.9 V are almost twice that of the commercial JM Pt/C catalyst and higher than those supported only on CB or RGO. In terms of mass activity, the Pt/RGO/CB remained triple the final mass activity of the commercial JM Pt/C. The ADT demonstrated that the hybrid supporting material can dramatically enhance the durability of the catalyst and retain the ECSA of Pt. In the hybrid structure, the flexible 2-D profile of RGO may function as a barrier that prevents leaching of dissolved Pt species into the electrolyte, while the CBs can serve as an active site for recapture or renucleation of small Pt clusters. Using Pt/G with different contents of CB in the catalyst layer, Park and coworkers investigated a serious of cathodes and tested them in a single PEMFC [57]. Carbon black was added as a spacer between G sheets in the catalyst layer to study its effect on electrochemical characteristics of the cathode in fuel cells. As confirmed by ECSA and double-layer capacitance (C_{dl}) of the cathodes with different CB contents, the addition of CB in Pt/G catalyst layer increased the ECSA of Pt nanoparticles dispersed on G sheets. The value of C_{dl} increased with the amount of CB in the catalyst layer, indicating that the interface between catalyst and ionomer increases with the CB added into the catalyst layer. The performance of the single PEMFC using Pt/G with 25 wt.% CB as cathode catalysts considerably increased with respect to that of a PEMFC using Pt/G without CB, and was comparable to that of a commercial Pt/C catalyst.

In recent work, the hybrid of G-CNT supports has also been proposed. The excellent properties of G emerge only in the planar direction. In CNTs, these properties emerge in the axial direction while providing high specific surface area, current density, and thermal conductivity. Thus, it is expected that the combination of 1-D CNT and 2-D G can be exceptionally advantageous when employed as a catalyst support material in fuel cells. On this basis, Aravind et al. reported an *in situ* method of preparation of a hybrid composite material of G and MWCNTs by solar exfoliation of the GO–MWCNT composite [58]. In the synthesis procedure, the MWCNTs were first refluxed with GO in HNO_3 by functionalization (defined as f-MWCNTs). Then, the GO-MWCNT composite was solar-exfoliated by harvesting and directing solar radiation (defined as sG). Finally, Pt nanoparticles were loaded over the catalyst support materials by an EG reduction method. The TEM image of Pt-dispersed sG-f MWCNT shows the uniform and full area distribution of Pt nanoparticles over G and MWCNT, as shown in Figure 7.5. In addition, the transparent nature of graphene in the G-MWCNT composite indicates that MWCNT can effectively prevent restacking of graphene, thus the ECSA and the Pt utilization of sG-f MWCNT was higher than both G- and MWCNT-supported P. All the above can be attributed to the uniform dispersion of Pt particles on the hybrid nanostructure and good accessibility of these sites for hydrogen adsorption and desorption reactions. Tests in PEMFCs showed that the performance of the cells with G-MWCNT-supported catalysts was better than that of the cells with catalysts supported on bare G and bare MWCNT. The enhanced performance of the G-MWCNT composite as a catalyst support compared to pure G has been attributed to the bridging of defects for the electron transfer and an increase in the basal spacing between graphene sheets with the incorporation of MWCNT.

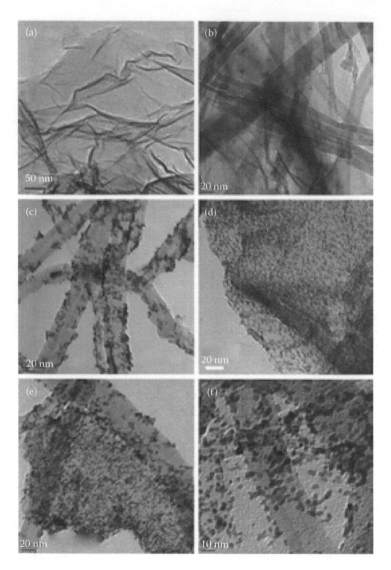

FIGURE 7.5 HRTEM images of (a) f-sG (b) sG-f-MWNT (c) Pt/f-MWNT (d) Pt/f-sG (e) low and (f) high magnification images of Pt/sG-f-MWNT composite. (From Aravind, S. J., Jafri, R. I., Rajalakshmi, N. and Ramaprabhu, S., Solar exfoliated graphene–carbon nanotube hybrid nano composites as efficient catalyst supports for proton exchange membrane fuel cells. *J. Mater. Chem.* 21(45):18199–18204, 2011. Reproduced by permission of The Royal Society of Chemistry.)

With MWCNT as the support mixture, Yun et al. synthesized Pt-graphene/ MWCNT composites with rodlike MWCNTs and studied their electrocatalytic activity [59]. Rodlike MWCNTs make a porous network structure, and therefore, Pt nanoparticles of 2–3 nm in diameter with little aggregation, even at a high metal loading of 50 wt.%, were incorporated homogeneously onto the G surfaces. Pt-graphene was bound homogeneously to the network structure of the MWCNTs. This makes the catalyst surface a very rough surface, which facilitates simultaneous access between the Pt electrocatalyst and reactant. In addition, the porous MWCNT network enabled the Pt-graphene electrode to overcome the deficiency caused by high electrical resistance by providing an electrical pathway for the ORR. From CVs, the ECSA of Pt-graphene/MWCNT composite cathode was calculated to be 38.2 m^2/g, which is slightly larger than the 34.5 m^2/g for Pt in the Pt-graphene cathode. These results indicate better utilization efficiency of the Pt-graphene/MWCNT composite cathode due to effects induced by MWCNTs. In the case of the ORR electrocatalytic activities evaluated by Nyquist plots, the ORR charge transfer resistance of the Pt-G/MWCNT composite cathode was found to be much smaller than that of the Pt/G cathode.

In a recent article, Rajesh et al. reported the fabrication and electrocatalytic activity of Pt nanoparticles decorated unique large-area 3-D-carbon electrode comprising of MWCNTs and a single graphene floor by one-step CVD method [60]. SEM analysis shows the flowerlike Pt nanostructures of about 60–90 nm in size with a cluster of needles that were electrochemically deposited over the entire hybrid nanostructure. The electrocatalytic activities of Pt/G-MWNTs and Pt/G were further investigated and compared toward the oxidation of methanol in different methanol concentrations from 0.5 to 3M. It was found that the methanol oxidation current density was highest at Pt/G-MWNTs at all methanol concentrations. Further kinetics study of MOR revealed a linear relationship between forward peak current density and $v^{1/2}$ for Pt/G-MWNTs and Pt/G. In contrast, Pt/G-MWNTs showed a larger slope (2.01) compared to Pt/G (0.40), indicating a faster diffusion process of methanol on the surface of G-MWNTs hybrid film than those of G catalyst supports. The enhanced performance of G-MWCNTs as catalyst support for electrooxidation of methanol is attributed to the large surface area of unique 3-D CNTs, which leads to an efficient charge transfer at the G-MWCNTs/electrolyte interface allowing higher dispersion of Pt, and good electrical properties of MWCNTs. Also, performance could be further improved by better controlling the Pt nanoparticle size with optimizing the electrodeposition process.

7.2.4 NITROGEN-DOPED GRAPHENE AS SUPPORT MATERIAL

Recently, it was widely reported in the literature that N-doped carbon nanostructures can modify electronic properties and show n-type or metallic behavior. They are expected to have greater electron mobility than their corresponding undoped carbon nanostructures. Consequently, NG is considered as the ideal support for metal nanoparticles. Incorporating different types of N into G would provide the NG with more functional groups and abundant active sites, which are very beneficial for uniform dispersion and small-size distribution of metal nanoparticles. More importantly, NG can increase the durability and catalytic activity of metal catalysts by improving graphene–metal (G-M) binding.

Jafri and colleagues synthesized NG nanoplatelet-supported Pt nanoparticles by nitrogen plasma treatment and a conventional chemical reduction technique [61]. Nitrogen plasma treatment created pyrrolic nitrogen defects revealed by Raman spectra and XPS survey spectrum, which acted as good anchoring sites for the deposition of Pt nanoparticles. The improved performance of fuel cells with NG as catalyst supports could be attributed to the structure formation of pentagons and heptagons in the catalyst surface, which increased in the reactivity of neighboring carbon atoms and thus the improved carbon–catalyst binding and increased electrical conductivity brought about by N doping.

NG was also successfully prepared by the same group through pyrolysis of G coated with PPy [62]. To efficiently dope the G-sheets with nitrogen while averting their agglomeration during wet-chemical synthesis, an anionic polyelectrolyte (sodium 4-styrenesulfonate [PSS]) is used to distribute negative charges over the surface of hydrogen-exfoliated graphene (HEG). These nitrogen-doped N-HEG sheets are used as catalyst supports for dispersing Pt and Pt-Co alloy nanoparticles synthesized by the modified-polyol reduction method. Pt and Pt_3Co nanoparticle sizes were estimated to have an average diameter of about 2.6 ± 0.4 and 2.3 ± 0.5 nm, and a uniform size distribution of Pt and Pt_3Co nanoparticles over the surface of NG was observed from SEM and TEM images. XPS analysis of the bonding configurations of N atoms and Pt or Pt–Co alloy nanoparticles revealed two important pieces of information: (1) in the N1s spectrum, pyridine-like N was the main component of the N-HEG specimen and the amount of N incorporated in N-HEG was found to be about 5.9 at%, and (2) the Pt species in the monometallic Pt and binary Pt–Co alloyed nanoparticles were predominantly in the zero valent state. By CV measurement, the ECSA of the Pt/C, Pt/N-HEG, and Pt_3Co/N-HEG are 42.1, 57.9, and 48.5 m^2/g, respectively. The larger ECSA for Pt/N-HEG and Pt_3Co/N-HEG compared with commercial Pt/C was attributed to the higher dispersion of Pt in the former sample because the N-incorporation increases the number of active Pt sites. The higher PEMFC performance of the Pt/N-HEG catalyst was attributed to the combined effects of the modified Pt state due to alloying and N doping, uniform dispersion of catalyst nanoparticles, and high catalytic activity of the N-HEG support itself. A further stability study showed that Pt_3Co/N-HEG cathode electrocatalyst is highly durable in acidic medium (longer than 100 h) due to the strong binding between metal nanoparticles and NG, which prevents the agglomeration of catalyst nanoparticles.

G and NG were synthesized by a solvothermal method and investigated as catalysts as well as catalyst supports for ORR [63]. Based on the work of Bai et al., NG exhibited enhanced electrocatalytic activity in both acidic and alkaline solutions in comparison to G. NG can act directly as a catalyst to facilitate $4e^-$ oxygen reduction in alkaline solution, whereas, in acidic solution NG activity for oxygen reduction was limited by a less efficient 2-electron pathway. However, when used as catalyst supports for Pt and Pt-Ru nanoparticles, the catalysts exhibited superior performance for the 4-electron pathway ORR even in acidic solution, indicated by higher peak current density, high peak potential, and better durability compared with those supported on G. In constrast, no obvious reduction peaks were observed for G- and NG-supported Pt and Pt-Ru nanoparticles in alkaline solution. The authors suggested that NG can be developed as an efficient catalyst for oxygen reductions to replace the use of precious Pt in alkaline solution and as catalyst supports in acidic solution.

Using NG as a catalyst support, Zhang et al. studied the catalytic activity of NG-supported Pt nanoparticles for MOR, where NG was obtained by thermal annealing of GO in NH_3 at different temperatures [64]. It was found that the amino groups and pyridinic species in NG sheets acted as the attaching sites for Pt nanoparticles and were favorable for uniformly distributing Pt nanoparticles on the G basal planes. A composite catalyst of Pt and N-rGO treated by annealing at 800°C showed excellent electrocatalytic activity toward methanol oxidation. The current density of methanol oxidation at Pt/N-rGO is three times higher than that of Pt/rGO catalysts or Pt/CB. The high methanol oxidation activity of the former catalyst was attributed to its increased conductivity upon thermal treatment at 800°C and the high dispersion state of the Pt nanoparticles facilitated by the pyridinic N doped in the carbon network. Using the same procedure, Xiong et al. prepared NG and used it as a conductive support for Pt nanoparticles (Pt/NG) [65]. After treatment in NH_3 and high temperature (from 500°C $\sim$ 800°C), the Pt nanoparticles grown on NG support were found to show smaller particle size, a narrower size distribution, and better dispersion. This may due to the fact that the surface microstructure might be destroyed and result in more edges and defects during the thermal reduction process. In addition, compared to the Pt/G, the Pt/NG showed a higher level of metallic (Pt^0) components, which lead to a higher electrochemical activity toward methanol oxidation. The onset potential for methanol oxidation on Pt/NG is 10 mV earlier than on Pt/G. At a scan rate of 50 mV/s, the forward peak current for Pt/NG was about four times higher than that of Pt/G. The presence of N functional groups in the G sheets yielded additional anchor sites for metal seeding and deposition. This lead to the uniform distribution of Pt particles on the doped G support, thus the electrochemical performance of Pt/N-G was greatly improved.

Very recently, Yang et al. synthesized Pt-Au/NG nanocomposites using an effective assembly strategy [50]. Alloy nanoparticles were well dispersed and anchored on the surface of NG sheets facilitated by the N-doped sites on the surface. The as-prepared Pt_3Au/NG nanocomposites exhibited much higher electrocatalytic activities for the MOR than the Pt_3Au/G and commercial Pt/C catalysts. This increase is attributed to both the alloying effect of Pt-Au and their synergistic interaction with the NG sheets. Later, a facile single-step approach for the simultaneous reduction of GO to G, functional doping of G with N, and loading of the doped G with well-dispersed Pt nanoparticles by a solvent mixture of EG and N-methyl-2-pyrrolidone (NMP) was reported by Zhao et al. [66]. As shown in Figure 7.6, TEM images show that the Pt particle size for the Pt/undoped G catalyst was $\sim$2.5–7.5 nm, while for the Pt/NG catalyst there was a significantly smaller particle mean size and narrower size distribution in the range of $\sim$1.5–4.5 nm. The ECSA of the Pt/NG catalyst was approximately 19% greater than that of the Pt/G catalyst. However, in the CV tests, the methanol oxidation peak current density for the Pt/NG catalyst was almost twice that of Pt/undoped G. It was suggested that at least some of the performance enhancement should be attributed to an intrinsic increase in the electrocatalytic activity rather than simple specific surface area scaling effects. From the above examples, it can be seen that G and NG have great potential as a high-performance catalyst support for fuel cell applications.

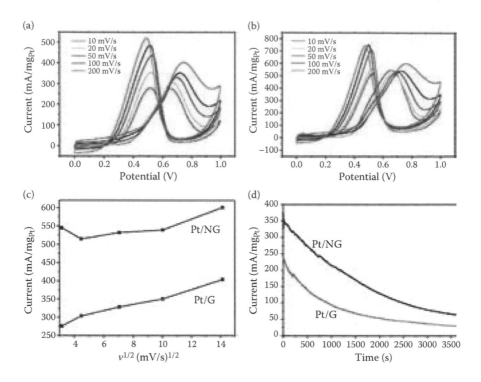

FIGURE 7.6 Cyclic voltammograms of (a) Pt/undoped graphene and (b) Pt/NG catalysts in 1 M CH_3OH+0.5 M H_2SO_4 electrolyte solution at T = 298 K. (c) The relationship of anodic peak current density vs. scan rate for both catalysts. (d) Chronoamperometry curves of methanol electrooxidation on Pt/G and Pt/NG electrodes at 0.6 V (vs. Ag/AgCl). (With kind permission from Springer Science + Business Media: Zhao, Y., Zhou, Y., Xiong, B. et al. 2013. Facile single-step preparation of Pt/N-graphene catalysts with improved methanol electrooxidation activity. *J. Solid-State Electrochem.* 17(4):1089–1098.)

7.3 GRAPHENE/NITROGEN-DOPED GRAPHENE AS SUPPORT MATERIAL FOR NONPRECIOUS METAL CATALYSTS

Although Pt and Pt-based alloys are most effective catalysts for ORR with excellent electrocatalytic activity, their high cost and scarcity in nature limit widespread application and commercialization of fuel cell technologies. Consequently, the search for higher activity, lower cost, and more durable electrocatalytic materials continues to be an active area of research. Generally, the research has taken to two different approaches. The first one is to reduce catalyst usage through increasing Pt utilization in the catalyst layers. An alternative approach is to develop nonprecious metal-based electrocatalyst materials. In comparison with noble metals, nonprecious metal catalysts (NPMCs), especially oxides are particularly attractive candidates because of the utilization of high abundant and inexpensive precursor materials. However, the performance of the NPMCs is still inferior due to their low intrinsic electrical conductivity. Recent studies revealed that blending metal compounds with G can greatly improve their catalytic performance. In particular, tremendous progress has been made with novel nanostructured G/NPMCs for ORR. These novel nanostructures possess high ORR activities as well as favorable durability, which can compete with commercial Pt and thus makes them highly promising electrode materials in fuel cells.

7.3.1 METAL OXIDES

Metal oxides are popular candidates as the most pursued alternative electrocatalysts for ORR. However, metal oxides frequently suffer from dissolution, sintering, and agglomeration during operation of the fuel cell, which results in catalyst degradation. Moreover, metal oxides prepared by the traditional ceramic route have limited electrocatalytic activity due to their large particle size and low specific surface areas. In searching for more robust and practical catalysts with comparative or even better catalytic performance than Pt, metal oxides supported on graphene have attracted much more attention. Low intrinsic electrical conductivity of metal oxides can be remarkably improved, which favors their electrocatalytic activities.

One notable contribution has been from Dai and coworkers, who explore cobalt oxide (Co_3O_4) nanocrystals supported on NG as an excellent catalyst for ORR in alkaline solution with different concentrations [67]. Controlled nucleation of Co_3O_4 on mildly oxidized graphene oxide (mGO) sheets was achieved by reducing the hydrolysis rate of cobalt acetate ($Co(OAc)_2$) by adjusting the ethanol/H_2O ratio and reaction temperature. Subsequent hydrothermal reaction at 150°C led to crystallization of Co_3O_4 and reduction of mGO. When an N-doped mGO was used as the support of a catalyst by adding NH_4OH in the synthesis steps, the particle sizes of Co_3O_4 (4–8 nm) were smaller than those grown on mGO (12–25 nm). XPS analysis revealed 4 at% N in Co_3O_4/N-reduced mildly oxidized GO (rmGO), but not in the Co_3O_4/rmGO sample made without NH_4OH. The authors demonstrated that Co_3O_4/rmGO or Co_3O_4/N-rmGO was one of the rare and highest performance bifunctional catalysts for ORR and water oxidation. The ORR kinetics of the obtained catalysts was studied by the rotating disk electrode (RDE) technique as well as the rotating ring-disk electrode (RRDE) technique in alkaline medium. The catalysts showed strong catalytic activities for ORR through a 4-electron oxygen reduction process similar to that of a high-quality commercial Pt/C catalyst and far exceeded Pt/C in stability and durability. The measured HO_2^- yields are below ~12% and ~6% for Co_3O_4/rmGO and Co_3O_4/N-rmGO, respectively, over the potential range of 0.45–0.80 V, giving an electron transfer number of ~3.9. XANES measurements demonstrated the formation of interfacial Co-O-C and Co-N-C bonds in the Co_3O_4/N-rmGO hybrid (i.e., the N-doping of GO could afford stronger coupling between Co and G in Co_3O_4/N-rmGO than in Co_3O_4/rmGO). N groups on reduced GO served as favorable nucleation and anchor sites for Co_3O_4 nanocrystals owing to the coordination with Co cations. In addition, in contrast with ORR-active sites in Fe- or Co-N/C catalysts prepared at temperatures much higher than 600°C with much lower metal loadings (<1~2 at% of metal), the Co in Co_3O_4/N-rmGO hybrid in low-temperature solution-phase synthesis was in the form of oxides with a high Co loading of ~20 at%. The $MnCo_2O_4$/N-rmGO as an electrocatalyst for ORR was also reported by the same authors [68]. XANES measurements indicated the formation of covalent interfacial M−O−C and M−N−C bonds in the hybrids.

Covalent coupling between spinel oxide nanoparticles and N-doped rmGO sheets afforded the resultant catalysts much higher activity and stronger durability than the physical mixture of nanoparticles and N-rmGO. Mn substitution increased the activity of catalytic sites of the hybrid materials, further boosting the ORR activity compared with the pure Co_3O_4/N-rmGO hybrid (Figure 7.7). At the same mass loading, the $MnCo_2O_4$/N-rmGO hybrid can outperform Pt/C in ORR current density with stability superior to Pt/C. The superior catalytic activity was ascribed to the increased catalytic sites of the spinel oxide hybrid. Their group also successfully synthesized a $Co_{1-x}S$/RGO hybrid via a low-temperature solution-phase reaction followed by a high-temperature annealing step [69]. Strong electrochemical coupling of the RGO support with the $Co_{1-x}S$ nanoparticles directly grown on top and the desirable morphology, size, and phase of the $Co_{1-x}S$ nanoparticles mediated by the RGO template afforded unprecedented high ORR catalytic performance among all cobalt chalcogenide based ORR catalysts in acid medium.

Wu et al. reported the preparation of Fe_3O_4 supported on 3-D N-doped graphene aerogel networks (N-GAs) as efficient catalysts for ORR [70]. The synthesis approach seems very simple. First, GO was dispersed in water by sonication. Next, iron acetate and PPy were added to the GO dispersion to form a stable aqueous suspension. Subsequently, these ternary components were hydrothermally assembled to form a G-based 3-D hydrogel. Fe_3O_4 nanoparticles could nucleate and grow on the graphene surface with simultaneous incorporation of nitrogen species into the graphene lattice. The as-prepared hydrogel was directly dehydrated via a freeze-drying process to maintain the 3-D monolithic architecture and then treated in high temperature at 600°C for 3 h under N_2 to obtain the final product. The G hybrid showed an interconnected macroporous framework of G sheets with uniform deposition of Fe_3O_4 nanoparticles. Because of the 3-D macroporous structure and high surface area, the resulting Fe_3O_4/N-GAs showed excellent electrocatalytic activity for the ORR in alkaline electrolytes, including a higher current density, lower ring current, lower H_2O_2 yield, higher electron transfer number (~4), and better durability.

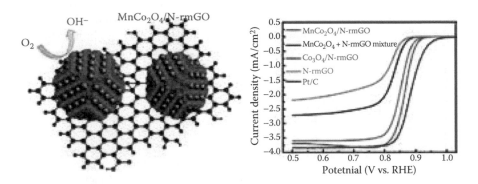

FIGURE 7.7 Rotating-disk electrode voltammograms of $MnCo_2O_4$/N-rmGO hybrid, $MnCo_2O_4$/N-rmGO mixture, Co_3O_4/N-rmGO hybrid, N-rmGO, and Pt/C in O_2-saturated 1 M KOH at a sweep rate of 5 mV/s at 1600 rpm. (Reprinted with permission from Ref. 68, 3517–3523. Copyright 2012 American Chemical Society.)

To control the desired size and morphology of the oxide nanoparticles and extend the synthesis strategies of G-based nanocomposites, Guo et al. reported that monodisperse Co/CoO NPs can be presynthesized and deposited on a G surface through the solution-phase self-assembly method [71]. The Co NPs tend to form a layer (ca. 1 nm) of natural CoO once they are exposed to an ambient environment. This CoO layer prevents Co from deep oxidation unless the Co/CoO NPs (1 nm shell) are heated at an elevated temperature (70°C). With this controlled oxidation, a series of G–Co/CoO NPs with tunable Co size and CoO thickness was obtained. Co in Co/CoO NPs can be completely oxidized by an excess of Me_3NO, thereby forming hollow CoO NPs. Compared to G and C–Co/CoO NPs, the G–Co/CoO NPs showed a great deal of enhanced catalytic activity for the ORR in alkaline solution, and this activity depends on the CoO thickness. The G–Co/CoO NPs with a 1-nm CoO shell showed maximum activity. The optimized G–Co/CoO NPs had comparative activity and better stability than the commercial Pt/C. Additionally, the combination of G with Cu_2O [72], MnO_2 [73], and Ni(Cu)-a-MnO_2 [74] was also reported as advanced electrocatalysts for ORR.

7.3.2 Pyrolized Metal Macrocyclic Compounds

Besides metal oxides, the class of nonprecious catalysts that has attracted the most attention over the years is pyrolized metal macrocyclic compounds, such as metal porphyrins and phthalocyanines (Pc) since Jasinski observed that cobalt phthalocyanine (CoPc) catalyzed ORR in 1964 [75]. Metal macrocycles after heat-treatment such as Fe- and Co-centered Pc complex are considered the most promising catalysts for ORR. In general, these catalysts can be synthesized by pyrolyzing metal macrocyclic complex under controlled conditions of temperature, pyrolysis time, and metal concentration in hybrids. The high-temperature treatments produces species of a nitrogen-coordinated metal (Me-N^x) type, which have been suggested to be the active ORR sites. Tang et al. developed a novel strategy to prepare the M-Nx-C catalysts, in which the molecular architecture of the cobalt porphyrin multilayers is incorporated onto the reduced graphite oxide sheets using the layer-by-layer assembly technique [76]. The electrochemical properties of the prepared catalysts were investigated by the RDE technique and the calculated electron transfer number was approximately 3.85 between −0.3 and approximately −0.5 V. This result was further confirmed by the RRDE measurements, demonstrating that ORR proceeds by an approximate 4-electron reduction pathway. Moreover, as compared to the commercial Pt/C catalysts, the catalyst showed comparable electrocatalytic activity but better stability and increased tolerance to the crossover effect.

7.3.3 Transition Metal-Nitrogen-Containing Complexes

In spite of significant progress achieved with heat-treated macrocylic complex toward ORR, the activity and durability of this family of catalysts are still insufficient for replacing Pt. Furthermore, the complex structure of macrocylic complex makes their synthesis expensive and potentially noncompetitive with precious-metal-based catalysts also from a materials cost point of view. A breakthrough in the synthesis of these catalysts was achieved by Yeager et al. who demonstrated that expensive macrocycles can instead be substituted by catalyst materials derived from individual metal and nitrogen [77]. Following this approach, proper metal-nitrogen (M-N) coordination is of great concern in pursuit of catalytically active catalysts with high performance. Accordingly, researchers have developed comparable active catalysts to the state-of-the-art Pt/C by rational design of unique nanostructures. Another key issue that contributes to the high electrocatalytic activity is related to the nature of the carbon support, which can assist in both dispersing the metal catalysts and dominating electron transfer, facilitating mass transport kinetics at the electrode surface. Based on these attractive properties, research groups have used a combination of G and Me-Nx to construct efficient catalysts for ORR. The typical candidate is from the word of Byon and coauthors, who successfully prepared a Fe-N-rGO electrocatalyst with high ORR activity and stability in acid for the applications in PEMFC [78]. They introduced Fe-N moieties into graphene (Fe-N-rGO) through the pyrolysis of chemically reduced graphene oxide (rGO), Fe salt, and graphitic carbon nitride (g-C_3N_4). After acid-leaching of the pyrolyzed product (in 2 M H_2SO_4 at 80°C for 3 h) in order to remove extraneous Fe species such as iron oxide and metallic iron, the ORR mass activity of Fe-N-rGO in the study (~0.5 and ~1.5 mA/mg Fe-N-rGO at 0.8 and 0.75 V vs. reversible hydrogen electrode [RHE], respectively) was a factor of three of the highest ORR activity reported for Fe-N-C catalysts synthesized from Fe-salt-based precursor in inert atmosphere. The enhanced catalytic activity was attributed to the pyridinic N and the intergration of Fe-N_x moieties in rGO sheets, which were proposed as active sites for ORR. The XPS N 1s spectrum revealed that the presence of pyridinic-N had the highest concentration (40 atom%) among all N components. This is different from N-containing G prepared by chemical vapor deposition growth with NH_3 gas and posttreatments of GO with either NH_3 gas annealing or N_2 plasma, all of which had shown predominantly quaternary-N. The ORR current of Fe-N-rGO remained constant over 70 h at 0.5 V vs. RHE in O_2-bubbled 0.5 M H_2SO_4 at 80°C after an initial drop of 30% in the first 3 h.

Kamiya et al. developed an instantaneous one-pot synthesis method of G modified with Fe and N (Fe-N-G) using GO with defined graphitic structures as a precursor to improve the carbon-based ORR catalysts for PMEFC [79]. A big difference was that the use of GOs as a precursor allowed the period of heat treatment to be drastically shortened (within 3 s) into an Ar atmosphere maintained at 900°C together with N and Fe sources. From the Raman spectra, the intensity ratio of the G-band (sp²-carbon, 1620 cm⁻¹) and D-bands (edge or defect sites, 1370 cm⁻¹), I_D/I_G, for Fe-N-G (1.14) and NG (1.15) were markedly higher than that for undoped-G (0.89). The Fe concentration calculated from the XPS spectra was 0.83 mol% for Fe-N-G and in the Fe^{3+} oxidation state. The N/C-ratio for Fe-N-G calculated from the XPS spectra was 0.059, which was larger than that for NG, 0.035, indicating that anchoring of N atoms on G sheets was reinforced by coordination bonds with Fe. The resulting catalysts possessed well-defined structures and functioned as an efficient electrocatalyst for ORR in acidic solutions with an onset potential of 850 mV vs. RHE. In this case, Fe did not serve as the catalyst for graphitization. Furthermore, the formation of metallic Fe particles and Fe carbides, whose existence implies the disappearance of M–N bonds, was not detected in TEM and XRD measurements of the synthesized catalysts. The results on XPS spectra suggested that Fe existed at the surface of G and the N atoms that coordinated with Fe remained. Although both graphite-like N and pyridinic-like N were present even in NG to some extent, NG exhibited much less ORR activity than Fe-N-G. Based on the above analysis, the authors concluded that atomic Fe coordinated with N is the active reaction center of the synthesized catalyst.

In another interesting study, Zhang et al. reported the facile synthesis and superior electrocatalytic activities of a Fe-N doped nanocarbon composite of CNTs grown *in situ* on/between G [80]. The in-between growth of CNTs prevented the GO sheets from restacking during the chemical doping, while the presence of G sheets as support resulted in efficient dispersion and reaction between the melamine and Fe ions, and therefore uniform nucleation and growth of Fe/N-doped CNTs. Such a synergistic coupling led to not only enhanced specific surface area but also large amounts of Fe-N coordination structures for ORR electrocatalysis. The ORR activity of the composite was nearly identical to that of the Pt/C in alkaline media and approaching that of Pt/C in acid fuel cells. When the catalyst-loaded electrode was subject to continuous polarization in O_2-saturated 0.1M $HClO_4$ solution at 0.62 V (vs. RHE), a typical working potential of the cathodes in PEMFC, the composite exhibited lower initial activity than Pt/C but were more stable with the evolution of time under the operating potential. Apart from the SEM and TEM images, Fe particles were found to be predominantly embedded inside the CNTs, in which Fe particles were protected by the relatively thick walls (ca. 5 nm) of the CNTs. Therefore, no obvious dissolution of Fe from the composite catalyst occurred in the electrolyte solution. In addition, the composite had remarkable selectivity toward ORR in a methanol-containing acid medium. For the large-scale synthesis of M-N-C catalysts with high active site density, Fu et al. demonstrated an effective approach for fabrication of G-based NPMCs for ORR through pyrolysis of a mixture of Fe, Co salt, PANI and rGO [81]. The transition metal nitrogen containing moieties (M-Nx, M=Fe or/and Co, x = 2 or 4) could be embedded into the G sheets. The embedded M-Nx moieties could act as catalytic sites and boost the activity of catalysts in both acidic and alkaline media. In comparison with commercial Pt/C catalyst, FeCo-N-rGO catalyst presented higher ORR onset potential and 46 mV more positive ORR half-wave potential in alkaline solution. For clarifying the mechanism of this improvement, a detailed analysis of the relative atomic ratios of each type of N species was made by XPS. The following facts were found: (1) the shape of N1s peaks significantly changed when the metal precursors were varied, (2) the total N content in M-N-rGO was higher than that of N-rGO, and (3) the pyridinic N is found to be the dominant component for all samples, especially in FeCo-N-rGO catalyst. Based on these facts, it seems like that the binary metal precursors could assist embedded FeCo-Nx moieties into G sheets, which would exert a large influence on their electrocatalytic performances for oxygen reduction. Additionally, the FeCo-N-rGO catalyst also demonstrated good tolerance to methanol and better stability than Pt/C.

Tsai et al. synthesized NG-supported carbon-containing iron nitride (FeCN/NG) by the chemical impregnation of iron and N-containing precursors in the presence of NH_3 under thermal treatment [82]. The resultant G-based material showed excellent performance in ORR applied to fuel cells as compared with that of FeN and that of

FeN/C. The steady-state catalytic current density at the FeCN/NG electrode is about six times higher than that at the NG electrode over a measured potential range. The number of transferred electrons obtained from RDE and RRDE are 3.91 and 3.93, which is extremely close to the theoretical 4-electron transferred process of the H_2O product, indicating that H_2O is a major product of this catalyzed ORR process. In other words, the presence of harmful H_2O_2 is almost not produced. The better performance can be attributed to the formation of active centers for graphite-like N positions and FeCN nanoparticles as well as the increase in electrical conductivity. As demonstrated in these examples, G and NG can be used as an effective support material for applications in the fabrication of high-performance electrocatalysts in the development of fuel cells.

7.4 GRAPHENE AS A METAL-FREE CATALYST FOR ORR AND MOR

Recent years have seen the development of metal-free electrocatalysts in the field of fuel cells. One of the significant advantages of metal-free catalysts is their enhanced durability against fuels like CH_3OH and CO, especially in an alkaline medium. Based on their unique electrical and thermal properties, wide availability, corrosion resistance, and large surface area, G-based nanomaterials are ideal candidates for metal-free electrocatalysts. Therefore, extensive research efforts have been made to synthesize G-based nanomaterials with enhanced performance toward ORR. Recent advances revealed that heteroatom-doped G shows an excellent ability to electrochemically catalyze ORR without the need for Pt or other metal catalysts. Both quantum mechanical calculations and experimental studies have revealed that heteroatom-doped G-nanomaterials can tailor their electronic property and chemical reactivity, as well as give rise to new functions, which were regarded to be a class of promising metal-free electrocatalysts toward ORR. Among these catalysts, NG has been reported to exhibit excellent ORR electrocatalytic activity [83].

7.4.1 NG AS A METAL-FREE CATALYST

Following an earlier discovery that N-containing vertically aligned CNTs with carbon atoms positively charged by N-doping could catalyze a more efficient $4e^-$ ORR process [84], various new metal-free NG catalysts have been developed for ORR in fuel cells. When a N atom is introduced into G, it usually has three different types of N-functional groups, including pyridine-like, pyrrole-like, and graphite-like nitrogen atoms, as shown in Figure 7.8. Specifically, pyridinic N refers to N atoms at the edges or defects of G-planes, where each N atom bonds with two carbon atoms and contributes one p-electron to the aromatic π system. Pyrrolic N atoms are incorporated into five-membered heterocyclic rings, which are bonded to two carbon atoms and donate two p-electrons to the π system. Graphtic N atoms refer to N atoms that incorporate into the G layer and substitute carbon atoms within the G-plane [85]. Among these nitrogen types, pyridinic N and quaternary N are sp^2-hybridized and pyrrolic N is sp^3-hybridized. Apart from these three common nitrogen types, N-oxides of pyridinic N, pyridinic-N^+–O^- have also been observed in NG. NG shows different properties compared with PG. The introduction of N atoms into

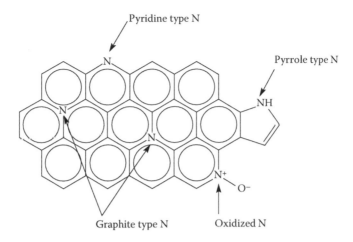

FIGURE 7.8 Bonding configurations for nitrogen atoms in N-graphene.

the sp²-hybridized carbon frameworks modulates the electronic surface state of G, which provides desirable chemical reactivity and electrocatalytic performance of G.

7.4.1.1 Synthesis Validation of NG

Similar to the synthesis of N-doped CNTs, various methods have been used to synthesize NG, and these methods can be categorized into two types: direct synthesis and posttreatment.

The posttreatment method is that in the preparation of NG, the reduction of GO and N-doping is simultaneously completed in these synthesis approaches. Considering that the reduction of GO is one of the most common methods to prepare G, the posttreatment of GO with N-containing precursors has been widely used in synthesizing NG. In this regard, Sheng et al. reported that NG could be synthesized by annealing GO in the presence of melamine at a high temperature [86]. Compared to other methods, such as CVD using NH_3 and pyridine as N source and arc discharge of graphite in the presence of pyridine vapor or G treated with nitrogen plasma, the thermal annealing of GO with melamine as a N source is simple, cost-effective, and safer. In the synthesis process, melamine molecules first adsorbed onto GO surfaces were condensed to carbon nitride with increasing temperature in a tubular furnace. At the same time, oxygen groups linked to GNSs in GO were removed at high temperature. Nitrogen atoms or other nitrogen species formed by decomposition of carbon nitride can attack these active sites and form NGs. The N content can be controlled by changing the mass ratio of GO and melamine, the annealing temperature, and time. The largest N content (10.1 at%) is obtained under the conditions of 700°C with a 0.2 mass ratio of GO to melamine. When used as cathode catalysts, NGs show similar electrochemical behavior. The onset potential of ORR is around −0.1 V and almost invariable with the increase of nitrogen content. The authors demonstrated that the pyridine-like N component in the NGs determines the electrocatalytic activity toward the ORR, but the N content in NGs does not significantly affect the electrocatalytic activity.

Recently, Hou's group successfully synthesized NG by hydrothermal reaction of GO with urea [87]. The introduction of urea leads to successful N insertion in the form of pyridinic, pyrrolic, and graphitic bonding configurations with enhanced reduction of GO, and an increase in the mass ratio between urea and GO gives higher N contents, which is accompanied by more defects in microstructure. Geng et al. reported that NG was synthesized by heat-treatment of G using NH_3 as a metal-free catalyst for ORR [88]. The highest N content was 2.8 at% at 900°C. The obtained catalysts had a very high ORR activity promoting the desired 4e⁻ ORR in alkaline medium by using RDE as well as RRDE techniques. However, although pyridinic-N and pyrrolic-N dominated in NG, no obvious dependence of the ORR activity on the content of these N species was observed. In contrast with other NG, they demonstrated that quaternary-N atoms seem to be the most important species for the ORR because of the matching relationship between activity and quaternary N contents. In comparison to commercial Pt/C, NG catalysts presented higher ORR onset potential (0.308 V) and 43 mV more positive ORR half-wave potential. More importantly, it demonstrated better stability than Pt/C. However, the N content in the NG prepared by these well-developed high-temperature routes is still relatively low, and high-temperature treatment has the problems of high-energy consumption and toxic gas emission. As one option to address these challenges, Zhang et al. prepared NG via a wet chemical reaction between GO and dicyandiamide (DCDA) at temperatures as low as 180°C [89]. Without any high-temperature treatments, the obtained NG catalysts showed a high ORR catalytic activity and favored a 4-electron pathway that was comparable to many previous N-doped carbon catalysts by high-temperature pyrolysis. They suggest that rational utilization of the graphitic carbon template, the N-containing molecules, and the wet chemical reactions may offer a low-temperature route to create interesting ORR electrocatalysts with easier surface property manipulation.

Apart from the posttreatment of GO and N-containing precursors, NG can also be prepared from direct synthesis including CVD [90], segregation growth [91], solvothermal [92], and arc-discharge approaches [93]. Among them, CVD is a process in which gaseous precursors are reactively transformed into a thin film, coating, or other solid-state material on the surface of a catalyst or substrate. It is widely used in controllable growth of various carbon nanomaterials like CNTs and G. Recently, it was successfully applied to prepare NG. Qu et al. reported that NG synthesized by CVD of CH_4 in the presence of NH_3 acted as a metal-free catalyst toward ORR [94]. The resultant NG consisted of only one or a few layers of the graphite sheets after HRTEM characterization. The N content in this method is calculated to be 4 at%. More importantly, the NG had a much better electrocatalytic activity, long-term operation stability, and tolerance to crossover effect than Pt for oxygen reduction via a 4e⁻ pathway in alkaline fuel cells (Figure 7.9). In addition, Luo et al. also synthesized single-layer G doped with pure pyridinic N by thermal CVD of hydrogen and ethylene on Cu foils in the presence of ammonia [90]. By adjusting the flow rate of NH_3, the atomic ratio of N and C can be modulated from 0% to 16%. The UV photoemission spectroscopy investigation demonstrated that the pyridinic N efficiently changed the valence band structure of G, including the raising of density of π states near the Fermi level and the reduction of work function. Such pyridinic N doping in carbon materials was generally considered to be responsible for their ORR activity.

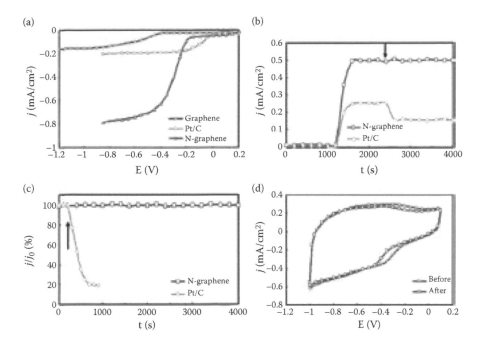

FIGURE 7.9 (a) RRDE voltammograms for the ORR in air-saturated 0.1 M KOH. Electrode rotating rate: 1000 rpm, scan rate: 0.01 Vs⁻¹, catalyst loading: 7.5 μg. (b) *j*-t curves obtained at the Pt/C and NG electrodes at −0.4 V in air-saturated 0.1 M KOH. The arrow indicates the addition of 2% (w/w) methanol into the air-saturated electrochemical cell. (c) *j*-t curves of Pt/C and NG electrodes to CO. The arrow indicates the addition of 10% (v/v) CO into air-saturated 0.1 M KOH at −0.4 V; j_o defines the initial current. (d) CVs of NG electrode in air saturated 0.1 M KOH before and after a continuous potentiodynamic swept for 200,000 cycles at 25°C. Scan rate: 0.1 Vs⁻¹. (Reprinted with permission from Ref. 94, 1321–1326. Copyright 2011 American Chemical Society.)

However, the 2-electron reduction mechanism of ORR determined by RDE on the NG indicated that the pyridinic N may not be an effective promoter for ORR activity. In the CVD approach, many factors, such as precursors, catalyst, gas flow, pressure, growth time, and growth temperature may all play a significant role in determining the morphologies and configuration of NG. Therefore, further research is still needed to prepare NG with enhanced activity for ORR.

7.4.1.2 Morphology Control of NG

Other than nitrogen doping, accurate control of the ideal morphologies of the final products is another effective way to enhance their electrochemical activities. Yang et al. reported an effective approach for the fabrication of nanoporous NG [95]. As shown in Figure 7.10, GO-based silica nanosheets were first fabricated by the hydrolysis of tetraethylorthosilicate (TEOS) on the surface of GO with the help of a cationic surfactant (cetyltrimethyl ammonium bromide [CTAB]). The resulting GO-based silica nanosheets were converted to rGO by heating under Ar. Further pyrolysis of rGO-based silica nanosheets impregnated with ethylenediamine and

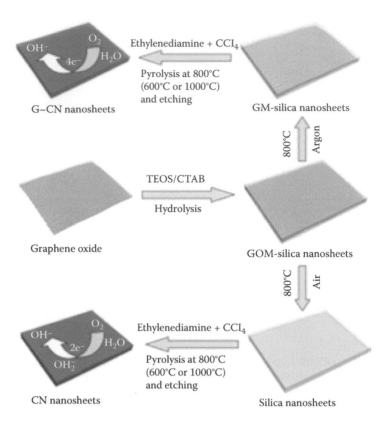

FIGURE 7.10 Fabrication of G-based carbon nitride (G–CN) and CN nanosheets for ORR. (Yang, S., Feng, X., Wang, X. and Müllen, K.: Graphene-based carbon nitride nanosheets as efficient metal-free electrocatalysts for oxygen reduction reactions. *Angew. Chem. Int. Ed.* 2011. 50(23). 5339–5343. Copyright Wiley-VCH Verlag GmbH & Co. KGaA, Weinheim. Reproduced with permission.)

carbon tetrachloride was carried out at different temperatures for the formation of carbon nitride (CN). Final etching of silica in 2 M NaOH solution generated porous NG. The resultant composite had high N content, thin thicknesses, high surface areas (542 m^2 g^{-1}), and enhanced electrical conductivity. When employed as metal-free catalysts for ORR, the porous NG nanosheets pyrolyzed at 800°C exhibited the highest electrochemical activity, good durability, and high selectivity. In association with the XPS analysis of G–CN nanosheets, with the increase of pyrolysis temperature from 800°C to 1000°C, the content of graphitic N is kept constant while that of pyridinic N is largely decreased. Since the pyridinic N atoms with strong electron-accepting ability can create a net positive charge on the adjacent carbon atoms in the G–CN sheets, they are favorable for the adsorption of oxygen atoms and can readily attract electrons from the anode, thus facilitating the ORR.

Unni et al. reported an efficient template-free synthetic route for the preparation of mesoporous NG containing a high weight percentage of pyrrolic N, good specific surface area, and comparable electrochemical oxygen reduction activity as that of the state-of-the-art 40 wt% Pt/C catalyst [96]. The desired coordination of N in the carbon framework of G has been conceived by a mutually assisted redox reaction between GO and pyrrole, followed by thermal treatment at elevated temperatures. NG exhibits a high surface area of 528 m^2 g^{-1} and a pore diameter of 3 to 7 nm. The heat-treatment temperature plays an important role in establishing the desired pyrrolic coordination of nitrogen in graphene for the electrochemical ORR. The NG sample obtained after heat treatment at 1000°C has 53% pyrrolic N content compared to the similar samples prepared by treating at low temperatures. Most importantly, NG displayed a significantly low overpotential for oxygen reduction with the onset potential comparable to that of the commercial 40 wt% Pt/C. In RDE measurements, the ORR involves the desired 4e$^-$ transfer as observed in the case of the Pt-based electrocatalysts, leading to a significantly high kinetic current density of 6 mAcm^2 at 0.2 V. Moreover, the methanol tolerance and durability under the electrochemical environment of the NG catalyst is found to be superior to the Pt/C catalyst.

7.4.1.3 Active Sites of NG for ORR

Since N incorporation can effectively enhance the electrocatalytic activity in NG, it is necessary to understand the influence of nitrogen content and its chemical configuration on the ORR activity. However, the assignments and exact role of active sites for ORR are still controversial. To study how the NG influences the process of ORR, Zhang et al. studied the electrocatalytic active sites and the processes of electron transformation of NG in acidic environment by using density functional theory (DFT) [97], by which the energy variations were calculated during each reaction step. The simulations demonstrate that the ORR on NG is a direct 4-electron pathway, which is consistent with the experimental observations. The energy calculated for each ORR step shows that the ORR can spontaneously occur on NG. The active catalytical sites on single NG have either high positive spin density or high positive atomic charge density. The N doping introduces asymmetry spin density and atomic charge density, making it possible for NG to show high electroncatalytic activities for the ORR. Also, quantum mechanical calculations reveal that pyridinic- or/and graphitic-N in the carbon frameworks plays an essential role in the highly electrocatalytic activity for ORR [97,98].

In additional to theoretical calculations, a variety of experiments have also been carried out for getting the real facts. Lai et al. presented two different ways to fabricate NG as a metal-free catalyst to study the catalytic active site for ORR [85]. One is annealing of GO under NH$_3$. Another is annealing of a N-containing polymer/rGO composite (PANi/rGO or PPy/rGO). They investigated the effects of the N precursors and annealing temperature on the performance of the catalyst for ORR and found that the bonding state of the N atom has a significant effect on the selectivity and catalytic activity. It was found that annealing of GO with NH$_3$ preferentially formed graphitic N and pyridinic N centers, while annealing of PANI/rGO or PPy/rGO tended to generate pyridinic and pyrrolic N moieties, respectively. Most importantly, the electrocatalytic activity of N-containing metal-free catalysts is highly dependent on the graphitic N content while pyridinic N species improve the onset potential for ORR. However, the total atomic content of N did not play an important role in the ORR process. In addition, graphitic N can greatly increase the limiting current density; while pyridinic N species might convert the ORR reaction mechanism from a 2-electron-dominated process to a 4-electron-dominated process. Zhang et al. reported that the total content of graphitic- and pyridinic-N atoms is the key factor to enhance the current density in the electrocatalytic activity for ORR [99]. Similar conclusions have been reported by other research groups [92,100]. In contrast with the previous observations, Luo et al. used a CVD method to produce NG with pyridinic-N as the dominant dopant [90]. They found that as-prepared NG does not show remarkable ORR catalytic ability as previously reported. Therefore, pyridinic N content may not be the key parameter determining the ORR catalytic activity. Although many studies have suggested the possible effect of N on ORR activity, the detailed mechanism is still unclear. Consequently, further research is still required to understand the real relationship between the catalytic activity and the microstructure for NG.

7.4.2 OTHER HETEROATOM-DOPED GRAPHENE AS A CATALYST FOR ORR AND MOR

Recent progress involving doping G with high electronegativity N atoms to transform G into metal-free catalysts for ORR has motivated researchers to pay more attention to the corresponding performance of G doped with other low electronegativity atoms, such as B, sulfur (S), and selenium (Se). Similar to NG, graphene doping with these atoms also showed much better electrocatalytic activity toward the ORR than the undoped one. Yang et al. successfully fabricated sulfur-doped graphene (SG) by directly annealing GO and benzyl disulfide in argon [101]. They studied the electrochemical activity of SG treated at various temperatures in the range from 600°C to 1050°C. All of the SG samples have more positive onset potentials and higher limiting current density than those of the corresponding graphene without dopants. The onset potential for SG-1050 is close to that of the Pt/C catalyst and its current density is larger than the Pt/C catalyst. Significantly, the SG-1050 was much more stable and had excellent tolerance to crossover effects than commercial Pt/C. Considering the fact that S has a close electronegativity to carbon and the C-S bonds are predominately at the edge or the defect sites, the change of atomic charge distribution for the SG is relatively smaller compared to N (B or P) doped carbon materials. They speculated that the spin density is another dominant factor to regulate the observed ORR activity in SG.

Sheng et al. prepared BG via annealing a mixture of GO and boron oxide (BO) [102]. B atoms could be successfully doped into a G framework with an atomic percentage of 3.2%. The electrochemical measurements indicated that the resultant catalysts had excellent electrocatalytic activity toward ORR in alkaline electrolytes and high CO tolerance, which can be ascribed to its particular structure and unique electronic properties. Se-doped graphene (SeG) can also be obtained through the same procedure by replacing BO with diphenyl diselenide [103]. Yao et al. also successfully fabricated I-doped graphene (IG) by annealing GO and iodine in Ar [104]. The iodine-bonding states were examined to be triiodide (I_3^-) and pentaiodide (I_5^-). The authors suggested that the I_3^- structure plays a crucial role for enhancement of the ORR activity of G.

Although these newly developed G-doped materials have shown better performance than the N-doped materials, improvement in the ORR catalytic activity has been very limited owing to the low content of the doped elements. It is believed that codoping with two elements; that is, one with higher and one with lower electronegativity than that of C ($\chi = 2.55$), can create a unique electronic structure with a synergistic coupling effect between heteroatoms, such as B ($\chi = 2.04$) and N (c = 3.04). This effect makes such dual-doped G catalysts much more catalytically active than singly doped G catalysts. As for the codoping method, CVD was reported early on [105]; however, using CVD for the synthesis of dual-doped G is too tedious and expensive for mass production. The recently developed chemical method involving the coannealing of solution-exfoliated GO in the presence of H_3BO_3 and NH_3 seems facile [106]. Unfortunately, it produced a covalent boron–carbon–nitride (BCN) but not a B- and N-codoped G (B,N-G). Because of this, almost all physical and chemical strategies inevitably lead to undesired by-products—for example, hexagonal boron nitride (h-BN), which is chemically inert and results in poor activity of the catalyst. To address this issue, a two-step B and N doping in G strategy was developed by Qiao's research group [107]. In their work, N was first incorporated by annealing with NH_3 at an intermediate temperature (e.g., 500°C), and then B was introduced by pyrolysis of the intermediate material (N-graphene) with boric acid (H_3BO_3) at a higher temperature (e.g., 900°C). This new method enabled the incorporation of heteroatoms at selected sites of the G framework to induce a synergistic enhancement of the activity of the B,N-graphene and at the same time could prevent the formation of inactive by-products. Thus, the resultant catalyst showed excellent activity in the ORR and perfect (nearly 100%) selectivity for the 4-electron ORR pathway in an alkaline medium. Its activity and 4-electron pathway selectivity in the ORR was much higher than that observed for single BG or NG. Additionally, the B,N-graphene also showed complete methanol tolerance and excellent long-term stability. These properties are superior to those observed for the commercial Pt/C catalyst (Figure 7.11). Through DFT calculations, they demonstrated that that in the case of NG, a twofold-coordinated pyridinic N dopant at the edge of the cluster has a tendency to suppress the ORR activity; only a graphitic N atom that is far from the G edge can induce HO_2^- adsorption (which is the rate-determining step of the ORR process) at an adjacent C atom. For dual-doped B,N-graphene, however, the pyridinic N group could be changed from "inactive" in N-graphene to "active" after B incorporation, which ensured a better activity of B,N-graphene owing to its higher content than that of the graphitic N group in the final product (2.03 atom% vs. 0.9 atom%, as based on experimental data). That is, the chemical coupling of B and N is the origin of a synergistic effect that boosts the ORR activity of the B,N-graphene through an alternative mechanism to that observed for singly doped G with B or N (Figure 7.11).

The same group also reported for the first time the design and one-step synthesis of N and S dual-doped mesoporous graphene (N-S-G) [108]. Both dopant elements in N-S-G are efficiently utilized for ORR catalysis and do not form inactive compositions. Moreover, the authors used commercially available colloidal silica nanoparticles to create large mesopores in the G catalyst. Melamine and benzyl disulfide (BDS) were selected as N and S precursors, and the doping process was carried out by heating a mixture of melamine/BDS/GO/SiO_2 in a weight ratio of 5:5:1:5 at 900°C

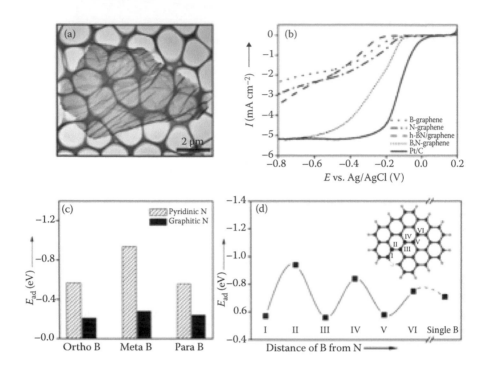

FIGURE 7.11 (a) TEM image of B,N-graphene. (b) Linear sweep voltammetry (LSV) of various electrocatalysts on a RDE (1500 rpm) in an O_2-saturated 0.1-m solution of KOH (scan rate: 10 mV s^{-1}). (c) The energies of HO_2 adsorption (E_{ad}) on various B,N-graphene models with pyridinic or graphitic N groups. (d) E_{ad} of HO_2 on various B,N-graphene models with B active sites as a function of the distance to a pyridinic N atom. (Zheng, Y., Jiao, Y., Ge, L., Jaroniec, M. and Qiao, S. Z.: Two-step boron and nitrogen doping in graphene for enhanced synergistic catalysis. *Angew. Chem.* 2013. 125(11). 3192–3198. Copyright Wiley-VCH Verlag GmbH & Co. KGaA, Weinheim. Reproduced with permission.)

in Ar to form N-S-G/SiO$_2$. This N and S dual-doped mesoporous graphene displayed a very high ORR onset potential of −0.06 V, which is close to that of commercial Pt/C (−0.03 V) and much more positive than that of N-G, S-G, or G (ca. −0.18 V). Also, compared to the fuel-sensitive and vulnerable Pt/C, the proposed N-S-G material was much more tolerant to fuel crossover and displayed long-term durability in an alkaline environment. From the above, it can be seen that these experimental results open an effective pathway to synthesize ideal low-cost metal-free catalysts with high electrocatalytic activity and practical durability for fuel cell applications.

7.5 GRAPHENE FOR MEA ASSEMBLY AND FUEL CELL PERFORMANCE

From a practical point of view, the single cell test is the ultimate evaluation criterion of novel electrocatalysts materials. MEA, which consists of catalyst layers, gas diffusion layers, and electrolyte membrane, is the heart of a single cell and determines the cell's performance and durability. Relative to the studies dealing with catalytic activity for the ORR or MOR, there are less reports of G-based catalysts in single fuel cells. Generally, the performance of single cells constructed with G-supported catalysts was better than that of conventional Pt/C in DMFC tests. For example, Jhan et al. prepared G-carbon nanotubes composite supported Pt nanoparticles and used them as anode electrocatalysts in DMFCs [109]. A large number of small and homogeneous Pt nanoparticles (from 3.4–4.2 nm) were deposited on this hybrid support. In contrast to catalyst supported on G (28.1 m^2g^{-1}), Pt/G-CNT showed a very high electrochemically active surface area (77.4 m^2g^{-1}). To fabricate MEA, the anode was the obtained catalyst composites while the standard Pt/C was used as the cathode. The catalyst loadings on the anode and cathode layers were both 1.0 mg Pt cm^{-2}. Nafion 117 (Du Pont) was used as the PEM. The MEA was fabricated via hot pressing at 140°C under 30.0 kg cm^{-2} of pressure and tested at 70°C. The performance of Pt/G-CNT composite was much higher than that of Pt/G and E-TEK. The maximum power density value for Pt/G-CNT was 32.0 mW cm^{-2} and was 74% higher than that of Pt/G (18.4 mW cm^{-2}) and 26% higher than that of E-TEK (25.4 mW cm^{-2}). In another study, Jha et al. prepared a series of nanostructured PtRu and Pt dispersed functionalized G-functionalized MWCNT nanocomposites by varying the mass ratio of G and MWCNT [110]. Electrochemical measurements showed that the best activity of these catalysts toward methanol oxidation was obtained in the G/MWCNT weight ratio of 1. With PtRu/(50 wt% MWCNT+ 50 wt% G) as the anode electrocatalyst and Pt/(50 wt% MWCNT+ 50 wt% G) as the cathode electrocatalyst, an MEA was fabricated using Nafion 117 membrane and hot-pressed at 130°C and 3 bar for 5 min. A power density of 68 mW cm^{-2} was achieved at 80°C at a loading of 2.5 mg cm^{-2} for both the anode and cathode with 1 M methanol feed. The high performance could be attributed to the good accessibility of catalyst particles for the methanol and oxygen reduction reactions due to the change in the morphology of the G-sheets by the addition of MWCNT.

Contrary to the catalysts test for DMFCs, the G-based catalysts conducted in PEMFC often get conflicting results. Park et al. prepared a Pt-rGO nanohybrid with 40% Pt/GO and used it as the cathode catalyst to carry out a single-cell test experiment for PEMFC application. They found that after three times of activation under 0.6 V, 1 h, and I–V curve conditions, the power density reached a maximum value of 480 mW cm^{-2} at 75°C at a relative humidity of 100%. The catalyst loading was 0.4 mg(Pt) cm^{-2} on both sides of the Nafion membrane and the MEA was 1 cm^{-2}. The Pt-RGO showed an open-circuit voltage of over 0.98 V even after activation. The activation process was found to be necessary to achieve better Pt catalyst activity and membrane proton conductivity, resulting in the higher performance of PEMFC [38]. Using Pt/f-sG as cathodes, Aravind et al. prepared MEA by sandwiching a pretreated Nafion 212 CS membrane between the anode (0.25 mg cm^{-2} loading) and cathode (0.5 mg cm^{-2} loading) [58]. The gas diffusion layer (GDL) was formed by coating a mixture of Vulcan XC-72 carbon and PTFE. Prior to the polarization studies, the electrodes were activated between OCV and high current densities for catalyst activation. A maximum power density of 260 mW cm^{-2} was obtained at 40°C and increased to 304 mW cm^{-2} and 355 mW cm^{-2} at 50°C and 60°C. When changing the Pt/sG-f MWCNT cathode, the maximum power densities with this cell were 592, 657, and 675 mW cm^{-2}. The cell performances were further increased to 704,

742, and 781 mW cm^{-2}, respectively, when a back pressure of 1 atm at the cathode was applied.

Du et al. developed a two-step growth of GNSs in which CNTs are grown directly on a carbon cloth [111]. Subsequently, GNSs are constructed on the CNT surface. Using 0.1 mg cm^{-2} Pt/GNS-CNT as a cathode catalyst, 0.25 mg cm^{-2} E-TEK Pt/C as an anode catalyst, and Nafion 212 as electrolyte membrane, they prepared an MEA and tested on a 2.25 cm^{-2} single cell. Under their testing conditions, MEA based on Pt/GNS-CNT outperformed those using Pt/C, which gave a maximum power density of 1072 mW cm^{-2} at 80°C under H$_2$/O$_2$, respectively. However, this is not the case in the study by Yun et al. [59]. Yun et al. synthesized porous Pt-G/MWCNT composite, and using this catalyst as a cathode for PEMFC, the fuel cell gave a maximum power density of 303 mW cm^{-2} at 70°C for a catalyst loading of 0.13 mg(Pt)cm^{-2} with an anode Pt loading of 0.2 mg cm^{-2}. This was found to increase fourfold compared to that of the Pt/G cathode, but was 17% lower than that of the Pt/CB cathode with a Pt loading of 0.2 mg cm^{-2}. Park et al. also observed a similar result in their study [57]. These results can be ascribed to the fact that G tended to horizontally stack, resulting in a decrease of active sites during MEA fabrication.

In fact, MEA performance was largely affected by many factors, including the ink preparation, hydrophobic or hydrophilic property of GDL and treatment, catalyst loading, MEA hot-press conditions, cell operating temperature, and activation conditions. To address this point, Jafri et al. constructed a full cell with Pt/NG nanoplatelets and compared it with Pt/G nanoplatelets [61]. In their MEA fabrications, a fixed Nafion to an electrocatalyst ratio of 1:1 (by weight) was applied. A mixture of Vulcan XC 72 and PTFE was coated on carbon cloth, which formed a gas-diffusion layer. The MEA was 11.56 cm^2. A Pt loading of 0.25 mg cm^{-2} and 0.5 mg cm^{-2} was maintained at the anode and cathode, respectively. A maximum power density of 440 and 390 mW cm^{-2} at 60°C under 15 psig back pressure was obtained with Pt/NG and Pt/G nanoplatelets as ORR catalysts, respectively. In another work, the authors used Pt/N-HEG and Pt$_3$Co/N-HEG as cathodes to study MEA performances [62]. The catalyst layer was prepared by coating catalyst ink over the GDL and the catalyst ink was prepared by ultrasonicating the required amount of catalyst in deionized water and 2-propanol with 5 wt% Nafion solution. Pt loadings of 0.25 and 0.4 mg cm^{-2} were maintained at the anode and cathode electrocatalysts, respectively. The MEA was prepared by sandwiching a pretreated Nafion 212 CS membrane between the anode and the cathode by hot-pressing at 130°C and 70 bar for 4 min. The performance of the fuel cell was then studied at three different temperatures (40°C, 50°C, and 60°C) and a relative humidity of 90% without any back pressure. When constructed as cathode catalysts in a single fuel cell, Pt/N-HEG and Pt$_3$Co/N-HEG showed high performances with current densities of 665 and 1110 mA cm^{-2} at 0.6 V, respectively, which are 2.41 and four times higher than that of the performance of commercial Pt/C cathode. The maximum power densities exhibited by commercial Pt/C, commercial Pt$_3$Co/C, Pt/N-HEG, and Pt$_3$Co/N-HEG cathode electrocatalysts were 241, 379, 512, and 805 mW cm^{-2} at 60°C, respectively (Figure 7.12). Consequently, further studies are highly desirable for optimizing the MEA fabrication process to improve fuel cell performance. Meanwhile, theoretical studies such as modeling and simulation are also needed to gain fundamental insight for the development of advanced MEAs.

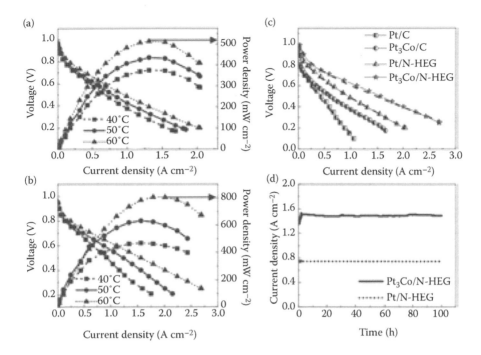

FIGURE 7.12 Polarization curves of cathode electrocatalysts. (a) Pt/N-HEG and (b) Pt₃ Co/N-HEG with Pt/C as the anode electrocatalyst at three different temperatures (40°C, 50°C, 60°C) without any back pressure. (c) Theoretical fit of the polarization data for Pt/C, Pt₃Co/C, Pt/N-HEG, and Pt₃Co/N-HEG using Equation 7.2 at 60°C. (d) Stability study of the PEMFC with cathode electrocatalyst containing Pt/N-HEG and Pt₃Co/N-HEG at 60°C without any back pressure. (Vinayan, B. P., Nagar, R., Rajalakshmi, N. and Ramaprabhu, S.: Novel platinum-cobalt alloy nanoparticles dispersed on nitrogen-doped graphene as a cathode electrocatalyst for PEMFC applications. *Adv. Funct. Mater.* 2012. 22(16). 3519–3526. Copyright Wiley-VCH Verlag GmbH & Co. KGaA, Weinheim. Reproduced with permission.)

7.6 GRAPHENE AS ADDITIVES FOR MEMBRANE PREPARATION AND MODIFICATIONS

PEM, which conducts ions generated at electrochemical interfaces, is a key functional component of the MEA. Currently, the most widely used PEMs are PEMFC fluoropolymer-based materials such as Nafion, developed by DuPont, which has dominated the fuel cell industry as the membrane material of choice. However, the practical uses of PEMFC for transportation and stationary applications have been hindered by several challenges, including the high cost, low conductivity at low humidity and/or high temperature, and loss of mechanical stability at high temperature. As well, for DMFCs, the permeation of methanol from the anode to the cathode not only decreases fuel cell efficiency but also the cathode performance. Therefore, extensive research efforts have been dedicated to improving the performance of presently available membranes and developing new ones during the past few decades. Recent studies have shown that the incorporation of G into Nafion membrane can improve the mechanical stability and the proton conductivity [112,113]. G has great potential as an additive in PEMs due to its fascinating properties. For example, Zarrin et al. reported sulfonic-acid-functionalized GO Nafion nanocomposites (F-GO/Nafion) as a potential PEM replacement for high-temperature PEMFC applications [114]. They observed that F-GO/Nafion composite membranes showed a fourfold higher proton conductivity than that of the unmodified one at a low humidity value of 30% and high temperature of 120°C. The enhanced proton conductivity can be explained by the Grotthuss mechanism, where protons diffuse through a hydrogen bond network of water molecules. The F-GO/Nafion composite membranes could hold more water due to an enormous amount of sulfonated functional groups on the extremely high-surface F-GO. As a result, the membrane could facilitate the transfer of more protons. Moreover, the increase of F-GO nanoadditive loading extends the number of available ion exchange sites per cluster, resulting in the increment of proton mobility in the membrane at high temperatures and low humidity values. The high proton conductivity contributes to the good cell performance. Under similar operating conditions, single-cell MEA testing revealed a significantly higher performance of 10 wt% F-GO/Nafion membrane than that of the recast Nafion, displaying peak power densities of 0.15 and 0.042 W cm^{-2}, respectively. Kumar et al. also prepared GO/Nafion composite membrane with a chemical strategy in which the GO was incorporated into the Nafion matrix to enhance water retention [115]. Nano-sized GO additives were successfully embedded in the composite polymer matrix. The GO/Nafion membrane showed an enhanced proton conductivity and tensile strength. The proton conductivity of GO (4 wt%)/Nafion composite was 0.17 S cm^{-1} at 80°C and 100% humidity, which is much higher than that of recast Nafion with a proton conductivity of 0.092 S cm^{-1}. The high proton conductivity of the GO/Nafion membrane can be ascribed to the chemical interactions between GO/Nafion and the presence of different oxygen functionalities on GO. A single-cell test with a PEMFC demonstrated that the maximum power density value of GO/Nafion membrane (212 mW cm^{-2}) is much higher than that of the Nafion 212 (56 mW cm^{-2}) at 100°C and 25% relative humidity.

Choi et al. synthesized a sulfonated GO (SGO)/Nafion membrane from a highly homogenous solution of Nafion and SGO by solution casting and used it for DMFC application [116]. Compared to the Nafion membrane, the obtained SGO/Nafion membrane showed higher proton conductivity and lower methanol permeability. They explained these results on the basis of the state of water confined in the composite. The bound water (b-H_2O) contributes to proton transfer through interaction with $-SO_3H$ groups of ionomers, while the free water (f-H_2O), which is not bound to the polymer, shows behavior similar to that of bulk water. The SGO/Nafion membrane showed a larger portion of b-H_2O than f-H_2O due to the functionality of $-SO_3H$ groups. MEA with SGO/Nafion membrane was conducted and tested under fuel cell operating conditions of 1 M methanol at the anode fuel and for O_2 gas at the cathode side. Single cell performance showed a maximum power density value of 132 mW cm^{-2} using SGO/Nafion membrane, which is higher than that of the Nafion 112 membrane (101 mW cm^{-2}) at 60°C. To further improve DMFC performance under higher methanol feed concentrations, Lin et al. prepared a dual-layer laminate membrane that used 2-D GO paper as a methanol barrier for a DMFC operating at higher methanol feed concentrations [117]. The GO paper was prepared by using a vacuum filtration method and was subsequently laminated with Nafion 115 membrane through transfer printing and hot-pressing to get the dual-layer membrane. The methanol permeability of the GO-laminated membrane was approximately 70% lower than that of Nafion 115, while their proton conductivity only decreased 22%. This membrane showed significantly higher DMFC performance than the pristine Nafion 115 at higher methanol feed concentration, giving a maximum power density of 33 mW cm^{-2} at 8 M methanol feed at 50°C, which is 100% higher than that of Nafion 115.

In addition to the modification of Nafion-based membrane, many efforts have been devoted to exploring alternative membrane materials [118–120]. Tseng et al. prepared sulfonated polyimide (SPI)/GO composite membranes and that found incorporation of GO in SPI can greatly enhance proton conductivity, reduce methanol permeability, and improve mechanical properties [119]. In particular, they discovered a remarkable utility of this composite membrane by incorporating G in the presence of poly(sodium 4-styrenesulfonate) (PSS-G). As shown in Figure 7.13, the SPI/GO (0.3 and 0.5 wt%) and SPI/PSS-G (0.5 wt%) showed proton conductivities at least approximately two- to fourfold greater than Nafion 117 at 90°C and 98% RH. The methanol permeability decreased with an increasing content of GO for all of the composite membranes and a 4.8-fold decrease compared to that of pristine SPI at 30°C and 6.4-fold decrease at 80°C because the GO dispersion prevented methanol from migrating through the membrane. In addtion, as the temperature was increased to 80°C the selectivity of the GO/SPI composite membranes was about 9.5-fold greater compared to Nafion 117 and 25-fold greater compared to pristine SPI. These results show that the addition of GO and PSS-G significantly inhibits methanol crossover and that high temperature has a significant impact on the reduction of methanol permeability. Among the various PEMs, sulfonated aromatic poly(ether ether ketone)s (SPEEKs) have attracted great interest because of their good mechanical strength and high chemical stability. Jiang et al. used SDBS-functionalized GO as an additive to modify SPEEK [121]. Due to the weak π–π and hydrophobic interactions, SDBS could be adsorbed onto the surface of GO to produce GO functionalized with sulfonic acid groups. With a well-controlled amount of SDBS-GO, the membranes and SPEEK/SDBS-GO exhibited even higher ion exchange capacity (IEC), water uptake, and proton conductivity but lower methanol permeability than the commercially available Nafion 112 membrane, as seen in Table 7.2. DMFCs with SPEEK/SDBS-GO membranes exhibited significantly improved performance compared to plain SPEEK.

In fact, GO not only has most of the properties of graphite, but also has improved characteristics, such as better dispersion capability, because of the increased number of O and N atoms. The presence of different acidic functional groups, like carboxylic acid and epoxy oxygen, could provide more facile hopping of protons. Based on the above conceptions, most recently, polybenzimidazole/graphite oxide (PBI/GO) and PBI/sulfonated graphite oxide (SGO) composite membranes has been reported by Scott's research group for high-temperature PEFC [122]. They prepared the PBI/SGO by immersing the membranes in aqueous H$_3$PO$_4$ solution (2 M). The membranes were loaded with H$_3$PO$_4$ to provide suitable proton conductivity. SGO was prepared using chlorosulfonic acid. The resultant PBI/SGO composite membrane

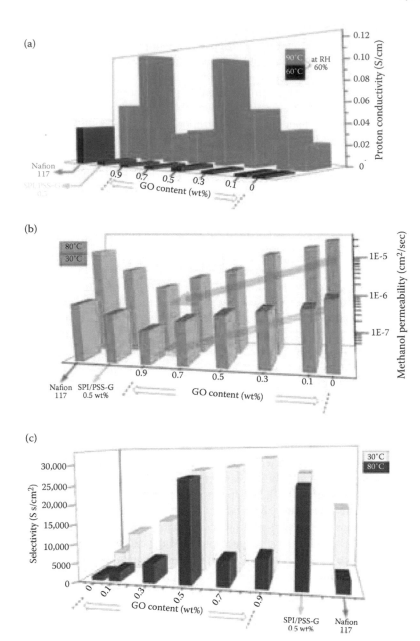

FIGURE 7.13 (a) Influence of 60% relative humidity on the proton conductivity of the composite membranes at 60°C and 90°C, (b) methanol permeability, and (c) selectivity of composite membranes at 30°C and 80°C. (Tseng, C. Y., Ye, Y. S., Cheng, M. Y. et al.: Sulfonated polyimide proton exchange membranes with graphene oxide show improved proton conductivity, methanol crossover impedance, and mechanical properties. *Adv. Energy Mater.* 2011. 1(6). 1220–1224. Copyright 2011 Wiley-VCH Verlag GmbH & Co. KGaA, Weinheim. Reproduced with permission.)

TABLE 7.2
1EC, Water Uptake, Swelling Ratio, Proton Conductivity (σ)

Membrane	IEC (meq. g⁻¹)	Water Uptake (%)		Swelling Ratio (%)		σ (mS cm⁻¹)		P (× 10⁶ cm² s⁻¹)	E_a (kJ mol⁻¹)
		25°C	65°C	25°C	65°C	25°C	65°C		
SPEEK	1.710	24.2	40.1	3.4	15.5	39.5	88.1	1.15	21.1
SPEEK/GO (5 wt%)	1.704	17.6	33.1	3.2	13.8	53.4	98.3	0.59	15.8
SPEEK/SDBS-GO (3 wt%)	1.768	24.7	47.7	3.5	16.3	60.8	101.7	0.63	16.4
SPEEK/SDBS-GO (5 wt%)	1.795	25.8	50.5	3.8	17.1	79.4	155.0	0.82	15.5
SPEEK/SDBS-GO (8 wt%)	1.830	27.9	53.2	4.2	21.7	93.8	162.6	0.95	13.0
Nafion® 112[a]	0.94	31.4	37.1	7.2	13.9	87.2	125.4	1.53	7.3

Note: Activation energy of proton conduction (E_a), and methanol permeability (P) of the plain SPEEK, GO-incorporated SPEEK (SPEEK/GO), SDBS-GO-incorporated SPEEK (SPEEK/SDBS-GO), and Nafion® 112 membranes.

[a] The values for Nafion® 112 were obtained from experiments conducted under the same conditions for comparison.

had an ionic conductivity of 0.052 S cm^{-1} at 175°C and 0% humidity and a peak power density of 600 mW cm^{-2} under H_2/O_2 conditions in fuel cell tests. The authors suggested the PBI composites with SGO loaded with H_3PO_4 may be potential membranes for high-temperature PEMFC. However, there are some aspects that need further study to improve the fuel cell performance. For example, from current densities 0 to 0.1 A cm^{-2}, there was an approximately 250-mV voltage loss, indicating that the catalyst compositions in the MEA were not optimal for the fuel cell. Overall, although it has been reported that GO/Nafion or GO/polymer composites are potential membranes for polymer membrane fuel cells, more studies are required, especially to optimize the electrode catalyst layer composition and to establish suitable MEA preparation in future works.

7.7   CONCLUSIONS AND OUTLOOK

Under the strong driving force of fuel cell commercialization, great progress in PEFCs has been made, especially to the key components including electrocatalysts and PEMs in the last two decades. At the present time, the most practical catalysts for fuel cell applications are still Pt-based catalysts supported on conventional carbon material. As for the future of carbon support, the combination of the high specific surface area, high electrical conductivity, unique 2-D basal plane structure, and the potential low cost and availability for mass production makes G a promising material in replacing conventional materials such as Valcun XC-72 fuel cells. Not only can it be used as a support to improve the performance of conventional catalysts, but it can also serve as a metal-free catalyst with fascinating electrochemical activities. Also, the membrane characteristics can be greatly improved using GO or SGO as additives because of the increased number of O and N atoms that result in better dispersion capability. The presence of different acidic functional groups, like carboxylic acid and epoxy oxygen, could also provide more facile hopping of protons, which would in turn improve the fuel cell power densities. In this chapter, achievements and major challenges, including catalyst material selection, design, synthesis, catalytic performance, catalytic mechanism, and even theoretical approaches have been reviewed, and technological developments for addressing these challenges in recent years discussed, with a focus on developing functional, nanostructured G as catalyst support material and metal-free catalysts as well as for membrane preparation and modifications.

In order to address these requirements, five aspects of the approaches have been reviewed: (1) G as a support material for Pt/Pt-alloy catalysts, (2) G/NG as a support material for NPMCs, (3) G as a metal-free catalyst for ORR and MOR, (4) graphene for MEA assembly and fuel cell performance, and (5) graphene as additives for membrane preparation and modifications. It was demonstrated that advanced G and doped G nanostructures showed great potential in solving issues of insufficient activity and low stability of the catalyst for PEM fuel cell technologies by combining the advantages of its unique structure and exceptional physical properties. However, in spite of the considerable progress that has been achieved, the following challenges are still remaining and should be clarified in future work:

1. A practical approach for the production of large-scale G with high quality is still lacking. As reviewed in this chapter, all the preparation of nanostructured G is only based at the lab level. In practical applications, either the amount or the synthesis procedure is large and time-consuming. Therefore, new methods of synthesizing highly conductive G materials with controllable nanostructures are strongly required.
2. It appears that there are no clear answers about the nature of electrocatalytically active sites facilitating ORR, and this area should be an ongoing research topic in catalyst development, especially for NG. Currently, achieving nitrogen doping at specific positions and with accurate control of doping content is still a challenge.
3. The performance of G composite catalysts strongly depends on their composition, including type of transition metal, doping elements and structure, and geometric and electronic effects, and also features such as size, shape, and morphology. Development of facile and effective approaches to preparing ideal morphologies and controllable nanoparticle sizes composite catalysts is highly desirable.
4. There are few studies that validate the activity and stability of these remarkable G catalysts using real fuel cell operating conditions, and future work should focus on this development.
5. The restacking of G during its processing due to the strong van der Waals interactions inevitably decrease electrochemical performance. Therefore, new approaches that can extensively prevent restacking of graphene sheets are obviously desired.

Additional approaches to the theoretical calculations such as using the DFT method and other modeling techniques are urgently needed for elevating improved catalytic mechanisms both for ORR and MOR from the theoretical basement into practical applications.

ACKNOWLEDGMENTS

The authors would like to thank the National Natural Science Foundation of China (grant no. 21173039), the Specialized Research Fund for the Doctoral Program of Higher Education, SRFD (20110075110001) of China, the Innovation Program of the Shanghai Municipal Education Commission (14ZZ074) and International Academic Cooperation and Exchange Program of Shanghai Science and Technology Committee (14520721900) for their financial support.

REFERENCES

1. Debe, M. K. 2012. Electrocatalyst approaches and challenges for automotive fuel cells. *Nature* 486(7401):43–51.
2. Antolini, E. 2009. Carbon supports for low-temperature fuel cell catalysts. *Appl. Catal. B* 88(1):1–24.

3. Sharma, S. and Pollet, B. G. 2012. Support materials for PEMFC and DMFC electrocatalysts—a review. *J. Power Sources* 208:96–119.
4. Wang, Y.-J., Wilkinson, D. P. and Zhang, J. 2011. Noncarbon support materials for polymer electrolyte membrane fuel cell electrocatalysts. *Chem. Rev.* 111(12):7625–7651.
5. Antolini, E. 2012. Graphene as a new carbon support for low-temperature fuel cell catalysts. *Appl. Catal. B* 123:52–68.
6. Geim, A. K. and Novoselov, K. S. 2007. The rise of graphene. *Nat. Mater.* 6(3):183–191.
7. Chen, D., Tang, L. and Li, J. 2010. Graphene-based materials in electrochemistry. *Chem. Soc. Rev.* 39(8):3157–3180.
8. Allen, M. J., Tung, V. C. and Kaner, R. B. 2009. Honeycomb carbon: A review of grapheme. *Chem. Rev.* 110(1):132–145.
9. Guo, S. and Dong, S. 2011. Graphene nanosheet: Synthesis, molecular engineering, thin film, hybrids, and energy and analytical applications. *Chem. Soc. Rev.* 40(5): 2644–2672.
10. Novoselov, K. S., Geim, A. K., Morozov, S. et al. 2004. Electric field effect in atomically thin carbon films. *Science* 306(5696):666–669.
11. Yoo, E., Kim, J., Hosono, E., Zhou, H.-S., Kudo, T. and Honma, I. 2008. Large reversible Li storage of graphene nanosheet families for use in rechargeable lithium ion batteries. *Nano Lett.* 8(8):2277–2282.
12. Wang, H., Cui, L.-F., Yang, Y. et al. 2010. Mn_3O_4–graphene hybrid as a high-capacity anode material for lithium ion batteries. *J. Am. Chem. Soc.* 132(40):13978–13980.
13. Wu, Q., Xu, Y., Yao, Z., Liu, A. and Shi, G. 2010. Supercapacitors based on flexible graphene/polyaniline nanofiber composite films. *ACS Nano* 4(4):1963–1970.
14. Dikin, D. A., Stankovich, S., Zimney, E. J. et al. 2007. Preparation and characterization of graphene oxide paper. *Nature* 448(7152):457–460.
15. Wang, H., Casalongue, H. S., Liang, Y. and Dai, H. 2010. Ni (OH) 2 nanoplates grown on graphene as advanced electrochemical pseudocapacitor materials. *J. Am. Chem. Soc.* 132(21):7472–7477.
16. Wang, X., Zhi, L. and Müllen, K. 2008. Transparent, conductive graphene electrodes for dye-sensitized solar cells. *Nano Lett.* 8(1):323–327.
17. Bonaccorso, F., Sun, Z., Hasan, T. and Ferrari, A. 2010. Graphene photonics and optoelectronics. *Nat. Photon.* 4(9):611–622.
18. Zhu, C. and Dong, S. 2013. Recent progress in graphene-based nanomaterials as advanced electrocatalysts towards oxygen reduction reaction. *Nanoscale* 5(5):1753–1767.
19. Guo, S., Dong, S. and Wang, E. 2009. Three-dimensional Pt-on-Pd bimetallic nanodendrites supported on graphene nanosheet: Facile synthesis and used as an advanced nanoelectrocatalyst for methanol oxidation. *ACS Nano* 4(1):547–555.
20. Huang, C., Li, C. and Shi, G. 2012. Graphene based catalysts. *Energy Environ. Sci.* 5(10):8848–8868.
21. Park, S. and Ruoff, R. S. 2009. Chemical methods for the production of graphenes. *Nat. Nanotechnol.* 4(4):217–224.
22. Luo, B., Liu, S., Zhi, L. 2012. Chemical approaches toward graphene-based nanomaterials and their applications in energy-related areas. *Small* 8(5):630–646.
23. Xu, C., Wang, X. and Zhu, J. 2008. Graphene–metal particle nanocomposites. *J. Phys. Chem. C* 112(50):19841–19845.
24. Zhang, Y., Gu, Y.-E., Lin, S. et al. 2011. One-step synthesis of PtPdAu ternary alloy nanoparticles on graphene with superior methanol electrooxidation activity. *Electrochim. Acta* 56(24):8746–8751.
25. He, W., Jiang, H., Zhou, Y. et al. 2012. An efficient reduction route for the production of Pd–Pt nanoparticles anchored on graphene nanosheets for use as durable oxygen reduction electrocatalysts. *Carbon* 50(1):265–274.

26. Liao, C.-S., Liao, C.-T., Tso, C.-Y. and Shy, H.-J. 2011. Microwave-polyol synthesis and electrocatalytic performance of Pt/graphene nanocomposites. *Mater. Chem. Phys.* 130(1):270–274.

27. Sharma, S., Ganguly, A., Papakonstantinou, P. et al. 2010. Rapid microwave synthesis of CO tolerant reduced graphene oxide-supported platinum electrocatalysts for oxidation of methanol. *J. Phys. Chem. C* 114(45):19459–19466.

28. Kundu, P., Nethravathi, C., Deshpande, P. A., Rajamathi, M., Madras, G. and Ravishankar, N. 2011. Ultrafast microwave-assisted route to surfactant-free ultrafine Pt nanoparticles on graphene: Synergistic co-reduction mechanism and high catalytic activity. *Chem. Mater.* 23(11):2772–2780.

29. Qiu, J.-D., Wang, G.-C., Liang, R.-P., Xia, X.-H. and Yu, H.-W. 2011. Controllable deposition of platinum nanoparticles on graphene as an electrocatalyst for direct methanol fuel cells. *J. Phys. Chem. C* 115(31):15639–15645.

30. Maiyalagan, T., Dong, X., Chen, P. and Wang, X. 2012. Electrodeposited Pt on three-dimensional interconnected graphene as a free-standing electrode for fuel cell application. *J. Mater. Chem.* 22(12):5286–5290.

31. Hung, T.-F., Wang, B., Tsai, C.-W. et al. 2012. Sulfonation of graphene nanosheet-supported platinum via a simple thermal-treatment toward its oxygen reduction activity in acid medium. *Int. J. Hydrogen Energy* 37(19):14205–14210.

32. Choi, S. M., Seo, M. H., Kim, H. J. and Kim, W. B. 2011. Synthesis and characterization of graphene-supported metal nanoparticles by impregnation method with heat treatment in H2 atmosphere. *Synth. Met.* 161(21):2405–2411.

33. Zhu, C., Guo, S., Zhai, Y. and Dong, S. 2010. Layer-by-layer self-assembly for constructing a graphene/platinum nanoparticle three-dimensional hybrid nanostructure using ionic liquid as a linker. *Langmuir* 26(10):7614–7618.

34. Guo, S. and Sun, S. 2012. FePt nanoparticles assembled on graphene as enhanced catalyst for oxygen reduction reaction. *J. Am. Chem. Soc.* 134(5):2492–2495.

35. Li, Y., Li, Y., Zhu, E. et al. 2012. Stabilization of high-performance oxygen reduction reaction Pt electrocatalyst supported on reduced graphene oxide/carbon black composite. *J. Am. Chem. Soc.* 134(30):12326–12329.

36. Ha, H.-W., Kim, I. Y., Hwang, S.-J. and Ruoff, R. S. 2011. One-pot synthesis of platinum nanoparticles embedded on reduced graphene oxide for oxygen reduction in methanol fuel cells. *Electrochem. Solid-State Lett.* 14(7):B70–B73.

37. Xin, Y., Liu, J.-G., Zhou, Y. et al. 2011. Preparation and characterization of Pt supported on graphene with enhanced electrocatalytic activity in fuel cell. *J. Power Sources* 196(3):1012–1018.

38. Park, D.-H., Jeon, Y., Ok, J. et al. 2012. Pt Nanoparticle-reduced graphene oxide nanohybrid for proton exchange membrane fuel cells. *J. Nanosci. Nanotechnol.* 12(7):5669–5672.

39. Seger, B. and Kamat, P. V. 2009. Electrocatalytically active graphene-platinum nanocomposites. Role of 2-D carbon support in PEM fuel cells. *J. Phys. Chem. C* 113(19):7990–7995.

40. Yoo, E., Okata, T., Akita, T., Kohyama, M., Nakamura, J. and Honma, I. 2009. Enhanced electrocatalytic activity of Pt subnanoclusters on graphene nanosheet surface. *Nano Lett.* 9(6):2255–2259.

41. Zhou, Y.-G., Chen, J.-J., Wang, F.-B., Sheng, Z.-H. and Xia, X.-H. 2010. A facile approach to the synthesis of highly electroactive Pt nanoparticles on graphene as an anode catalyst for direct methanol fuel cells. *Chem. Commun.* 46(32):5951–5953.

42. Liu, X.-W., Mao, J.-J., Liu, P.-D. and Wei, X.-W. 2011. Fabrication of metal-graphene hybrid materials by electroless deposition. *Carbon* 49(2):477–483.

43. Liu, S., Wang, J., Zeng, J. et al. 2010. "Green" electrochemical synthesis of Pt/graphene sheet nanocomposite film and its electrocatalytic property. *J. Power Sources* 195(15):4628–4633.

44. Hu, Y., Zhang, H., Wu, P., Zhang, H., Zhou, B. and Cai, C. 2011. Bimetallic Pt–Au nanocatalysts electrochemically deposited on graphene and their electrocatalytic characteristics towards oxygen reduction and methanol oxidation. *Phys. Chem. Chem. Phys.* 13(9):4083–4094.
45. Zhang, H., Xu, X., Gu, P., Li, C., Wu, P. and Cai, C. 2011. Microwave-assisted synthesis of graphene-supported Pd1Pt3 nanostructures and their electrocatalytic activity for methanol oxidation. *Electrochim. Acta* 56(20):7064–7070.
46. Huang, Z., Zhou, H., Li, C., Zeng, F., Fu, C. and Kuang, Y. 2012. Preparation of well-dispersed PdAu bimetallic nanoparticles on reduced graphene oxide sheets with excellent electrochemical activity for ethanol oxidation in alkaline media. *J. Mater. Chem.* 22(5):1781–1785.
47. Liu, J., Zhou, H., Wang, Q., Zeng, F. and Kuang, Y. 2012. Reduced graphene oxide supported palladium–silver bimetallic nanoparticles for ethanol electro-oxidation in alkaline media. *J. Mater. Sci.* 47(5):2188–2194.
48. Lee, S. H., Kakati, N., Jee, S. H., Maiti, J. and Yoon, Y.-S. 2011. Hydrothermal synthesis of PtRu nanoparticles supported on graphene sheets for methanol oxidation in direct methanol fuel cell. *Mater. Lett.* 65(21):3281–3284.
49. Rao, C. V., Reddy, A. L. M., Ishikawa, Y. and Ajayan, P. M. 2011. Synthesis and electrocatalytic oxygen reduction activity of graphene-supported Pt3Co and Pt3Cr alloy nanoparticles. *Carbon* 49(3):931–936.
50. Yang, G., Li, Y., Rana, R. K. and Zhu, J.-J. 2013. Pt–Au/nitrogen-doped graphene nanocomposites for enhanced electrochemical activities. *J. Mater. Chem. A* 1(5):1754–1762.
51. Hu, Y., Wu, P., Zhang, H. and Cai, C. 2012. Synthesis of graphene-supported hollow Pt–Ni nanocatalysts for highly active electrocatalysis toward the methanol oxidation reaction. *Electrochim. Acta* 85:314–321.
52. Luo, B., Yan, X., Xu, S. and Xue, Q. 2012. Polyelectrolyte functionalization of graphene nanosheets as support for platinum nanoparticles and their applications to methanol oxidation. *Electrochim. Acta* 59:429–434.
53. Wietecha, M. S., Zhu, J., Gao, G. et al. 2012. Platinum nanoparticles anchored on chelating group-modified graphene for methanol oxidation. *J. Power Sources* 198:30–35.
54. Zhao, Y., Zhan, L., Tian, J., Nie, S. and Ning, Z. 2011. Enhanced electrocatalytic oxidation of methanol on Pd/polypyrrole–graphene in alkaline medium. *Electrochim. Acta* 56(5):1967–1972.
55. Wang, Z., Gao, G., Zhu, H., Sun, Z., Liu, H. and Zhao, X. 2009. Electrodeposition of platinum microparticle interface on conducting polymer film modified nichrome for electrocatalytic oxidation of methanol. *Int. J. Hydrogen Energy* 34(23):9334–9340.
56. Shi, Q. and Mu, S. 2012. Preparation of Pt/poly (pyrogallol)/graphene electrode and its electrocatalytic activity for methanol oxidation. *J. Power Sources* 203:48–56.
57. Park, S., Shao, Y., Wan, H. et al. 2011. Design of graphene sheets-supported Pt catalyst layer in PEM fuel cells. *Electrochem. Commun.* 13(3):258–261.
58. Aravind, S. J., Jafri, R. I., Rajalakshmi, N. and Ramaprabhu, S. 2011. Solar exfoliated graphene–carbon nanotube hybrid nano composites as efficient catalyst supports for proton exchange membrane fuel cells. *J. Mater. Chem.* 21(45):18199–18204.
59. Yun, Y. S., Kim, D., Tak, Y. and Jin, H.-J. 2011. Porous graphene/carbon nanotube composite cathode for proton exchange membrane fuel cell. *Synth. Met.* 161(21):2460–2465.
60. Rajesh, Paul, R. K. and Mulchandani, A. 2013. Platinum nanoflowers decorated three-dimensional graphene–carbon nanotubes hybrid with enhanced electrocatalytic activity. *J. Power Sources* 223:23–29.
61. Jafri, R. I., Rajalakshmi, N. and Ramaprabhu, S. 2010. Nitrogen doped graphene nanoplatelets as catalyst support for oxygen reduction reaction in proton exchange membrane fuel cell. *J. Mater. Chem.* 20(34):7114–7117.

62. Vinayan, B. P., Nagar, R., Rajalakshmi, N. and Ramaprabhu, S. 2012. Novel platinum-cobalt alloy nanoparticles dispersed on nitrogen-doped graphene as a cathode electro-catalyst for PEMFC applications. *Adv. Funct. Mater.* 22(16):3519–3526.

63. Bai, J., Zhu, Q., Lv, Z., Dong, H., Yu, J. and Dong, L. 2013. Nitrogen-doped graphene as catalysts and catalyst supports for oxygen reduction in both acidic and alkaline solutions. *Int. J. Hydrogen Energy* 38(3):1413–1418.

64. Zhang, L.-S., Liang, X.-Q., Song, W.-G. and Wu, Z.-Y. 2010. Identification of the nitrogen species on N-doped graphene layers and Pt/NG composite catalyst for direct methanol fuel cell. *Phys. Chem. Chem. Phys.* 12(38):12055–12059.

65. Xiong, B., Zhou, Y., Zhao, Y. et al. 2013. The use of nitrogen-doped graphene supporting Pt nanoparticles as a catalyst for methanol electrocatalytic oxidation. *Carbon* 52:181–192.

66. Zhao, Y., Zhou, Y., Xiong, B. et al. 2013. Facile single-step preparation of Pt/N-graphene catalysts with improved methanol electrooxidation activity. *J. Solid-State Electrochem.* 17(4):1089–1098.

67. Liang, Y., Li, Y., Wang, H. et al. 2011. Co_3O_4 nanocrystals on graphene as a synergistic catalyst for oxygen reduction reaction. *Nat. Mater.* 10(10):780–786.

68. Liang, Y., Wang, H., Zhou, J. et al. 2012. Covalent hybrid of spinel manganese–cobalt oxide and graphene as advanced oxygen reduction electrocatalysts. *J. Am. Chem. Soc.* 134(7):3517–3523.

69. Wang, H., Liang, Y., Li, Y. and Dai, H. 2011. $Co_{1-x}S$–graphene hybrid: A high-performance metal chalcogenide electrocatalyst for oxygen reduction. *Angew. Chem. Int. Ed.* 50(46):10969–10972.

70. Wu, Z.-S., Yang, S., Sun, Y., Parvez, K., Feng, X. and Müllen, K. 2012. 3D nitrogen-doped graphene aerogel-supported Fe_3O_4 nanoparticles as efficient electrocatalysts for the oxygen reduction reaction. *J. Am. Chem. Soc.* 134(22):9082–9085.

71. Guo, S., Zhang, S., Wu, L. and Sun, S. 2012. Co/CoO nanoparticles assembled on graphene for electrochemical reduction of oxygen. *Angew. Chem. Int. Ed.* 124(47):11940–11943.

72. Yan, X.-Y., Tong, X.-L., Zhang, Y.-F. et al. 2012. Cuprous oxide nanoparticles dispersed on reduced graphene oxide as an efficient electrocatalyst for oxygen reduction reaction. *Chem. Commun.* 48(13):1892–1894.

73. Qian, Y., Lu, S. and Guo, F. 2011. Synthesis of manganese dioxide/reduced graphene oxide composites with excellent electrocatalytic activity toward reduction of oxygen. *Mater. Lett.* 65(1):56–58.

74. Lambert, T. N., Davis, D. J. and Lu, W. et al. 2012. Graphene–Ni–α-MnO_2 and–Cu–α-MnO_2 nanowire blends as highly active non-precious metal catalysts for the oxygen reduction reaction. *Chem. Commun.* 48(64):7931–7933.

75. Jasinski, R. 1964. A new fuel cell cathode catalyst. *Nature* 201(4925):1212–1213.

76. Tang, H., Yin, H., Wang, J., Yang, N., Wang, D. and Tang, Z. 2013. Molecular architecture of cobalt porphyrin multilayers on reduced graphene oxide sheets for high-performance oxygen reduction reaction. *Angew. Chem. Int. Ed.* 125(21):5695–5699.

77. Gupta, S., Tryk, D., Bae, I., Aldred, W. and Yeager, E. 1989. Heat-treated polyacrylonitrile-based catalysts for oxygen electroreduction. *J. Appl. Electrochem.* 19(1):19–27.

78. Byon, H. R., Suntivich, J. and Shao-Horn, Y. 2011. Graphene-based non-noble-metal catalysts for oxygen reduction reaction in acid. *Chem. Mater.* 23(15):3421–3428.

79. Kamiya, K., Hashimoto, K. and Nakanishi, S. 2012. Instantaneous one-pot synthesis of Fe–N-modified graphene as an efficient electrocatalyst for the oxygen reduction reaction in acidic solutions. *Chem. Commun.* 48(82):10213–10215.

80. Zhang, S., Zhang, H., Liu, Q. and Chen, S. 2013. Fe–N doped carbon nanotube/graphene composite: Facile synthesis and superior electrocatalytic activity. *J. Mater. Chem. A* 1(10):3302–3308.

81. Fu, X., Liu, Y., Cao, X., Jin, J., Liu, Q. and Zhang, J. 2013. FeCo–Nx embedded graphene as high performance catalysts for oxygen reduction reaction. *Appl. Catal. B* 130–131:143–151.

82. Tsai, C.-W., Tu, M.-H., Chen, C.-J. et al. 2011. Nitrogen-doped graphene nanosheet-supported non-precious iron nitride nanoparticles as an efficient electrocatalyst for oxygen reduction. *RSC Adv.* 1(7):1349–1357.

83. Parvez, K., Yang, S., Hernandez, Y. et al. 2012. Nitrogen-doped graphene and its iron-based composite as efficient electrocatalysts for oxygen reduction reaction. *ACS Nano* 6(11):9541–9550.

84. Gong, K., Du, F., Xia, Z., Durstock, M. and Dai, L. 2009. Nitrogen-doped carbon nanotube arrays with high electrocatalytic activity for oxygen reduction. *Science* 323(5915):760–764.

85. Lai, L., Potts, J. R. and Zhan, D. et al. 2012. Exploration of the active center structure of nitrogen-doped graphene-based catalysts for oxygen reduction reaction. *Energy Environ. Sci.* 5(7):7936–7942.

86. Sheng, Z.-H., Shao, L., Chen, J.-J., Bao, W.-J., Wang, F.-B. and Xia, X.-H. 2011. Catalyst-free synthesis of nitrogen-doped graphene via thermal annealing graphite oxide with melamine and its excellent electrocatalysis. *ACS Nano* 5(6):4350–4358.

87. Wu, J., Zhang, D., Wang, Y. and Hou, B. 2013. Electrocatalytic activity of nitrogen-doped graphene synthesized via a one-pot hydrothermal process towards oxygen reduction reaction. *J. Power Sources* 227:185–190.

88. Geng, D., Chen, Y., Chen, Y. et al. 2011. High oxygen-reduction activity and durability of nitrogen-doped graphene. *Energy Environ. Sci.* 4(3):760–764.

89. Zhang, Y., Fugane, K., Mori, T., Niu, L. and Ye, J. 2012. Wet chemical synthesis of nitrogen-doped graphene towards oxygen reduction electrocatalysts without high-temperature pyrolysis. *J. Mater. Chem.* 22(14):6575–6580.

90. Luo, Z., Lim, S. and Tian, Z. et al. 2011. Pyridinic N doped graphene: Synthesis, electronic structure, and electrocatalytic property. *J. Mater. Chem.* 21(22):8038–8044.

91. Zhang, C., Fu, L., Liu, N., Liu, M., Wang, Y. and Liu, Z. 2011. Synthesis of nitrogen-doped graphene using embedded carbon and nitrogen sources. *Adv. Mater.* 23(8):1020–1024.

92. Deng, D., Pan, X., Yu, L. et al. 2011. Toward N-doped graphene via solvothermal synthesis. *Chem. Mater.* 23(5):1188–1193.

93. Panchakarla, L., Subrahmanyam, K., Saha, S. et al. 2009. Synthesis, structure, and properties of boron-and nitrogen-doped graphene. *Adv. Mater.* 21(46):4726–4730.

94. Qu, L., Liu, Y., Baek, J.-B. and Dai, L. 2010. Nitrogen-doped graphene as efficient metal-free electrocatalyst for oxygen reduction in fuel cells. *ACS Nano* 4(3):1321–1326.

95. Yang, S., Feng, X., Wang, X. and Müllen, K. 2011. Graphene-based carbon nitride nanosheets as efficient metal-free electrocatalysts for oxygen reduction reactions. *Angew. Chem. Int. Ed.* 50(23):5339–5343.

96. Unni, S. M., Devulapally, S., Karjule, N. and Kurungot, S. 2012. Graphene enriched with pyrrolic coordination of the doped nitrogen as an efficient metal-free electrocatalyst for oxygen reduction. *J. Mater. Chem.* 22(44):23506–23513.

97. Zhang, L. and Xia, Z. 2011. Mechanisms of oxygen reduction reaction on nitrogen-doped graphene for fuel cells. *J. Phys. Chem. C* 115(22):11170–11176.

98. Ikeda, T., Boero, M., Huang, S.-F., Terakura, K., Oshima, M. and Ozaki, J.-I. 2008. Carbon alloy catalysts: Active sites for oxygen reduction reaction. *J. Phys. Chem. C* 112(38):14706–14709.

99. Zhang, C., Hao, R., Liao, H. and Hou, Y. 2013. Synthesis of amino-functionalized graphene as metal-free catalyst and exploration of the roles of various nitrogen states in oxygen reduction reaction. *Nano Energy* 2(1):88–97.

100. Jeon, I.-Y., Yu, D., Bae, S.-Y. et al. 2011. Formation of large-area nitrogen-doped graphene film prepared from simple solution casting of edge-selectively functionalized graphite and its electrocatalytic activity. *Chem. Mater.* 23(17):3987–3992.

101. Yang, Z., Yao, Z., Li, G. et al. 2011. Sulfur-doped graphene as an efficient metal-free cathode catalyst for oxygen reduction. *ACS Nano* 6(1):205–211.
102. Sheng, Z.-H., Gao, H.-L., Bao, W.-J., Wang, F.-B., and Xia, X.-H. 2012. Synthesis of boron doped graphene for oxygen reduction reaction in fuel cells. *J. Mater. Chem.* 22(2):390–395.
103. Jin, Z., Nie, H., Yang, Z. et al. 2012. Metal-free selenium doped carbon nanotube/ graphene networks as a synergistically improved cathode catalyst for oxygen reduction reaction. *Nanoscale* 4(20):6455–6460.
104. Yao, Z., Nie, H., Yang, Z., Zhou, X., Liu, Z. and Huang, S. 2012. Catalyst-free synthesis of iodine-doped graphene via a facile thermal annealing process and its use for electro-catalytic oxygen reduction in an alkaline medium. *Chem. Commun.* 48(7):1027–1029.
105. Wang, S., Iyyamperumal, E., Roy, A., Xue, Y., Yu, D. and Dai, L. 2011. Vertically aligned BCN nanotubes as efficient metal-free electrocatalysts for the oxygen reduction reaction: A synergetic effect by co-doping with boron and nitrogen. *Angew. Chem. Int. Ed.* 50(49):11756–11760.
106. Wang, S., Zhang, L., Xia, Z. et al. 2012. BCN graphene as efficient metal-free electro-catalyst for the oxygen reduction reaction. *Angew. Chem. Int. Ed.* 51(17):4209–4212.
107. Zheng, Y., Jiao, Y., Ge, L., Jaroniec, M. and Qiao, S. Z. 2013. Two-step boron and nitrogen doping in graphene for enhanced synergistic catalysis. *Angew. Chem. Int. Ed.* 125(11):3192–3198.
108. Liang, J., Jiao, Y., Jaroniec, M. and Qiao, S. Z. 2012. Sulfur and nitrogen dual-doped mesoporous graphene electrocatalyst for oxygen reduction with synergistically enhanced performance. *Angew. Chem. Int. Ed.* 51(46):11496–11500.
109. Jhan, J.-Y., Huang, Y.-W., Hsu, C.-H., Teng, H., Kuo, D. and Kuo, P.-L. 2013. Three-dimensional network of graphene grown with carbon nanotubes as carbon support for fuel cells. *Energy* 53:282–287.
110. Jha, N., Jafri, R. I., Rajalakshmi, N. and Ramaprabhu, S. 2011. Graphene-multi walled carbon nanotube hybrid electrocatalyst support material for direct methanol fuel cell. *Int. J. Hydrogen Energy* 36(12):7284–7290.
111. Du, H.-Y., Wang, C.-H., Hsu, H.-C. et al. 2012. Graphene nanosheet–CNT hybrid nano-structure electrode for a proton exchange membrane fuel cell. *Int. J. Hydrogen Energy* 37(24):18989–18995.
112. Ansari, S., Kelurakis, A., Estevez, L. and Giannelis, E. P. 2010. Oriented arrays of gra-phene in a polymer matrix by in situ reduction of graphite oxide nanosheets. *Small* 6(2):205–209.
113. Feng, K., Tang, B. and Wu, P. 2013. "Evaporating" graphene oxide sheets (GOSs) for rolled up GOSs and its applications in proton exchange membrane fuel cell. *ACS Appl. Mater. Interfaces* 5(4):1481–1488.
114. Zarrin, H., Higgins, D., Jun, Y., Chen, Z. and Fowler, M. 2011. Functionalized gra-phene oxide nanocomposite membrane for low humidity and high temperature proton exchange membrane fuel cells. *J. Phys. Chem. C* 115(42):20774–20781.
115. Kumar, R., Xu, C. and Scott, K. 2012. Graphite oxide/Nafion composite membranes for polymer electrolyte fuel cells. *RSC Adv.* 2(23):8777–8782.
116. Choi, B. G., Hong, J., Park, Y. C. et al. 2011. Innovative polymer nanocomposite electro-lytes: Nanoscale manipulation of ion channels by functionalized graphenes. *ACS Nano* 5(6):5167–5174.
117. Lin, C. W. and Lu, Y. S. 2013. Highly ordered graphene oxide paper laminated with a Nafion membrane for direct methanol fuel cells. *J. Power Sources* 237:187–194.
118. Scott, K. 2012. Freestanding sulfonated graphene oxide paper: A new polymer electro-lyte for polymer electrolyte fuel cells. *Chem. Commun.* 48(45):5584–5586.
119. Tseng, C. Y., Ye, Y. S., Cheng, M. Y. et al. 2011. Sulfonated polyimide proton exchange membranes with graphene oxide show improved proton conductivity, methanol cross-over impedance, and mechanical properties. *Adv. Energy Mater.* 1(6):1220–1224.

120. Cao, Y.-C., Xu, C., Wu, X., Wang, X., Xing, L. and Scott, K. 2011. A poly (ethylene oxide)/graphene oxide electrolyte membrane for low temperature polymer fuel cells. *J. Power Sources* 196(20):8377–8382.
121. Jiang, Z., Zhao, X., Fu, Y. and Manthiram, A. 2012. Composite membranes based on sulfonated poly(ether ether ketone) and SDBS-adsorbed graphene oxide for direct methanol fuel cells. *J. Mater. Chem.* 22(47):24862–24869.
122. Xu, C., Cao, Y., Kumar, R., Wu, X., Wang, X. and Scott, K. A. 2011. Polybenzimidazole/ sulfonated graphite oxide composite membrane for high temperature polymer electrolyte membrane fuel cells. *J. Mater. Chem.* 21(30):11359–11364.

Index

Page numbers followed by f and t indicate figures and tables, respectively.

A

Accelerated durability testing (ADT), 254
Acetonitrile, 53, 70, 96, 188
Acetylene black (AB), 191
Acrylonitrile, 153
Activated carbon (AC), 171, 187f
Activated carbon nanofibers (ACN), 200
Adsorption isotherms, 175f
Aerogels, 108, 123, 172
Aerosol spray process, 97
Air electrode, 157, 160f, 166f
Alkaline fuel cell (AFC), 246
Allotrope of carbon materials, 1, 2
Aluminum chloride, 42
Aluminum iodide, 42
Aluminum zinc oxide (AZO), 218
Ambient pressure chemical vapor deposition
    (APCVD), 221, 230
Ambipolar electric field effect, 8
Aminophenyl groups, 96
Ammonia, 47
    as doping agent, 161
Ammonium ferrous sulfate, 118
Ammonium hydroxide, 57
Ammonium vanadate, 91
Amorphous carbon, 153
Amphiphilic nonionic surfactants, 39
Amylopectin, 154
Anatase, 115
Anisotropic silicon nanowires, 99
Annealing, 30, 46, 72
Anode materials, graphene. *See also* Composite
    anode materials, graphene-based;
    Lithium ion batteries
    doped, 70–73, 71f
    graphene paper as flexible anode, 73–77, 73f,
        75f, 76f
    pure, 66–70, 67f, 69f
Anodic electropolymerization (AEP), 195
Aqueous electrolytes, graphene-based hybrid
    supercapacitor in, 200–202, 201f,
    202f
Aqueous Li-air batteries, 165–168, 166f. *See also*
    Graphene in Lithium-air battery
Arc discharge exfoliation method, 59
Arc-discharge method, 3

Argon (Ar), 24
Ascorbic acid, 83, 125
Asymmetric supercapacitor, 199, 200, 202f
    in organic electrolyte, 203–205, 204f, 206f.
    See also Hybrid supercapacitor
Atmospheric pressure chemical vapor deposition
    (APCVD), 32, 32f, 223–224
Atomic force microscopy (AFM), 10, 23, 40f, 45,
    45f, 46
Atomic layer deposition (ALD) technique, 108
Auger electron spectroscopy, 110

B

Back electrodes, graphene materials as. *See also*
    Solar cells
    applications in PV solar cells, 231–233, 232f,
        233f, 234f
    conductive 3-D graphene, 230–231, 231f
    pristine graphene/B-doped graphene
        powders, 228–230, 229f
Barrett–Joyner–Halenda (BJH) method, 98
Bath sonication, 50
B-doped graphene powders, 228–230. *See also*
    Back electrodes, graphene materials
    as
Benzyl disulfide (BDS), 277
Bilayer graphene, 7
Bimodal porous structure, 158f
Binary materials, 254
Bioapplications, graphene, 15–16
Bis(trifluoromethylsulfonyl)amide anion, 189
Borane, 198
Borane-THF adduct, 198f
Boric acid, 277
Boron–carbon–nitride (BCN), 277
Boron chloride, 71
Boron-doped graphene, 228
Boron nitride (BN), 30
Boron oxide (BO), 276
Boron tribromide, 229
Bottom-up process, 22
Bragg grating, 9
B-rG-O nanoplatelets, 198
Brodie method, 38, 185
Broken egg structure, 157, 158
Brunauer–Emmett–Teller (BET) method, 46, 98

Buckminsterfullerene, 4f
1-butyl-3methylimidazoliu tetrafluoroborate (BMIMBF4), 178

C

Cadmium sulfide (CdS), 225
Cadmium telluride (CdTe), 224, 232f
Calcination, 107, 123, 124, 124f
Carbonaceous gases, 31
Carbon black (CB), 140, 251, 258–261
    commercial, 154f, 156
Carbon-black-based catalyst, 166
Carbon–catalyst binding, 262
Carbon disulfide, 105
Carbon isotope labeling, 29, 29f
Carbonization, 159
    of oleylamine, 100
Carbon materials, 1–4, 2f, 4f, 94, 138
Carbon nanofibers (CNF), 150
    S-coated, 151f
Carbon nanofoams, 4f
Carbon nanomaterials, 251
Carbon nanotubes (CNT), 3, 25, 161, 251
    helicoidal, 4f
    unzipping of, 57, 58f
Carbon nitride, 274, 274f
Carbon precipitation process, 219
Carbon precursor, 31–33, 32f. See also Chemical vapor deposition (CVD)
Carboxyl groups, 143, 150, 159
Carburization, 27
Catanionic reverse-micelle method, 88
Cathode electrocatalysts, 281f
Cathode materials (composite), graphene-based. See Composite cathode materials, graphene-based
Cellulose, 42
Cetyltrimethyl ammonium bromide (CTAB), 273
Charcoal, 1
Charge/discharge measurement, 153
Charge transfer, 14
    resistance, 81, 112
Chemical activation, 181, 207
Chemical-converted graphene (CCG), 182
Chemical doping, 70, 199
Chemical exfoliation
    of graphenes, 47f
    of graphite, 17
    of graphite oxide, 230
Chemically modified graphene (CMG), 198
Chemical reduction, defined, 44
Chemical vapor deposition (CVD), 3, 11, 15, 22, 81, 219, 272. See also Graphene, synthesis of
    carbon precursor for controlled morphologies of graphene, 31–33, 32f

formation of graphene, 27–30, 28f, 29f
    graphene film production in roll-to-roll way, 35–37, 36f
    graphene on Cu substrate, 30–31
    overview, 25–26, 26f
    transfer technique for CVD graphene film, 33–35, 34f
Chloroauric acid, 37
Chronoamperometry, 264f
CNT. See Carbon nanotubes (CNT)
Cobalt acetate, 265
Cobalt oxide (Co_3O_4)-graphene composite anode materials, 124–127, 124f, 126f. See also Metal oxide-graphene composite anode materials
Cobalt phthalocyanine (CoPc), 267
Coelectrodeposition, 127
Composite anode materials, graphene-based. See also Lithium ion batteries
    lithium titanate-graphene composite anode materials, 128–130, 129f
    metal-graphene composite anode materials
        germanium/graphene composite anode materials, 100–102, 101f
        Si/graphene composite anode materials, 94–99, 95f, 98f, 100f
        tin (Sn)/graphene composite anode materials, 102–105, 103f
    metal oxide-graphene composite anode materials
        cobalt oxide (Co_3O_4)-graphene composite anode materials, 124–127, 124f, 126f
        iron-oxide–graphene composite anode materials, 115–124, 116f, 120f, 121f
        tin dioxide (SnO_2)-graphene composite anode materials, 105–110, 106f, 108f
        titanium dioxide (TiO_2)-graphene composite anode materials, 110–115, 111f, 113f
Composite cathode materials, graphene-based. See also Lithium ion batteries
    Li-metal-oxide graphene composite cathode materials, 86–90, 87f
    Li metal phosphate-graphene composite cathode materials, 77–86, 79f, 82f, 83f, 84f
    metal-oxide-graphene composite cathode materials, 90–94, 91f, 93f
Composite materials and coatings, 13–14. See also Graphene
Conduction band (CB), 8, 235
Conductive 3-D graphene, 230–231, 231f. See also Back electrodes, graphene materials as
Conductive graphene films, 222–223, 223f, 224f. See also Transparent conducting front contact, graphene materials as
Conductive graphite, 231

Conductive polymer-functionalized GNS as
    support material, 256–258, 257f
Copper, melting point of, 31
Copper substrate, graphene on, 30–31. *See also*
    Chemical vapor deposition (CVD)
Coprecipitation method, 78
Coulombic efficiency, 68, 93, 99, 101, 102, 107,
    109, 110, 141, 142
Coulomb repulsion, 51
Coulomb scattering, 44
Counterelectrode (CE), 223
Covalent bond, 183
Cross-linking, 109
Crystallization process, 111
Cuprous oxide, 35
Curved graphene sheets (CGN), 184, 185f
CV curves, 191, 192f, 201f, 202f
CVD. *See* Chemical vapor deposition (CVD)
Cyclic voltammetry (CV), 47, 196, 254, 264f
Cycling tests, 197f

D

Delamination, 35
Density functional theory (DFT), 275
Density of states (DOS), 218
Deprotonation, 41
Depth of discharge (DOD), 150
Diamond, 1, 2
    atomic structure of, 2f
Diamond lattice, 2
Dicyandiamide (DCDA), 272
Dielectric substrates. *See also* Transparent
    conducting front contact, graphene
    materials as
    few-layer graphene on, 221, 221f, 222f
    vertically-erected graphene walls on, 222
Dimethylformamide (DMF), 41
Dirac point, 8, 218
Direct methanol fuel cell (DMFC), 246, 250, 279
Discharge process, 156
DMFC. *See* Direct methanol fuel cell (DMFC)
DNA bases, 15
Dopamine, 41
    self-polymerization of, 109
Doped graphene anode materials, 70–73, 71f
Dye sensitized solar cell (DSSC), 218, 223, 226f,
    227, 227f, 228f, 234

E

EDLC. *See* Electrical double-layer capacitance
    (EDLC)
Electrical conductivity, 88, 164
    of cathode materials, 84
    of GO, 140
    of lithium titanate, 128

Electrical double-layer capacitance (EDLC), 12,
    173–190. *See also* Supercapacitors
    electrode materials
        GNR-based, 183–185, 183f, 184f, 185f,
            186f
        graphene-based, 174–176, 175f, 176f, 177f
        graphene hybrid composites-based,
            177–180, 178f, 179f
        porous graphene-based, 180–182, 181f
    electrolyte, effect of, 186–190, 187f, 188f,
        189f
Electrical transport, 41
Electric vehicles, 137
Electroactive materials, 172
Electrocatalysts, 250
    Pt-based, 252
Electrochemical capacitors, 12
Electrochemical deposition, 253
Electrochemical exfoliation, 54–56, 55f. *See also*
    Graphene, synthesis of
Electrochemical impedance spectroscopy (EIS),
    72, 74, 109
Electrochemical measurements, 127, 252, 279
Electrochemical processes, 105
Electrochemical reduction of GO, 47–49, 48f,
    49f
Electrochemical surface area (ECSA), 253, 259
Electrochemical tests, 197f, 253
Electrode materials, 65, 173
    with pseudocapacitive properties. *See also*
        Supercapacitors
    based on graphene with heteroatoms
        in carbon network, 196–199,
        197f, 198f
    graphene-conductive polymers-based,
        193–196, 194f, 195f
    graphene-metal-oxide composites-based,
        190–193, 192f, 193f
Electrodeposition process, 127
Electrode quantum capacitance, 196
Electrode rotating rate, 273f
Electrolyte dielectric constant, 173
Electron beam irradiation, 68
Electron conductivity, 222
Electron diffraction, 53, 54f
Electronegativity atoms, 276
Electron energy-loss spectroscopy, 97
Electron gas, 9
Electronic properties, graphene, 8–9
Electronics, 10–11. *See also* Graphene
Electron microscopy, 119, 124
Electron transfer, 266
Electrophoretic deposition (EPD) method, 48
Electrospinning, 130
Electrostatic repulsion, 143
Elemental mapping, 78, 97, 98f, 155
Energy barrier, 162

Energy conversion
  efficiency, 223
  and storage, 12–13. *See also* Graphene
Energy density, 203
Energy-dispersive spectroscopy, 145, 145f
Energy dispersive x-ray (EDX), 110
  mapping, 109, 162
Epitaxial growth of graphene on SiC, 24–25, 25f.
      See also Graphene, synthesis of
Etching process, 35
  metal catalysed, 180
Ethanol, 42, 52, 59
Ethylene, 104
Ethylenediamine, 174
Ethylenediamine triacetic acid (EDTA), 256, 257,
      257f, 258
Ethylene glycol (EG), 126, 127, 252
1-ethyl-3-methyl imidazolium tetrafluoroborate
      (EMIMBF$_4$) electrolyte systems, 180
Evaporation process, 123

F

Fabrication process, 26, 44, 124, 149f, 151f
  of G-based carbon nitride (G–CN), 274f
Fast electron transport, 102, 200
Fermi level, 8, 13, 272
Ferric oxide hydrate, 116
FESEM images
  of graphene film, 224f
  of RGO, 175f
Fillers, graphene-based, 13
Fill factor, 232
Flexibility
  2-D graphene sheets, 145
  graphene paper, 77
Floating graphene film, 34f
Fluid–liquid–solid process, 99
Fluorine tin oxide (FTO), 218
Formic acid, 140
Fossil fuels, 12
Fourier transfer (FT)-IR, 142
Fourier-transformed x-ray absorption fine
      structure spectra, 87
Freeze-drying technique, 95, 159
Fresnel equations, 9
Fuel cells
  graphene as additives for membrane
      preparation/modifications, 281–286,
      284f, 285t
  graphene as metal-free catalyst for ORR/MOR
      active sites of NG for ORR, 275–276
      heteroatom-doped graphene as catalyst
          for ORR/MOR, 276–278, 278f
      morphology control of NG, 273–275, 274f
      NG as metal-free catalysts, 270, 271f
      synthesis validation of NG, 271–273, 273f

graphene as support material for nonprecious
  metal catalysts, 264–270
  metal oxides, 265–267, 266f
  pyrolized metal macrocyclic compounds,
      267
  transition metal-nitrogen-containing
      complexes, 267–270
graphene as support material for Pt/Pt-alloy
  catalysts
  conductive polymer-functionalized GNS,
      256–258, 257f
  hybrid supporting material by
      intercalating carbon black and CNT,
      258–261, 260f
  nitrogen-doped graphene, 261–263, 264f
  pristine single-layer graphene, 252–255,
      255f
graphene for MEA assembly/fuel cell
  performance, 278–280, 281f
polymer electrolyte fuel cells
  history of fuel cells, 246–247, 247f
  operating principles of/components of,
      247–252, 248f
Fullerenes, 3

G

Galvanostatic charge/discharge curve, 192f,
      201f
Galvanostatic charge/discharge measurement,
      101
Galvanostatic charge/discharge tests, 147, 159
Gas diffusion layer (GDL), 279, 280
Gas sensors, 14
Germanium dioxide, 102
Germanium (Ge)/graphene composite anode
      materials, 100–102, 101f. *See also*
      Metal-graphene composite anode
      materials
Germanium tetrachloride, 101
Glucose-derived carbon, 104
G nanosheets (GNS), 252, 253, 280
  as support material, conductive polymer-
      functionalized, 256–258, 257f
G-noble metal catalysts, 252
GNR, properties of, 183
GNR-based electrode materials, 183–185, 183f,
      184f, 185f, 186f. *See also* Electrical
      double-layer capacitance (EDLC)
GO. *See* Graphite oxide (GO)
Gold nanoparticles, 99
Graphene
  as additives for membrane preparation/
      modifications, 281–286, 284f, 285t
  application
      composite materials and coatings, 13–14
      electronics, 10–11

energy conversion and storage, 12–13
photonic, 11–12
sensors, 14
thermal management, 14–15
bioapplications, 15–16
carbon material for lithium iron phosphate
modification, 78
commercial production of, 23
Cu/Ni sphere-wrapped, 222–223, 223f,
224f. *See also* Transparent
conducting front contact, graphene
materials as
defined, 5
electronic properties, 8–9
few-layer, on dielectric substrates, 221, 221f,
222f
formation mechanism of, 229f
heteroatom-doped, 276–278, 278f
history of, 5–6
layered structure, 5
for MEA assembly/fuel cell performance,
278–280, 281f
mechanical properties, 10
as metal-free catalyst for ORR/MOR.
See also Fuel cells
active sites of NG for ORR, 275–276
heteroatom-doped graphene as catalyst
for ORR/MOR, 276–278, 278f
morphology control of NG, 273–275,
274f
NG as metal-free catalysts, 270, 271f
synthesis validation of NG, 271–273,
273f
optical properties, 9
research directions, 16–18
structure, 6–8, 7f
as support material for nonprecious metal
catalysts, 264–270
metal oxides, 265–267, 266f
pyrolized metal macrocyclic compounds,
267
transition metal-nitrogen-containing
complexes, 267–270
as support material for Pt/Pt-alloy catalysts
conductive polymer-functionalized GNS
as support material, 256–258, 257f
hybrid supporting material by
intercalating carbon black and CNT,
258–261, 260f
nitrogen-doped graphene as support
material, 261–263, 264f
pristine single-layer graphene as support
material, 252–255, 255f
thermal properties, 9–10
Graphene, synthesis of
arc discharge exfoliation method, 59
CVD method

carbon precursor, manipulation of, 31–33,
32f
formation of graphene, 27–30, 28f, 29f
graphene film production in roll-to-roll
way, 35–37, 36f
graphene on Cu substrate, 30–31
overview, 25–26, 26f
transfer technique for CVD graphene
film, 33–35, 34f
electrochemical exfoliation, 54–56, 55f
epitaxial growth on SiC, 24–25, 25f
graphite oxide (GO), exfoliation/reduction of
electrochemical reduction, 47–49,
48f, 49f
green chemical reduction methods,
40–44, 42f, 43f
overview, 37–38
reduction of GO by hydrazine, 38–40, 40f
thermal exfoliation/reduction of GO,
44–47, 45f, 47f
igniting magnesium in dry ice, 59
liquid-phase exfoliation
of graphite, 50–52, 50f
of graphite intercalation compounds
(GIC), 52–53, 54f
overview, 49–50
solvothermal exfoliation, 53–54
mechanical exfoliation, 23–24
organic synthesis, 56–57, 56f
overview, 21–23
solvothermal method, 59
unzipping of CNT, 57, 58f
Graphene anode materials. *See* Anode materials,
graphene
Graphene-based composite anode materials.
See Composite anode materials,
graphene-based
Graphene-based composite cathode materials.
See Composite cathode materials,
graphene-based
Graphene-based electrode materials, 174–176,
175f, 176f, 177f. *See also* Electrical
double-layer capacitance (EDLC)
Graphene cluster, 4f
Graphene-conductive polymers-based electrode
materials, 193–196, 194f, 195f
Graphene-encapsulated sulfur (GES), 145f
Graphene film
floating, 34f
low-defect, 218–220, 219f, 220f
patterned, 34f
production in roll-to-roll way, 35–37, 36f
Graphene hybrid composites-based electrode
materials, 177–180, 178f, 179f.
See also Electrical double-layer
capacitance (EDLC)
Graphene hybrids, 164

Graphene in lithium-air battery
  overview, 156
  progress
    aqueous Li-air batteries, 165–168, 166f
    nonaqueous Li-air batteries, 156–165,
      157f, 158f, 160f, 161f, 163f, 165f
Graphene in lithium-sulfur (Li-S) battery
  overview, 138–139
  progress, 139–156, 139f, 141f, 144f, 145f,
    146f, 147f, 149f, 151f, 152f, 154f
Graphene–metal (G-M) binding, 261
Graphene-metal-oxide composites-based
  electrode materials, 190–193, 192f,
  193f. *See also* Electrode materials
  with pseudocapacitive properties
Graphene nanoribbons (GNR), 7, 56, 56f
Graphene nanosheets, 67f, 74, 139
  TEM images of, 87f
Graphene oxide, 8, 10, 15, 16, 113f, 114
  in propylene carbonate (PC), 189f
Graphene paper as flexible anode, 73–77, 73f,
  75f, 76f
Graphene/polyaniline composite paper (GPCP),
  195
Graphene quantum dots (GQD), 7
Graphene-sulfur-amylopectin composite, 154f
Graphene with heteroatoms in carbon
  network, 196–199, 197f, 198f.
  See also Electrode materials with
  pseudocapacitive properties
Graphite
  allotrope of carbon materials, 1
  atomic structure of, 2f
  chemical exfoliation of, 17
  exfoliation process, 6
  in flakes or lumps, 2
  oxidation of, 22
  Slonczewski–Weiss–McClure band model
    of, 9
Graphite intercalation compounds (GIC), 52–53,
  54f. *See also* Liquid-phase exfoliation
Graphite onions, 4f
Graphite oxide (GO), 37–49, 271, 272
  chemical exfoliation of, 230
  chemical reduction of, 73
  electrochemical reduction, 47–49, 48f, 49f
  green chemical reduction methods, 40–44,
    42f, 43f
  immobilizing sulfur, 141f
  in lithium ion batteries, 89
  optical photos of, 176f
  reduction of GO by hydrazine, 38–40, 40f
  thermal exfoliation of, 70
  thermal exfoliation/reduction of, 44–47, 45f,
    47f
Graphitization, 3, 44
Gravimetric capacitances, 197f

Green chemical reduction methods, 40–44, 42f,
  43f. *See also* Graphite oxide (GO)
Grove, William
  father of fuel cells, 246
  sketch of fuel cells, 247f

H

Haeckelite surface, 4f
Halogenation, 43f
Hansen solubility, 51
Hard-template-assisted method, 111
Hard template method, 146
Helium–neon (He–Ne) laser, 9
Hematite, nanocomposites of, 120f
Heteroatom-doped graphene, 276–278, 278f
Heteroatoms, 71
Hexabromobenzene precursors, 56
Hexagonal boron nitride (h-BN), 277
Hierarchically porous air electrode,
  graphene-based, 159, 160f
Highly corrugated graphene sheets (HCGS), 176,
  177f
Highly oriented pyrolytic carbon (HOPG)
  platelets, 23
High-resolution transmission electron
  microscopy (HRTEM), 52–53, 80,
  120f, 220f, 260f
Hildebrand solubility, 51
Holey GO sheets, 74, 75f
Hummers method, 38, 41, 185f
Hybrid electric vehicle (HEV), 205
Hybrid supercapacitor, 207
  graphene-based, 199–206. *See also*
    Supercapacitors
  in aqueous electrolytes, 200–202, 201f,
    202f
  asymmetric supercapacitor in organic
    electrolyte, 203–205, 204f, 206f
Hybrid supporting material by intercalating
  carbon black/CNT, 258–261, 260f
Hydrated vanadium pentoxide, 93
Hydrazine, 174
  reduction, 68, 69f
  reduction of GO by, 38–40, 40f
Hydrazine hydrate, 232
Hydrazine hydrate vapor, 130
Hydride reduction process, 42f
Hydriodic acid-acetic acid, 44
Hydrocarbon gas, 27
Hydrochloric acid (HCl), 76
Hydrogel, 266
Hydrogen, 32, 248
Hydrogen bonding, 193f
Hydrogen-exfoliated graphene (HEG), 262, 280,
  281f
Hydrogen fluoride (HF), 142

Hydrogen iodide (HI), 43, 144
Hydrogen oxygen reaction (HOR), 250
Hydrogen peroxide, 57
Hydrohalic acids, 43f
Hydroiodic acid, 43
Hydrolysis, 116, 273
Hydroquinone, 174
Hydrothermal method, 40, 94, 117, 119
Hydrothermal process, 145
Hydroxyl titanium oxalate (HTO), 113, 114

I

Indium tin oxide (ITO), 218
Infrared (IR) spectroscopy, 51
Inorganic nonmetallic material, 1
Intercalation compounds (LixCy), 55
Interdiffusion, 219
Interfacial capacitance, 196
Ion exchange capacity (IEC), 283
Iron chloride, 33
Iron-doped manganese oxide, 83
Iron nitride, 269
Iron-oxide-graphene composite anode materials,
        115–124, 116f, 120f, 121f. See also
        Metal oxide-graphene composite
        anode materials
Isopropyl alcohol, 124f

J

Jahn-Teller distortion, 87
Johnson noise, 14

K

Klein tunneling, 9
Knudsen diffusion, 46

L

Laser ablation, 3
Laser irradiation, 2
Linear sweep voltammetry (LSV), 278f
Liquid phase exfoliation, 49–54, 207. See also
        Graphene, synthesis of
    of graphite, 50–52, 50f
    of graphite intercalation compounds (GIC),
        52–53, 54f
    solvothermal exfoliation, 53–54
Lithium, 12
Lithium-air battery. See also Graphene in
        Lithium-air battery
    aqueous, 165–168, 166f
    nonaqueous, 156–165, 157f, 158f, 160f, 161f,
        163f, 165f
Lithium hydroxide (LiOH), 55

Lithium ion batteries (LIB), 199
    graphene anode materials
        doped graphene anode materials, 70–73,
            71f
        graphene paper as flexible anode, 73–77,
            73f, 75f, 76f
        pure graphene anode materials, 66–70,
            67f, 69f
    graphene-based composite anode materials
        lithium titanate-graphene composite
            anode materials, 128–130, 129f
        metal-graphene composite anode
            materials, 94–105, 95f, 98f, 100f, 101f,
            103f
        metal oxide-graphene composite anode
            materials, 105–127, 106f, 108f, 111f,
            113f, 116f, 120f, 121f, 124f, 125f
    graphene-based composite cathode
            materials
        Li-metal-oxide graphene composite
            cathode materials, 86–90, 87f
        Li metal phosphate-graphene composite
            cathode materials, 77–86, 79f, 82f,
            83f, 84f
        metal-oxide-graphene composite cathode
            materials, 90–94, 91f, 93f
    overview, 65–66
    performance of, 66
Lithium iron phosphate (LiFePO$_4$), 77
Lithium manganese phospate (LiMnPO$_4$), 81
Lithium-metal-oxide graphene composite
        cathode materials, 86–90, 87f
Lithium metal phosphate-graphene composite
        cathode materials, 77–86, 79f, 82f,
        83f, 84f
Lithium rechargeable batteries, 55
Lithium-sulfur (Li-S) battery, graphene in. See
        Graphene in lithium-sulfur (Li-S)
        battery
Lithium superion-conductive glass film
        (LISICON), 166
Lithium titanate-graphene composite anode
        materials, 128–130, 129f. See also
        Composite anode materials,
        graphene-based
Lithium vanadium phosphate, 85
Low-energy electron diffraction, 25f
Low pressure chemical vapor deposition
        (LPCVD), 31, 32, 32f

M

Macropores, 182
Magnesium in dry ice, igniting, 59. See also
        Graphene, synthesis of
Magnesium oxide, 97
Manganese oxide, 200

Manganese oxide nanowire/graphene (MGC), 200
Mechanical exfoliation, 23–24
Mechanical properties, graphene, 10
Melamine, 277
Membrane electrode assembly (MEA), 247, 278–280. *See also* Fuel cells
Mesoporous carbon, 151–152, 172
Mesoporous graphene, 147f, 277, 278
Mesoporous silica, 125
Metal-catalyzed etching, 180
Metal-free catalyst for ORR/MOR, graphene as. *See also* Fuel cells
    active sites of NG for ORR, 275–276
    heteroatom-doped graphene as catalyst for ORR/MOR, 276–278, 278f
    morphology control of NG, 273–275, 274f
    NG as metal-free catalysts, 270, 271f
    synthesis validation of NG, 271–273, 273f
Metal-graphene composite anode materials. *See also* Composite anode materials, graphene-based
    germanium/graphene composite anode materials, 100–102, 101f
    silicon (Si)/graphene composite anode materials, 94–99, 95f, 98f, 100f
    tin (Sn)/graphene composite anode materials, 102–105, 103f
Metal organic framework (MOF), 153, 154
Metal oxide-graphene composite anode materials. *See also* Composite anode materials, graphene-based
    cobalt oxide (Co_3O_4)-graphene composite anode materials, 124–127, 124f, 126f
    iron-oxide-graphene composite anode materials, 115–124, 116f, 120f, 121f
    tin dioxide (SnO_2)-graphene composite anode materials, 105–110, 106f, 108f
    titanium dioxide (TiO_2)-graphene composite anode materials, 110–115, 111f, 113f
Metal-oxide-graphene composite cathode materials, 90–94, 91f, 93f
Metal oxides, 172, 203, 265–267, 266f
Methane, 222
Methanol, 246, 248
    as anode fuel, 249t
    oxidation, 257
    permeability, 283, 284f
Methanol oxidation reaction (MOR), 250. *See also* Fuel cells
    heteroatom-doped graphene as catalyst for, 276–278, 278f
Methylene blue dye adsorption method, 46
Microelectromechanical systems (MEMS) technology, 33
Microexplosion reaction, 55

Micropores, 182
Micro-Raman characterization, 29
Microwave-assisted hydrothermal method, 80
Microwave-exfoliated and reduced GO (MEGO), 175, 176f
Mildly oxidized graphene oxide (mGO), 265
Molten carbonate fuel cell (MCFC), 246
MOR. *See* Methanol oxidation reaction (MOR)
Multilayer graphene platelets, 16
Multiwalled carbon nanotubes (MWCNT), 3, 149, 149f, 150, 259, 260f, 279f

N

Nafion, 155
Nafion membrane, 279, 282, 283
Nanocones/nanohorns, 4f
Nanocrystals (NC), 107, 146, 258
Nanomesh graphene, 180, 181f
Nanopores, 148, 150
Nanotoroids, 4f
Nanotube unzipping, 57, 58f
Nanowires, 233
National Institute of Advanced Industrial Science and Technology, Japan, 67
Near-edge x-ray absorption fine structure (NEXAFS), 148
Negative electrode, 192, 203, 204f
Nested giant fullerenes, 4f
NG. *See* Nitrogen-doped graphene (NG)
Nickel, 25, 27, 218
    formation of graphene on, 28f
    graphene films on, 34f
Nickel oxide, 159
Nickel oxyhydroxide (NiOOH), 159
Nitric acid, 37
Nitrogen atoms in N-graphene, 271f
Nitrogen-coordinated metal, 267
Nitrogen cry-adsorption method, 46
Nitrogen-doped graphene (NG), 92, 110, 122, 160, 161f, 167, 196, 197
    active sites of, for ORR, 275–276
    as metal-free catalysts, 270, 271f
    morphology control of, 273–275, 274f
    as support material, 261–263, 264f
    synthesis validation of, 271–273, 273f
Nitrogen-doped graphene aerogel networks (N-GA), 266
Nitrogen-doped graphene sheets, 71f, 72, 161f
Nitrogen-doped graphene-supported carbon-containing iron nitride, 269
Nitrogen plasma treatment, 261
N-methylpyrrolidone (NMP), 50, 51, 55, 263
Noble metals, 164, 254
Nonaqueous Li-air batteries, 156–165, 157f, 158f, 160f, 161f, 163f, 165f. *See also* Graphene in Lithium-air battery

Nonlocal density functional theory (NLDFT), 175
Nonnoble metal catalyst, 167
Nonprecious metal catalysts (NPMC), support material for, 264–270. *See also under* Graphene
Non-Sanger reaction-based technologies, 15
Normal hydrogen electrode (NHE), 193, 200
Nucleation-growth mechanism, 29
Nyquist plots, 85, 121

O

Oleum, 41
Open circuit voltage (OCV), 248
Optical images of graphene, 222f
Optical modulator, graphene-based, 11–12
Optical properties, graphene, 9
Optical transmittance, 218, 222f
Organic PV (OPV), 224
Organic synthesis, 56–57, 56f
ORR. *See* Oxygen reduction reaction (ORR)
Oxalic acid (OA), 113f
Oxidation method, 44
Oxidation of graphite, 38
Oxygen reduction reaction (ORR), 156, 157, 163, 252, 254, 268. *See also* Fuel cells
    active sites of NG for, 275–276
    heteroatom-doped graphene as catalyst for, 276–278, 278f
    polarization curves, 255f

P

Patterned graphene film, 34f
PDDA, 256
Phonons, 10
Phosphoric acid fuel cell (PAFC), 246
Photoanode applications in solar cells, 234–237, 235f, 236f. *See also* Solar cells
Photodetector, 11
Photonic applications, graphene, 11–12
Photovoltaic (PV) absorption spectrum, 217
Photovoltaic (PV) solar cells. *See also* Solar cells
    graphene materials as back electrodes in, 231–233, 232f, 233f, 234f
    transparent conducting front contact, 224–227, 225f, 226f, 227f, 228f
Phthalocynine, 167
Plasma enhanced chemical vapor deposition (PE-CVD), 31, 222
Platinum, particle size of, 13
Platinum foil, 35
Polarization curves of cathode electrocatalysts, 281f
Polarizers, 12
Poly(acrylic acid) (PAA), 109

Poly(ethyleneimine) (PEI), 127
Poly(ionic liquid) (PIL), 189, 189f
Poly(pyrogallol) (PPG), 258
Poly(sodium 4-styrenesulfonate) (PSS-G), 283
Polyacrylonitrile, 155
Polyaniline (PANI), 161f, 172, 193, 194, 194f, 275
Polybenzimidazole/graphite oxide (PBI/GO), 283
Polydimethylsiloxane (PDMS) films, 25, 33
Polydopamine (PD), 109
Polydopamine (PD) adlayer, 41
Polydopamine-capped reduced GO, 41
Polyelectrolyte (sodium 4-styrenesulfonate) (PSS), 262
Polyethylene glycol (PEG), 41, 140
Polyethylene terephthalate (PET), 37, 155
Polygonal nanomesh graphene, 181f
Polymer, 154
Polymer composites, 13
Polymer electrolyte fuel cells (PEFC), 246, 248f. *See also* Fuel cells
    electrode reactions for, 249t
    history of, 246–247, 247f
    operating principles of/components of, 247–252, 248f
Polymer electrolyte membrane (PEM), 247, 281
Polymethyl methacrylate (PMMA), 13, 33, 35
Polypyrrole (PPy), 155, 195, 196
Polysaccharide, 42
Poly(1-vinyl-3-ethylimidazolium) salts, 189
Polysulfides, 141, 143, 146, 148, 151
Polytetrafluoroethylene (PTFE), 157
Polythiophene (PTH), 172
Pore size distribution, 90, 114
Porous anodic alumina (PAA), 33
Porous carbon materials, 156
Porous graphene, 76, 180
Porous graphene-based electrode materials, 180–182, 181f. *See also* Electrical double-layer capacitance (EDLC)
Positive electrode, 192, 203, 204f
Potassium hydroxide (KOH), 147, 148, 181, 273f
Potassium permanganate, 38, 57
Potassium persulfate, 35
Power conversion efficiency (PCE), 224, 232
Precipitation method, 191
Pristine graphene, 14, 38, 72
Pristine graphene/B-doped graphene powders, 228–230, 229f. *See also* Back electrodes, graphene materials as
Pristine single-layer graphene as support material, 252–255, 255f
Propylene carbonate (PC), 174, 188
Proton conductivity, 284f, 285t
Proton-exchange membrane fuel cell (PEMFC), 246, 247, 250, 259, 279, 281f, 282
Pseudocapacitance, 12, 190
Pseudocapacitive faradic process, 205

Pseudocapacitors, 172
Pt nanoparticle-reduced graphene oxide
    (Pt-rGO), 253
Pulse-microwave-assisted synthesis, 126
Pulverization, 107, 117
Pure graphene anode materials, 66–70, 67f, 69f
Pyrolized metal macrocyclic compounds, 267
Pyrolytic carbon, 82f, 83f
Pyrolyzed polyacrylonitrile (PAN), 153

Q

Quantum dot sensitized solar cell (QDSSC), 218,
    235, 236, 236f, 237
Quantum Hall effect, 9, 14

R

Raman modes, 29
Raman spectra, 50, 68, 184, 223f
    of graphene films, 36f, 38
Raman spectroscopy, 8, 70, 79, 101
Reaction temperature, 31
Redox behavior, 201
Redox (faradic) reactions, 199
Reduced graphene oxide (rGO), 268, 269
Reduced graphite oxide (RGO), 38, 39, 44, 49,
    174, 177, 178, 178f, 206f, 257
    FESEM inages of, 175f
Reoxidation, 194
Reprotonation, 194
Resistivity, 35
Reversible hydrogen electrode (RHE), 268, 269
Rotating disk electrode (RDE), 265, 267, 278f
Rotating-disk electrode voltammograms, 266f,
    273f
Rotating ring-disk electrode (RRDE), 265, 267
Rutile, 115

S

Scanning electron microscope (SEM), 38, 40f,
    48f, 50f, 73f, 188f
    3-D graphene network, 231f
    of discharged sulfur/graphene electrodes,
        163f
    of GNS electrodes, 157f
    of graphene coated on sulfur particles, 144f
    of graphene films directly grown on SiO$_2$
        substrates, 221f
    of graphene grown on Ni-Cu substrate, 220f
    of graphite oxide composite, 91f
    of lithium iron phosphate, 82f
    of Ni foam template for graphene CVD
        growth, 231f
    of Si/graphene composite anode materials,
        95f

Scanning transmission electron microscopy
    (STEM), 71, 71f
    of Si/graphene composite microsphere, 98f
Scanning tunneling microscope (STM), 56
Schrodinger equation, 8
Selenium (Se)-doped graphene (SeG), 276
Self-assembly method, 267
Self-polymerization of dopamine, 109
Self-supporting electrode, 146f
Semiconductor material, 218
Sensors, 14. *See also* Graphene
Sheet resistance, 36f, 37, 44, 52
Short carbon chains, 4f
Short-circuit photocurrent density, 234
SiC, epitaxial growth of graphene on, 24–25, 25f
Silicate-anion-assisted hydrothermal method,
    120
Silicon carbide (SiC), 5
Silicon (Si)/graphene composite anode materials,
    94–99, 95f, 98f, 100f. *See also* Metal-
    graphene composite anode materials
Silicon nanowires, anisotropic, 99
Silicon-on-insulator, 15
Silicon oxide, 97
Single-crystal, 26, 30, 32, 105
Single-walled CNT (SWCNT), 3, 177, 178f, 187
Slonczewski–Weiss–McClure band model of
    graphite, 9
Smart janus substrate, 218–220. *See also* Solar
    cells
Sodium borohydride, 42
Sodium dodecyl benzene sulfonate (SDBS), 51,
    113f, 114, 283
Sodium hydride, 42
Sodium nitrate, 127
Sodium thiosulfate, 139
Solar cells
    back electrodes, graphene materials as
        applications in PV solar cells, 231–233,
        232f, 233f, 234f
    conductive 3-D graphene, 230–231, 231f
    pristine graphene/B-doped graphene
        powders, 228–230, 229f
    overview, 217–218
    photoanode applications in, 234–237, 235f,
        236f
    transparent conducting front contact,
        graphene materials as
    applications in PV solar cells, 224–227,
        225f, 226f, 227f, 228f
    Cu/Ni sphere-wrapped graphene on
        conductive graphene films, 222–223,
        223f, 224f
    few-layer graphene on dielectric
        substrates, 221, 221f, 222f
    low-defect graphene films via smart janus
        substrate, 218–220, 219f, 220f

vertically-erected graphene walls on
dielectric substrates, 222
Sol-gel process, 80, 85
Solid copper, 31
Solid electrolyte interphase (SEI), 102
Solid oxide fuel cell (SOFC), 246
Solvothermal exfoliation, 53–54. *See also*
Liquid-phase exfoliation
Solvothermal method, 59, 78, 164, 228
Specific surface area (SSA), 172, 173
Spindle-shaped ferric oxide hydrate, 116
Spintronics, 11
Spray-drying process, 79
Sputtering, 227
Staudenmaier method, 38
Substoichiometric hydrochloric acid, 76
Sulfonated aromatic poly(ether ether ketone)
(SPEEK), 283
Sulfonated graphite oxide (SGO), 282, 283
Sulfonated polyimide (SPI), 283
Sulfur cathode material, graphene-wrapped, 139f
Sulfuric acid, 38, 57
Sulphur doping, 162, 163f
Supercapacitor-battery hybrids, 199
Supercapacitors, 12
electrical double-layer capacitance (EDLC),
graphene-based, 173–190
electrolyte, effect of, 186–190, 187f, 188f,
189f
GNR-based electrode materials, 183–185,
183f, 184f, 185f, 186f
graphene-based electrode materials,
174–176, 175f, 176f, 177f
graphene hybrid composites-based
electrode materials, 177–180, 178f,
179f
porous graphene-based electrode
materials, 180–182, 181f
graphene-based electrode materials with
pseudocapacitive properties
based on graphene with heteroatoms
in carbon network, 196–199,
197f, 198f
graphene-conductive polymers-based,
193–196, 194f, 195f
graphene-metal-oxide composites-based,
190–193, 192f, 193f
graphene-based hybrid supercapacitor,
199–206
in aqueous electrolytes, 200–202, 201f,
202f
asymmetric supercapacitor in organic
electrolyte, 203–205, 204f, 206f
overview, 171–173
Support material
for nonprecious metal catalysts, graphene as,
264–270. *See also* Fuel cells

metal oxides, 265–267, 266f
pyrolized metal macrocyclic compounds,
267
transition metal-nitrogen-containing
complexes, 267–270
for Pt/Pt-alloy catalysts, graphene as
conductive polymer-functionalized GNS
as support material, 256–258, 257f
hybrid supporting material by
intercalating carbon black and CNT,
258–261, 260f
nitrogen-doped graphene as support
material, 261–263, 264f
pristine single-layer graphene as support
material, 252–255, 255f
Surface energy, 143
Symmetrical supercapacitor cells, 181

T

Tensile strength, 10
Ternary composite cathode material, 153
Tetrabutylammonium, 41
Tetrabutylammonium hydroxide (TBAOH), 44,
53, 54f
Tetrabutyl titanium, 113
Tetrachloromethane, 229
7,7,8,8-tetracyanoquinodimethane (TCNQ), 109,
122
Tetraethylammonium tetrafluoroborate
(TEABF4), 174, 179–180
Tetraethyl orthosilicate (TEOS), 97, 273
Tetrahydrofuran (THF), 198, 198f
Thermal annealing, 103f
Thermal evaporation, 103f
Thermal exfoliation/reduction of GO, 44–47,
45f, 47f
Thermal management, 14–15. *See also* Graphene
Thermal properties, graphene, 9–10
Thermal reduction, 44, 263
of GO, 175
Thermogravimetric analysis (TGA), 96, 121f, 150
THF. *See* Tetrahydrofuran (THF)
3-D graphene, conductive, 230–231, 231f
3-D graphite crystal, 4f
3-D nanotube networks, 4f
3-D Schwarzite crystals, 4f
Tin dioxide (SnO$_2$)-graphene composite anode
materials, 105–110, 106f, 108f.
See also Metal oxide-graphene
composite anode materials
Tin (Sn)/graphene composite anode materials,
102–105, 103f. *See also* Metal-
graphene composite anode materials
Titanium dioxide (TiO$_2$)-graphene composite
anode materials, 110–115, 111f, 113f
Top-down process, 22

Touch screens, 17
Toxic hydrazine hydrate, 40
Transfer technique for CVD graphene film,
        33–35, 34f
Transistors, 18
Transition metal-nitrogen-containing complexes,
        267–270
Transition metal oxides, 190
Transmission electron microscopy (TEM), 5,
        49, 79f
    of B,N-graphene, 278f
    and charge/discharge performance of
            Ru-based electrocatalyst, 165f
    of GNS electrodes, 157f
    of graphene composite, 93f
    of graphene dispersed by Nafion, 235f
    of graphene-encapsulated sulfur particles, 145f
    of graphene nanosheets, 87f
    of graphene wrapped mesoporous C/S
            composite, 152f
    of lithium iron phosphate, 82f
    of MEGO, 176f
    of mesoporous graphene paper, 147f
    of Si/graphene composite anode materials, 95f
Transparent conducting front contact, graphene
            materials as. See also Solar cells
    applications in PV solar cells, 224–227, 225f,
            226f, 227f, 228f
    Cu/Ni sphere-wrapped graphene on
            conductive graphene films, 222–223,
            223f, 224f
    few-layer graphene on dielectric substrates,
            221, 221f, 222f
    low-defect graphene films via smart janus
            substrate, 218–220, 219f, 220f
    vertically-erected graphene walls on
            dielectric substrates, 222
Transparent conducting oxide films, 217
Transparent conductive films (TCF), 221, 224,
            225, 226
Transparent conductive properties, 225f
Tris(hydroxymethyl)aminomethane, 152
Triton X-100, 140
Tungsten (W), 31

U

Ultracapacitors, 12, 197f
Ultrahigh vacuum (UHV), 24

Ultrasonication, 39, 46, 149
Ultrasonic coassembly process, 191
Ultrathin carbon films, 5
Ultrathin graphite flakes, 23
Ultraviolet (UV) adhesive, 37
Unzipping of CNT, 57, 58f, 68. See also
            Graphene, synthesis of
Urea, 174
UV photoemission spectroscopy, 272

V

Vacuum drying, 193
Vacuum filtration, 39, 43
Vacuum thermal treatment, 191
Vanadium isopropoxide, 92
Vanadium oxide, 90, 91f
Vanadium pentoxide, 93
Van der Pauw method, 48
Van der Waals forces, 2
Vapor carbon sources, 33

W

Wafer-scale graphene film, 28f
Wettability, 189
    of tin, 104
Wiedemann–Franz law, 10
Wormlike graphite (WEG), 52
Wurtz-type reductive coupling (WRC), 228

X

Xerogel, 93
X-ray absorption near-edge structure (XANES),
            85, 110, 265
X-ray absorption spectroscopy (XAS), 141
X-ray emission spectroscopy (XES), 148
X-ray photoelectron spectroscopy (XPS), 38, 46,
            121, 122, 142, 262
XRD, 121f, 129f

Y

Young's modulus, 10

Z

Zigzag edge graphene, 183, 183f

T - #0395 - 071024 - C318 - 234/156/14 - PB - 9780367783716 - Gloss Lamination